Anonymous

Rivers of Great Britain

The Thames, from Source to Sea

Anonymous

Rivers of Great Britain
The Thames, from Source to Sea

ISBN/EAN: 9783337203658

Printed in Europe, USA, Canada, Australia, Japan

Cover: Foto ©berggeist007 / pixelio.de

More available books at **www.hansebooks.com**

RIVERS OF GREAT BRITAIN.

The Thames, from Source to Sea.

DESCRIPTIVE, HISTORICAL, PICTORIAL.

CASSELL & COMPANY, LIMITED:

LONDON, PARIS & MELBOURNE.

1891.

[ALL RIGHTS RESERVED.]

CONTENTS.

CHAPTER I.
ABOVE OXFORD.—*By W. SENIOR.*

The Source of the Thames—Early Names of the River—Seven Springs—Thames Head—The Churn and its Course—Thames and Severn Canal—Cricklade—Castle Eaton—Inglesham—Fairford and the Coln—Lechlade—The First Lock—Some Thames Flowers—Old Buscot—Hart's Weir—Bird Life—Radcot Bridge—Eddying Pools and Golden Shallows—Canal-like Reaches—Tadpole Bridge—Bampton—Duxford Ferry—Canute's Country—The Windrush—The Oldest Bridge—Old Father Thames—Disused Weir-pools—Bablock Hythe, Stanton Harcourt, and Cumnor—Skinner's Weir and Pinkhill Lock—Eynsham Weir, Bridge, and Cross—The Evenlode—Witham Hill—Thames Angling—Godstow—King's Weir—Port Meadow—Folly Bridge 1

CHAPTER II.
OXFORD TO ABINGDON.—*By D. MACCOLL.*

Oxford, from the Upper River; the New Town—The Courses of the River, from Medley Weir to Folly Bridge—The Houses of the Regulars and Friars—The University and Parish Churches—The Halls and Colleges of the Seculars, from the Thirteenth Century to the Reformation—Jacobean Oxford—Classic Oxford—Convenient Oxford—The Architectural Revival—The Undergraduate Revival—The River below Folly Bridge, and the Invention of Rowing—The Navigation Shape of the River—Floods—The Barges—Iffley—Littlemore—Kennington—Radley—Sandford—Nuneham 33

CHAPTER III.
ABINGDON TO STREATLEY.—*By J. PENDEREL-BRODHURST.*

Abingdon—The Abbey—St. Nicholas' Church—The Market Cross—The Ancient Stone Cross—St. Helen's Church—Christ's Hospital—Culham—First View of Wittenham Clump—Clifton Hampden—The "Barley Mow"—A River-side Solitude—Day's Lock—Union of the Thames and the Isis—Dorchester—The Abbey Church—Sinodun Hill—Shillingford Bridge—Bensington—The Church—Crowmarsh Giffard—Wallingford—Mongewell—Newton Murren—Moulsford—The "Beetle and Wedge"—Cleeve Lock—Streatley 62

CHAPTER IV.
STREATLEY TO HENLEY.—*By W. SENIOR.*

Streatley, the Artists' Mecca—Goring *versus* Streatley—Goring from the Toll-gate—Streatley Mill—Weirs and Backwaters—Antiquity of Streatley and Goring—Goring Church—Common Wood—Basildon Ferry and Hart's Wood—A Thames Osier Farm—Whitchurch Lock—Pangbourne—Hardwicke House and Mapledurham—Caversham Bridge—Reading and its Abbey—A Divergence to the Kennet, with calls at Marlborough, Hungerford, and Newbury—The Charms of Sonning—"The Loddon slow, with verdant alders crowned"—St. Patrick's Stream—Shiplake Weir—Wargrave and Bolney Court—Park Place—Marsh Lock—Remarks on Thames Angling—The Approach to Henley 85

CHAPTER V.
HENLEY TO MAIDENHEAD.—*By the Rev. PROFESSOR BONNEY, F.R.S.*

The Best Bit of the River—Henley—The Church—The "Red Lion"—Shenstone's Lines—Henley Regatta—The First University Boat-race—Fawley Court—Remenham—Hambledon Lock—Medmenham Abbey and the Franciscans—Dissolution of the Order—Hurley—Lady Place and its History—A Strange Presentiment—Bisham Abbey and its Ghost—Bisham Church—Great Marlow—The Church and its Curiosities—"Puppy Pie"—Quarry Woods—The Thames Swans and the Vintners' Company—Cookham and Cliefden—Hedsor—Cliefden Woods—The House—Raymead—The Approach to Maidenhead 113

CONTENTS.

CHAPTER VI.
MAIDENHEAD TO WINDSOR.—*By H. SCHÜTZ WILSON.* PAGE

Maidenhead—Bray—Jesus Hospital—The Harbour of Refuge—Frederick Walker—A Boat-race—Monkey Island—The River—Surley Hall—Boveney Lock—Eton—Windsor—St. George's Chapel—The Castle—Mr. R. R. Holmes—James I.—Surrey—The Merry Wives of Windsor 143

CHAPTER VII.
WINDSOR TO HAMPTON COURT.—*By GODFREY WORDSWORTH TURNER.*

Leaving Windsor—Eton, its History and its Worthies—The College Buildings—Windsor Park—The Long Walk—The Albert Bridge—Datchet and Falstaff—Old Windsor—"Perdita's" Grave—The Tapestry Works—The "Bells of Ouseley"—Riverside Inns—The Loves of Harry and Anne Boleyn—Magna Charter Island—Runnymede—The Poet of Cooper's Hill—Fish at Bell Weir—A Neglected Dainty—Egham and Staines—John Emery—Penton Hook—Laleham—Dr. Arnold—Chertsey—The Lock and Bridge—Albert Smith and his Brother—Chertsey Abbey—Black Cherry Fair—Cowley the Poet—A Scene from "Oliver Twist"—St. Ann's Hill—Weybridge—Oaklands and the Grotto—Shepperton Lock and Ferry—Halliford—Walton—The Scold's Bridle—Sunbury—Hampton—Moulsey Hurst and its Sporting Associations—Hampton Court Bridge 161

CHAPTER VIII.
HAMPTON COURT TO RICHMOND.—*By J. PENDEREL-BRODHURST.*

Hampton Court—Thames Ditton; The "Swan"—The Church—Surbiton—Kingston; The Coronation Stone—Teddington—Twickenham—Eel Pie Island—Petersham—Richmond Park—Approach to Richmond 201

CHAPTER IX.
RICHMOND TO BATTERSEA.—*By the Rev. PROFESSOR BONNEY, F.R.S.*

The River at Richmond—A Spot for a Holiday—The Old Palace of Sheen—The Trumpeters' House—Old Sad Memories—Richmond Green—The Church—Kean's Grave—Water Supply—The Bridge—The Nunnery of Sion and Convent of Sheen—Sir William Temple—Kew Observatory, Isleworth—Sion House and its History—Kew Palace and the Georges—Kew Gardens—Kew Green—Brentford—Mortlake—Barnes—Chiswick—The Boat-race—Hammersmith—Putney—Barn Elms—Putney and Fulham—The Bishops of London—Hurlingham—The Approach to a Great City 229

CHAPTER X.
BATTERSEA TO LONDON BRIDGE.—*By EDMUND OLLIER.*

The Scene Changes—A City River—Battersea—Chelsea—The Old Church—Sir T. More and Sir Hans Sloane—Cheyne Walk—Don Saltero's Coffee-house and Thomas Carlyle—The Botanical Gardens—Chelsea Hospital—The Pensioners—Battersea Park—The Suspension Bridge—Vauxhall—Lambeth—The Church and Palace—Westminster Palace and the Abbey—Its Foundation and History—Westminster Hall—Westminster Bridge—The Victoria Embankment—York Gate—Waterloo Bridge and Somerset House—The Temple—Blackfriars Bridge—St. Paul's—Southwark Bridge—The Old Theatres—Cannon Street Bridge—London Bridge and its Traffic 258

CHAPTER XI.
LONDON BRIDGE TO GRAVESEND.—*By AARON WATSON.*

Hogarth's Water Frolic—Billingsgate—Salesmen's Cries—The Custom House—Queen Elizabeth and the Customs—The Tower, and Tower Hill—The Pool—The Docks—Ratcliff Highway—The Thames Tunnel—In Rotherhithe—The Isle of Dogs—The Dock Labourer—Deptford and Greenwich—Woolwich Reach and Dockyard—The *Warspite* 288

CHAPTER XII.
GRAVESEND TO THE NORE.—*By J. RUNCIMAN.*

Morning on the Lower Thames—Gravesend—Pilots and Watermen—A Severe Code—Tilbury and its Memories—The Marshes—Wild-fowl Shooting—Eel Boats—Canvey Island—Hadleigh Castle—Leigh, and the Shrimpers—Southend and the Pier—Sailing—Sheerness—The Mouth of the Medway—The Dockyard—The Town and its Divisions—The Nore—A Vision of Wonder—Shoeburyness—Outward Bound 337

TABLE OF DISTANCES ... 368

LIST OF ILLUSTRATIONS.

Frontispiece.—CLIEFDEN WOODS.

ON TITLE-PAGE, HEAD OF THAMES (*from bas-relief on Temple Pier*).

MAP OF THE THAMES . . . . *To face page* . . . 1

ABOVE OXFORD:—

The Seven Springs — Thames Head — The Sources of the Thames (*Map*) — The First Bridge over the Thames — Cricklade — Inglesham Round House — Lechlade: the First Lock — Radcot Bridge — The Ferry, Bablock Hythe — Cumnor Churchyard — Stanton Harcourt Church — Eynsham Weir — Cross at Eynsham — Oxford from Godstow — The Thames from Lechlade to Oxford (*Map*) 1–32

OXFORD TO ABINGDON:—

The Barges — Oxford, from Headington Hill — New College, from the Gardens — St. Mary's, from the High Street — Magdalen Tower, from the Cherwell — Stone Pulpit, Magdalen — "Tom" Gateway — The Dome of the Radcliffe, from Brasenose — The 'Varsity Barge — A "Bump" at the Barges — Iffley Mill — Iffley Church — Littlemore Church and Kennington Island — Oxford to Abingdon (*Map*) — A Picnic to Nuneham — The Bridge and Cottage, Nuneham — Distant View of Abingdon 33–61

ABINGDON TO STREATLEY:—

Abingdon, from the River — Abingdon Bridge — Culham Church — Clifton Hampden Church — Dorchester, from Little Wittenham — Sinodun Hill and Day's Lock — Shillingford Bridge — Wallingford Church and Bridge — Moulsford Ferry — Abingdon to Streatley (*Map*) — Streatley Mill 62–84

STREATLEY TO HENLEY:—

The Thames at Streatley — Streatley to Henley (*Map*) — Goring, from the Tollgate — Whitchurch Church and Mill — Mapledurham, the Church and the Mill — Flooded Meadows, from Caversham Bridge — The Thames at Reading, from the Old Clappers — Sonning-on-Thames — Sonning Weir Shiplake — A Camping-out Party — Backwater at Wargrave — A Pool of Water-lilies 85–112

HENLEY TO MAIDENHEAD:—

Henley Regatta — Henley, from the Towing-path — Regatta Island — Fawley Court — Aston Ferry — Medmenham Abbey — Below Medmenham — Bisham Abbey — Bisham Church — Great Marlow, from Quarry Woods — Henley to Maidenhead (*Map*) — A Picnic at Quarry Woods — A Group of Swans — Cookham — A Crowd in Cookham Lock — The Landing-Stage, Ray Mead — Taplow Woods 113–142

MAIDENHEAD TO WINDSOR:—

Bray Church — Maidenhead to Windsor (*Map*) — Surley — Boveney Lock — Windsor Castle, from Boveney Lock — St. George's Chapel, Windsor 143–160

LIST OF ILLUSTRATIONS.

WINDSOR TO HAMPTON COURT:—

Procession of the Boats, Eton—Eton, from the Playing-fields—The Albert Bridge—Old Windsor Lock—The "Bells of Ouseley"—Magna Charta Island—Runnymede—Windsor to Hampton Court (*Map*)—London Stone—Staines Bridge—Laleham Ferry—Laleham Church—Chertsey Bridge—Shepperton Lock—Shepperton—Halliford—Sunbury Weir—Sunbury Church—Between Hampton and Sunbury—Garrick's Weir, Hampton . 161—200

HAMPTON COURT TO RICHMOND:—

The Approach to Hampton Court—Entrance Porch—The First Quadrangle—Fountain Court—In the Reach below Hampton Court—The "Swan," Thames Ditton—Thames Ditton Church—Hampton Court to Richmond (*Map*)—Kingston, from the River—The Market-place, Kingston—The Coronation Stone—The Royal Barge—The "Anglers," Teddington—Strawberry Hill—Pope's Villa at Twickenham—Twickenham Ferry—Richmond: the Meadows and the Park—Richmond: The Terrace from the River . . 201—228

RICHMOND TO BATTERSEA:—

Richmond Bridge—Between Richmond and Kew—Sion House—The River at Kew—The Pagoda in Kew Gardens—Kew Gardens—Cambridge Cottage—High Water at Mortlake—Hogarth's Tomb—The University Boat-race—Richmond to Battersea (*Map*)—Old Hammersmith Bridge—Old Putney Bridge and Fulham Church . . . . 229—257

BATTERSEA TO LONDON BRIDGE:—

Battersea Bridge—Cheyne Walk—Vauxhall Bridge, from Nine Elms Pier—Lambeth Palace and Church—The Victoria Tower—The Abbey, from Lambeth Bridge—York Gate—The Embankment—The River at Blackfriars—St. Paul's, from the Thames—Southwark Bridge—Cannon Street Station—Battersea to London Bridge (*Map*) . . . . 258—287

LONDON TO GRAVESEND:—

In the Pool—St. Magnus' Church and the Monument—London Bridge to Woolwich (*Map*)—Billingsgate: Early Morning—The Tower, from the River—Limehouse Church—The River below Wapping—Entrance to the East India Docks—The West India Docks—Millwall Docks—Millwall—Greenwich Hospital—View from Greenwich Park—The Albert Docks—Woolwich Reach—Woolwich Arsenal—Woolwich—Plumstead—Dagenham Marshes—Barking Abbey—Barking Reach—At Purfleet—Erith Pier—Tilbury Fort—Gravesend—At Gravesend—Woolwich to Gravesend (*Map*) . . . . 288—336

GRAVESEND TO THE NORE:—

At Canvey Island—The Fringe of the Marshes—Hadleigh Castle—Leigh—Southend and the Pier—Sheerness Dockyard, looking up the Medway—Sheerness Dockyard, from the River—Mouth of the Thames: Low Water—Artillery Practice at Shoeburyness—Gravesend to the Nore (*Map*)—Outward Bound: Passing the Nore Light . . . . 337—362

We are indebted to Messrs. Taunt, of Oxford, for permission to use their photographs for the views on pages 15, 22, 56, 60, 65, 69, 79, 91, 149, 169, 174, 195, 198, 200, 209, 220, and 334; to Messrs. Hill and Saunders for that on page 58; to Messrs. G. W. Wilson and Co., of Aberdeen, for that on page 54; to Messrs. W. H. Beer and Co. for those on pages 71, 80, and 85; to Messrs. Marsh Bros., of Henley, for those on pages 63, 77, 113, and 132; to Messrs. Poulton and Son, of Lee, for that on page 335; to Mr. F. H. Scourable, of Southend, for that on page 351; to Mr. S. Cole, of Gravesend, for those on pages 354 and 355; and to Mr. F. G. O. Stuart, of Southampton, for those on pages 83 and 153.

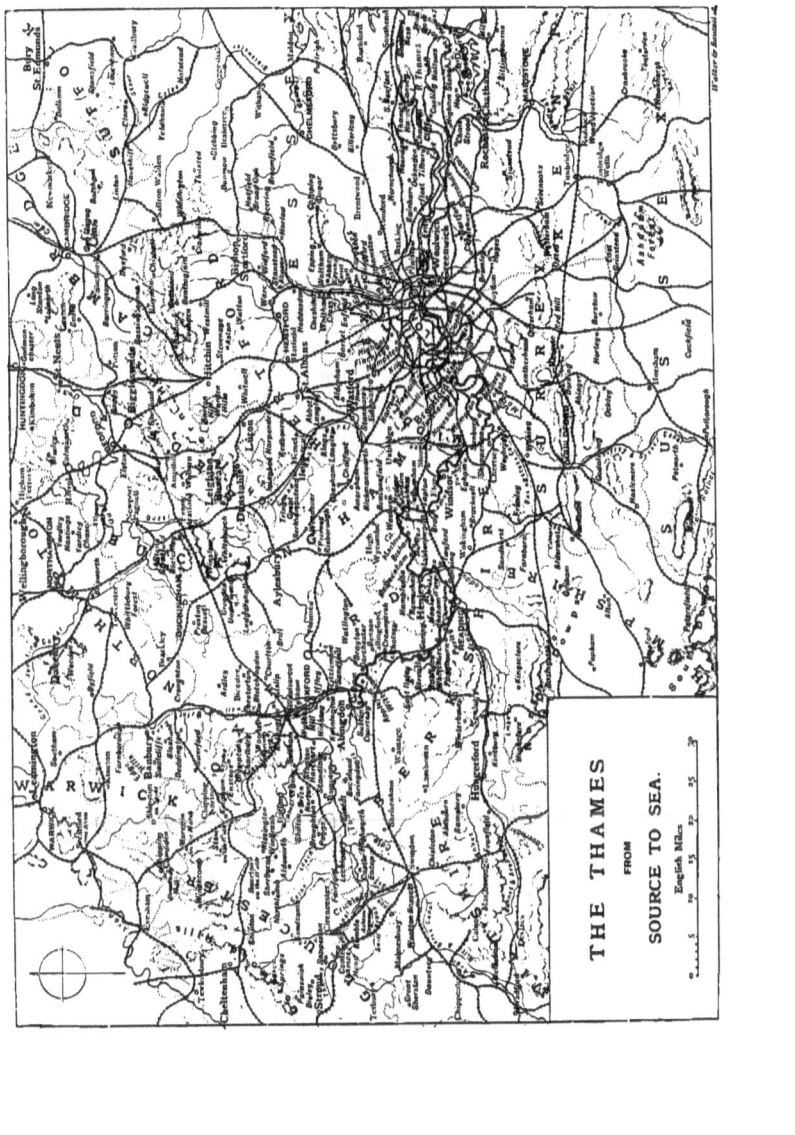

RIVERS OF GREAT BRITAIN.

THE SEVEN SPRINGS.

THE THAMES.
CHAPTER I.
ABOVE OXFORD.

The Source of the Thames—Early Names of the River—Seven Springs—Thames Head—The Churn and its Course—Thames and Severn Canal—Cricklade—Castle Eaton—Inglesham—Fairford and the Coln—Lechlade—The First Lock—Some Thames Flowers—Old Buscot—Hart's Weir—Bird Life—Radcot Bridge—Eddying Pools and Golden Shallows—Canal-like Reaches—Tadpole Bridge—Bampton—Duxford Ferry—Canute's Country—The Windrush—The Oldest Bridge—Old Father Thames—Disused Weir-pools—Bablock Hythe, Stanton Harcourt, and Cumnor—Skinner's Weir and Pinkhill Lock—Eynsham Weir, Bridge, and Cross—The Evenlode—Witham Hill—Thames Angling—Godstow—King's Weir—Port Meadow—Folly Bridge.

HE birds, flowers, and bees around are, doubtless, in their several ways, rejoicing with me in the balmy May morning radiant with warm sunshine. Down the unsullied emerald of the little slope yonder, carpeted with nodding cowslips, daisies, and buttercups, and faintly azured in sheltered spaces with wild hyacinths, I have descended into a rustic glade, not, at its widest, more than fifty yards across, and running, roughly reckoning, north and south. The slope is easy, springing as it does from a verdant bottom to the foot of a low wall; this, pushing aside the glossy sycamore branches, I have leaped from the Canal path, at a gap where the village children, on their recent half-holiday, wastefully cast aside the surplus of their cowslip harvest to wither and die. But from my present standpoint the low wall is nearly hidden in undergrowth, and by a plentiful intermixture of hawthorn, holly, and ash flourishing on the bank top. The sweet-smelling grass is spangled with daisies and buttercups, though not so profusely as in the field adjacent, which is destined for a crop of hay; and the grove resounds with bird-music set in the rapturous key of the bridal season. And there, a few paces athwart the sward, under the shadow of trembling foliage, is the spot which for centuries was said to be the birthplace of the River Thames. We are at Thames Head, in Trewsbury Mead, in the parish of Cotes, in the county of Gloucestershire, three miles south-west of Cirencester.

The mossy trunk, lying prostrate under the wall on the side of the glade opposite the sylvan slope by which entrance has been effected, invites the opportunity of a more minute observation. Seated thereupon, far from the noisy world, we may make a fair and leisurely start upon that long and interesting voyage from Source to Sea, upon which, in this and succeeding chapters, the reader is invited to embark with confidence and hope. Here, probably, is the identical spot which Peacock, author of the "Genius of the Thames," had in his mind when he wrote—

> "Let fancy lead from Trewsbury mead.
> With hazel fringed, and copsewood deep;
> Where, scarcely seen, through brilliant green,
> Thy infant waters softly creep."

The friendly branches of a wild rose hustle my elbow, or, rather, would do so, but that a sturdier bramble bough interposes. On the other side of me there is a charming tangle of hazel and blackberry bushes. There is also a more than commonly bushy hawthorn overspreading the wall at a portion where thick ivy covers it. A spreading wild rose is established in the very middle of the glade, which is graced with quite an unusual quantity of large and old hawthorn trees. A strong west wind soughs and sighs in the trees; blackbirds and thrushes, by their liquid notes, blithe and merry, seem to protest against the melancholy undertone, as does a grand humble bee, in magnificent orange-velvet smallclothes, who contributes a sympathetic bass solo as he drones by. But the object to be chiefly noticed at this moment is the aged ash-tree yonder. It is of medium size and no particular shape, though the ivy covering its bole and lower limbs gives it an air of picturesque importance. Ragged hawthorns and brambles surround it. The importance of the tree lies in the circumstance that it marks the spot which the old writers, and many modern authorities following in their footsteps, have pronounced to be the source of the Thames. The supposition is that in former times a perennial spring of water issued forth here, forming Thames Head. The well, however, out of which the water might once have gushed, and miscellaneously overspread the pasturage on its way to form a brook, has in these days lost potency. For a long time past it has ceased to yield water, and, as a matter of prosaic fact, from one end of the glade to the other there is no sign of water in any shape or form. The inhabitants of the countryside say that in the winter-time the waters, provoked by long rains, still well forth in copious flood; but, even granting this, we may not conclude that a spring so uncertain as this in Trewsbury Mead is the source of the Thames. The obvious reflection is that before the erection of the ugly pump-house which disfigures the locality, and before the neighbouring springs began to be drained for the service of the Canal, the supply of water was permanent and strong, albeit there is ground for supposing that Thames Head was never thoroughly to be relied upon. I have thus pictured for the reader the source which appears to be favoured by topographers and antiquaries; but there are other springs besides. Half a mile lower down there is, near the Roman way, a basin—another Thames Head—which is sometimes filled by a spring, and which is pictured on the next page in the precise condition in which it appeared to our artist during the rains of early spring. Yet another rill issues from a hill-side; and a fourth, lower still, is perhaps the most clearly defined and strongest of the group, and best entitled to the honour claimed on behalf of the dried-up well in the green glade just described. The Thames Head district seems, indeed, to abound in springs, and in wet weather the level ground is probably freely intersected by brooklets, forming the stream which is the undoubted head of Isis, and which has been called the Thames from time immemorial.

On the very threshold of our task we are confronted, indeed, with two some-time-disputed points which it will be necessary to clear away, or come to terms with, if we would proceed upon our voyage of some two hundred miles from source to Nore with a clear conscience. They relate, first, to the name of the river; and second, to the precise spot in the Cotswold country where it starts upon its wanderings. Neither of these controversial subjects shall, however, detain us long from an intimate acquaintance with the "mighty king of all the British rivers, superior to most in beauty, and to all in importance," setting forth on its career in humble smallness, gathering tranquil volume as it flows in succession through the fertile counties of Gloucestershire, Wiltshire, Berkshire, Oxfordshire, Buckinghamshire, Surrey, Middlesex, Kent, and Essex, and finally delivering its full tribute to the Northern Ocean. What rare historical memories it evokes, what varieties of landscape it touches and creates, let the following lines describe:—

"The blood-stain'd scourge no tyrants wield,
No groaning slaves enrich the field,
But Health and Labour's willing train
Crowns all thy banks with waving grain;
With beauty decks thy sylvan shades,
With livelier green invests thy glades;
And grace, and bloom, and plenty pours
On thy sweet meads and willowy shores.
The field where herds unnumber'd rove,
The laurell'd path, the beechen grove,
The oak, in lonely grandeur free,
Lord of the forest and the sea;

The spreading plain, the cultured hill,
The tranquil cot, the restless mill,
The lonely hamlet, calm and still;
The village spire, the busy town,
The shelving bank, the rising down,
The fisher's punt, the peasant's home,
The woodland seat, the regal dome,
In quick succession rise to charm
The mind, with virtuous feelings warm;
Till where thy widening current glides,
To mingle with the turbid tides,
Thy spacious breast displays unfurl'd
The ensigns of th' assembled world."

It is now generally accepted that, from times as remote as those which preceded the Conquest, the highest portion of the river was called the Thames. The Saxon Chronicles so refer to it, and there is no reason to suppose that the river crossed by the armies of Ethelwold and Canute on their expeditions into the land of Mercia was ever known by other name. How, and when, the river from Cricklade to Oxford acquired the local name of Isis is not clear; but the idea was probably fairly started, though not invented, by Camden, who had pretty visions of the "marriage of the Tame and

"THAMES HEAD."

Isis." "This," wrote he, "is that Isis which afterwards joining with Tame, by adding the names together, is called Tamisis, chief of the British rivers, of which we may truly say, as ancient writers did of Euphrates in the East, that it both plants and waters Britain." It is sufficient for us now to recognise the fact that above Oxford the river is impartially spoken of, now as the Thames, and now as the Isis; and it is rather as a matter of convenience than of dogmatic purpose that I shall elect henceforth to use the older and more reasonable name—the Tameses of the Romans, the Temese of the Saxons, and the Thames of modern days.

Equally fruitful of controversy has been the source of the Thames. It has long been a question whether this grassy retreat, in which we are supposed to be lingering, to wit, Thames Head, in the parish of Cotes, near Cirencester, in the county of Gloucestershire, or Seven Springs, near Cheltenham, should be regarded as the actual starting-point of the river. Gloucestershire and Wiltshire, and different portions of each, have occasionally contended for the honour. Many pages might be filled with rehearsals of learned argument, and quotations from ancient authorities, to support conflicting contentions; but I shall presently invite the reader to follow the course suggested in regard to the name, and make an arbitrary law unto himself. It is not to be denied that the balance of acceptation by topographers of olden times pointed to Thames Head as the generally received source. Leland, sometimes called the Father of English Antiquaries, settles it thus:—"Isis riseth at 3 myles from Cirencestre, not far from a village cawlled Kemble, within half a mile of the Fosse-way, wher the very hed of Isis ys. In a great somer drought there appereth very little or no water, yet is the streame servid with many of springes resorting to one botom."

THE SOURCES OF THE THAMES.

Ill, therefore, will fare the visitor to Thames Head, who seeks it, as I have done, full of poetical fancies and pretty conceits about the source of rivers in general, and the birthplace of the famous English stream in particular. However charming he may find the place to be, and charming it certainly is, he will be doomed to disappointment if he thinks he has reached the source of our royal stream. I was bound for the identical spot, as I congratulated myself, where

"From his oozy bed
Old Father Thames advanced his rev'rend head,
His tresses dropped with dews, and o'er the stream
His shining horns diffused a golden gleam."

As we have seen, the explorer will, at first, experience failure in his endeavour to find, with any satisfactory clearness, either old Father Thames or his oozy bed. Arrived at the ancient Akeman Street, or Fosse-way, "3 myles" from Cirencester, a choice of no fewer than four springs is presented. The village of Cotes, the Roman mound known as Trewsbury Castle, Trewsbury Mead, and the unromantic chimney of the Thames and Severn Canal Engine-house are plain enough, here and there—landmarks, all of them, for the industrious searcher; but there is no sign of flowing water, or, indeed, of water in repose. You will look in vain for semblance of a bed which might be that of a river. It was only after considerable trouble that I obtained any information, and was guided to this well, named by tradition as the original and primary source of the Thames, and reached by proceeding for a quarter of a mile from the high road (where it crosses the railway) along the walk bordering the Canal.

The reader, however, is hereby invited to regard, not Thames Head, but Seven Springs, near Cheltenham, as the natural and common-sense source of the River Thames. Some three miles south of the town, in the parish of Cubberley, or Coberley, to quote the words of Professor Ramsay, "the Thames rises not far from the crest of the oolitic escarpment of the Cotswold Hills that overlook the Severn."

After pausing on the shoulder of Charlton Hill, and admiring—as who can fail to do?— the magnificent panorama of hill and valley receding into the mist of distance north and north-east, you proceed from Cheltenham along the Cirencester road to the crossways. A short divergence to the right, and a dip in the road brings you to a piece of wayside turf, with, beyond, a corner shaped like an irregular triangle. One side of this might be, perhaps, seven yards in length, another four yards, and the third something between the two. The triangular depression is reached by one of those little green hillocks so often to be found on English waysides. The bottom is covered with water, which, in spite of the place being no-man's-land, is clear as crystal, and in its deepest part there was not, at the time of my visit, more than six inches of water. The bed of this open shallow reservoir is not paved with marble, or even concrete, but is liberally provided with such unconsidered trifles as the weather or playful children would cast there. When the wind sets that way a good deal of scum will gather in the farther corner, formed by two walls. The turf near the water's edge is worn away, and the green hillock has been trodden into a mere clay bank by the feet of cattle and men, for it is, as I have said, a patch of common land abutting upon the road. Overhead, stretched from the telegraph posts, you may count nine unmistakable wires parallel with the wall which forms the base of our triangle. On the side farthest from the road the bank is high. A venerable hawthorn has become wedded apparently to an equally venerable ash, whose topmost boughs coquette at close quarters with the telegraph wires. Another ash-tree, at the outer point of the triangle, leans over the water. Between the trees a little sloe bush keeps sturdy foothold. You may mark, moreover, a few straggling briars, bits of silver-weed, a root or two of the meadow cranesbill, a clump of poverty-stricken meadow-sweet, some fool's parsley, wild strawberry plants, and a good deal of bold and always flourishing dandelion. This is the environment of the true source of the great River Thames. We are at Seven Springs.

Hence multitudinous initials are rudely carved upon the old trees and on the stone walls; hence strangers, during summer, drive hither and pay homage. Clear away the scum from the water at the foot of the wall and a small iron grating explains how the waters, always bubbling clear and cool from the Seven Springs, pass away. On the other side of the wall the inflow forms a pond in private grounds. Thence it descends by a homely fall into a smaller pond, and by yet another insignificant fall into what for some distance is sometimes little better than a stagnant ditch. A lower fall, however, of more determined character than the others, sets in motion a clear rill, which, though tiny in volume and unpretentious in present aims, sets off upon its gravelly course as if it knew that by-and-by it would form an estuary upon which the navies of the world might ride in safety. Just now a child might leap across. It is a mere thread of water, yet the streamlet begins at once to proceed in a business-like way under the solid hedgerows separating the fields, and soon becomes a decided brook. This is a tangible beginning, at all events. The Seven Springs are on evidence in a convenient enclosure; they may be recognised as, silently sparkling, they gush from the bank which gives roothold to the hawthorn and ash; and the infant river is always in sight from the moment it assumes the form of a tiny streamlet.

It is difficult to conceive how it has come about that Thames Head on the one part and Seven Springs on the other have been considered rival claimants for the honour of being the cradle of the Thames. It is true that both streams (for Thames Head eventually, by sundry means, becomes a stream) rise from the eastern slopes of the Cotswolds; but they are many miles apart, and Thames Head is nearly fifteen miles nearer the sea than Seven Springs. The rivulet issuing from Seven Springs, and which presently becomes the River Churn is, in the present day at least, the distinct stream which continues its unbroken course to the Nore, and it is the source which is farthest from the mouth of the Thames.

Leland, nevertheless, writing at the time of Henry VIII., fixes, as we have seen, upon Thames Head as the source. Stow, with less detail, adopts the same locality; Camden does likewise; Atkins declares that the river riseth in the parish of Cotes; Rudder that it has been reputed "to rise in the parish of Cotes, out of a well." Modern tourists regularly visit both places, and in great numbers, during the summer season, and in the case of Thames Head are probably taken now to the uppermost glade, which I have described, and now to the spring nearer the engine-house of the Thames and Severn Canal, represented by the illustration. The neglect of the alleged sources by the local authorities of both Cirencester and Cheltenham is to be explained, probably, on the old principle, that what is everybody's is nobody's business. Since, however, people go in full faith to both Seven Springs and Thames Head, some record, however simple, might surely be upraised at both for the enlightenment of the wayfaring man.

Dealing with this question at more length perhaps than the subject requires, I may be allowed to repeat that in these days there ought to be no manner of doubt that the natural and legitimate source of the Thames is that shallow, neglected, triangular pool formed by the Seven Springs. The Cotswold Hills are, in any case, above dispute as the cradle-ground of the river, and may be happy with either claimant.

> "But Cotswold, be this spoke to th' onely praise of thee,
> That thou, of all the rest, the chosen soyle should bee,
> Faire Isis to bring forth, the mother of great Tames,
> With whose delicious brooks, by whose immortal streames
> Her greatnesse is begun."

Following the fortunes of the Seven Springs, you naturally enter with some degree of zeal into an expedition down the River Churn, and this you are able to do without losing sight of the excellent road between Cheltenham and Cirencester. The pretty little dancing trout stream runs hard by the highway, mostly through a succession of beautiful estates, and generally thickly overhung with alders and other bushes. Drayton hit off the character of the stream most accurately in calling it the "nimble-footed Churn;" and its picturesqueness, and musical flow between the wooded hills and through the fat meadows, as we near Cirencester, appeal to us, even on the score of sentiment. Surely it is more pleasant to identify this as the Thames than that commonplace current proceeding from the Thames Head series of springs. There is no necessity, however, to trace in detail the course of the beck-like Churn, by wooded uplands teeming with game, and through rustic villages and sequestered grounds. It runs through Rendcombe to North Cerney, down by Baunton, and through the once famous and still interesting town of Cirencester. The Fosse-way mentioned in connection with Thames Head was one of three great Roman roads which met here. Mentioned by Roman historians as Corinium and Cornovium, the strongly fortified city of Cirencester, the metropolis of a Roman province was, there is reason for believing, a considerable British town before it became a Roman centre. In the time of Henry VIII. the Roman wall surrounding the city might yet be traced, and, as the histories of Gloucestershire show, many Roman remains have from time to time been discovered here. The Churn sustains its brook-like character alongside the Cricklade road by Addington, South Cerney, and Hailstone Hill, and then within a mile of the town of Cricklade it unites with the other branch issuing from Thames Head, to which it is necessary briefly to return, in order to administer to it the justice already bestowed upon what we have agreed to regard as the rightful heir, namely, the Seven Springs stream, or River Churn.

The Thames and Severn Canal is so intimately associated with Thames Head, and so dominates that particular part of the country, that a few words respecting it may be spared. Indeed, it has dealings, directly or indirectly, with the Churn as well as with the Thames Head stream. Not far from Trewsbury Mead it gives a position to Thames Head Bridge, and the Canal lies within a few yards of the traditional spring. The first tributary is formed by a spring issuing from beneath the aqueduct, and not far from the Canal stands the single-arch watercourse, here illustrated as practically the first bridge over the Thames. The course of the Canal, however, almost immediately bears eastward, until it strikes the Churn, near

THE FIRST BRIDGE OVER THE THAMES.

which it keeps during the remainder of its independent career. The Thames and Severn Canal is an interesting fact which the present generation is in danger of forgetting. For many years the junction of "fair Sabrina" with "lordly Thames" was a burning question in the commercial worlds of London and Bristol. The merchants were much fascinated with the speculations in which they indulged. The Canal scheme was launched in a Bill in the reign of Charles II., and Mr. Hydrographer Moxon was engaged to survey the ground and prove to what extent the project was practicable. Pope, in the grandiloquent language of the time, in a famous letter written at Oakley, Lord Bathurst's country house at Cirencester, said, "I could pass whole days in describing the future and as yet visionary beauties that are to rise in these scenes: the palace that is to be built, the pavilions that are to glitter, the colonnades that are to adorn them; nay, more, the meeting of the Thames and Severn, which, when the noble owner has finer dreams than ordinary, are to be led into each other's embraces, through secret caverns of not above twelve to fifteen miles, till they rise and celebrate their marriage in the midst of an immense amphitheatre, which is to be the admiration of posterity a hundred years hence."

The Canal was completed sixty-eight years after this dream was indulged in, and in December, 1790, the first Canal boat, laden with coals, passed through. The Canal is a continuation of the Stroudwater system from the Severn to Wallbridge, near Stroud, and runs in a devious course from that point to Lechlade. It is thirty miles long, forty-two feet broad at the top, and thirty feet at the bottom. Between Stroud and Sapperton the water is raised 241 feet in less than eight miles, by means of locks.

Returning now to the lower spring of the Thames Head group, the course of this branch of the river may be traced from the expanded water giving growth to the ancient watercress bed, and receiving its first modest tributary rill from the spring proceeding from under the Canal aqueduct. Hence the brook meanders through meadows, and, near the railway, passes under the roadway. The village of Kemble lies half a mile back, and the stream passes under and alongside the road from Kemble to Ewen, beneath a considerable culvert, or trio of culverts. The first mill on the Thames was, in former times, at Ewen; but a cosy farmhouse now occupies the site, and the water which in former days set the drowsy wheel in motion is turned aside for sheep-washing purposes. The first mill now is Somerford Upper Mill, with its pretty setting of elm-trees, and charming rural surroundings. Somerford Keynes, on the elevated ground to the left, was bestowed as a marriage gift upon Ralph de Kaineto by Henry I., and an ancient charter granted to the Abbot Aldelm of Malmesbury contains the following incidental reference to the river as the Thames:—"Cujus vocabulum Temis juxta vadum qui appellatur Somerford."

Throughout the varied and interesting voyage upon which we are embarked the spires and towers of churches will be ever present, graceful and welcome features of many a landscape, now set upon a hill like a city which cannot be hid, now half concealed by mantling ivies, and shunning observation amongst the rugged elms which shelter their roofs and windows. The square tower of venerable All Saints, Somerford Keynes, is one of the earliest to claim attention as a typical parish church of rural England, very dignified in its age, and in its maternal relation to the cottages around the churchyard. The stream not far below this point serves another rustic mill, and a noticeable object later on is a homely foot-bridge supported by upright slabs of stone. At Ashton Keynes there are sundry small bridges spanning the current, soon to be sensibly increased in depth and width by Swill Brook, whose proportions have for the last mile of its course been not inferior to those of the Thames Head stream. The young River Thames was once, as is supposed, navigable to Water Hay Bridge for boats of moderate size; but this must have been before the aqueduct of the North Wilts Canal crossed it, or West Mill was built.

The ancient town of Cricklade reconciles any differences, and effectually ends all disputes as to individual claimants, by affording the two branches an opportunity of uniting their forces a short distance below the bridge. Here the Churn, from its north-western source, merges into the stream which has been always apparently called by the name of Thames. In this district it formed the boundary of the forest of Braden in the time of Canute. James Thorne, in his accurately written "Rambles by Rivers," does no injustice to the town of Cricklade when he speaks of it as dull to look at, dull to live in, and no less dull to talk about. There is, indeed, little about which to talk in connection with it, though we may, in passing, smile on recalling Drayton's words in his "Polyolbion"—

> "Cricklade whose great name yet vaunts that learned tongue
> Where to Great Britain first the sacred Muses sung,
> Who first were seated here at Isis' bounteous head,
> As telling that her fame should through the world be spread."

It has been alleged by certain authorities that a college here, founded by a school of ancient philosophers, became famous for its Greek learning, and hence the name

CRICKLADE.

of the town. It has also been insisted that a few miles down the river a rival college, maintained with similar success in the Latin interests, gave a name to the community which lived under its learned shadow; or, as Fuller said, "The Muses swam down the stream of the River Isis to be twenty miles nearer to the rising sun." In this manner fanciful writers have sought to explain the origin of the words Cricklade and Lechlade.

Cricklade is important to lovers of the Thames as being the first definite station on its upper waters. From the southern watershed come, besides the Swill Brook, the Dance and the Rey. The last-named, a contribution from the range of hills around Swindon, has been thought worthy by a few enthusiasts of the distinction of contesting with Thames Head and Seven Springs the responsibilities of parentage.

The Thames passes under a plank bridge at Cricklade, and becomes very shallow before receiving the tributaries above indicated. Thames tourists rarely push their explorations so high as Cricklade, which, other than two well-preserved specimens of fourteenth-century crosses (as is conjectured), and the prominent share in the landscape taken by St. Mary's Church and churchyard, as seen from Eisey foot-bridge, offers few attractions to the visitor. The scenery of the river hereto, and in truth for many a mile to come, is of a pleasing order, yet on a small and unpretentious scale. Farmhouses, with their surroundings of rick-yard and orchard; hamlets and villages in sleepy remove from the noisy world; a country house set in blooming gardens at odd intervals; pasture land and grain-fields, separated by old-fashioned hedges that are gay with flowers in spring and summer, with deeply-hued berries in the mellow autumn;—on every hand and at every turn these form the landscape. The river itself, so far, claims no particular notice, calls for no warmth of admiration. It makes no noise, performs no astonishing feats, inspires no terrors, but steals tranquilly through the meadows, and silently flows by the plentiful rushes which, unmolested, protect its banks in these remote reaches.

Castle Eaton Bridge, over four miles below Cricklade town, is perhaps the centre of the best of the rural scenery of this district, but it demands no special pause. The church tower, which shows boldly above the meadow, two miles farther on, belongs to Kempsford. King Harold was once a property-owner here. William the Conqueror subsequently gave the manor to one of his Norman soldiers, and, as was not uncommon in early days, it ultimately fell into possession of Mother Church, by whom, at the time of dissolution, it was disgorged and granted to the Thynne family. The edifice upon the river-side was probably built in the fourteenth century. There was also a castle at Kempsford, of which a fragment of window and a bit of wall remain, and a portion of the tower known as the Gunners' Room. The occupants of the Gunners' Room, when the building was habitable, had the advantage of looking out upon the river. A horseshoe nailed to the church-door long sustained the legend that when powerful Henry, Duke of Lancaster, the builder of the church, was quitting the place for ever, his horse cast a shoe, which the inhabitants nailed up in proud remembrance of the honour. The traces of an old weir, as we proceed downwards with such speed as the thickets of reeds and weeds will allow, if we are attempting the passage in a boat, remind us that in days when even inconsiderable streams were valuable as highways for barge and boat traffic the channel of the Thames was not so neglected as it will in these days be found. In other respects also its character has doubtless changed, as indicated by the stepping-stones across the foundation of the weir-sill, upon which the passenger, during summer level, may step dryshod from bank to bank. Below Hannington Bridge (with Highworth Church in the distance), the rushy pool, almost choked up with aquatic growths, still bears the name of Ham Weir. A sharp northerly bend in the river opposite the village of Upper Inglesham marks the separating-point of Berks and Wilts. Born in Gloucestershire, the Thames has latterly diverged for a brief excursion into Wilts, but now returns again, and from Kempsford until, a few miles onward, Oxfordshire is entered, it is the boundary between Gloucestershire and the southern counties of Wilts and Berks. The River Cole joins the Thames on the eastern bank. The little stream, in a charming Berkshire valley lying south, has given a name to Lord Radnor's mansion, Coleshill, famous as a perfect specimen of the style of Inigo Jones, and it arrives, accordingly, with some degree of repute on its own account.

For many years the highest weir upon the Thames was at Inglesham, known for

INGLESHAM ROUND HOUSE.

the picturesque church, with its bell-turret overlooking the river, and the remarkable piece of carved stone in the porch wall, but of more interest to us as the meeting-place of the Coln, the Thames and Severn Canal, and the Thames.

The Canal we have already glanced at. The Coln is a trout stream of some value, but it receives its fame principally from association with the town of Fairford; and Fairford is famous because of the painted church windows supposed to have been designed by Albert Dürer. The Round House at Inglesham is the final lock-house of the Thames and Severn Canal, and it indicates the stage at which the Thames becomes a river of importance. It is broader and deeper than heretofore, and was once navigable for barges of from thirty to seventy tons burden, drawing four feet of water; but a channel of such proportions has not existed for many years, and the river threatens to lose all pretensions to a waterway before long, unless its guardians dredge to more purpose than they have done in recent years. Once upon a time, when the ports of Bristol and London were not connected by railway, there was constant traffic through Inglesham Lock, and the Round House was a conspicuous beacon for the barges of the period; but the lock-keeper's berth has, it is needless to explain, in modern days become a sinecure. The angler at this meeting of the waters is the gainer, and his practised eye will mark the juncture as a probable haunt of the voracious pike.

The Highworth road is carried by a substantial one-arch bridge (to the left of the compass of the illustration on the next page) over the river at Lechlade, and in the fields, half a mile below, we arrive at the first lock on the Thames. There are a lock-house and garden to rest in, Thames Conservancy notices to be read, and ancient lock-keeping folk to talk with. It is a very old lock. In the natural order of things it cannot last much longer, and at no distant date, no doubt, it will give place to one of the more useful, but infinitely more prosaic, affairs of iron, with modern improvements in the machinery, which the Conservancy supplies when it is necessary to replace the original structures. The partly-decayed boards, the hand-rail rising from their outer edge, the lock gates patched many a time, and thinned in regard to their outer casing by many a winter flood, have done their work, and stand in weather-worn picturesqueness, all awry, doing their remaining duty as best they may. Looking westward, the spire of Lechlade Church, and, indeed, its tower and the greater part of the body of the building, shaded with ivy, make a very harmonious object of middle distance. The village, neat, substantial, and mature, rallies round it, with woods extending on either hand, and at its left flank stands the well-built, arched structure, which may be said to be the first bridge worthy of the name upon the river. It will be, perhaps, half a mile from the lock to the bridge, as the crow flies, but the Thames winds among the flat meadows in serpentine twists. Still farther on the line of the horizon, as, seated on the lever-beam used for opening the lock, we look westward, is a little picket of six poplars, marking the whereabouts of the solitary Round House, which substantially marks the limit of the navigable part of the Thames. The scene, thus comprising woods, village, church, bridge, and the long line of low trees terminated by poplars, is peculiarly English, and of a character that we shall see reproduced in endless variety—every prospect pleasing—until the last lock is reached at Teddington. But this is the first lock, 144 miles from London Bridge, and 125 from Teddington weir. Between the two there are many subjects of interest; but there is only one first lock, and upon this we may bestow closer attention than common. In the meadow is a big hawthorn on which the hips are already forming, and on a hot summer day the dairy kine will find shelter, lazily flicking the flies from their hides. Haycocks are plentiful on all sides. Yonder the men are hoisting a load of sweet-smelling hay upon the rick. Farther in the distance a late crop is falling in regular swathes; and when the gurgle of the water escaping from the dilapidated lock-gates moderates for a moment we can hear the mower whetting his scythe. The meadow upon which the first Thames Conservancy notice board has been erected, opposite the neat lock cottage, has not, it seems, been laid down for hay this year, and so offers a variety for the satisfaction of the artistic eye—to wit, masses of newly-shorn sheep lying on the grass, and dappled kine steadily feeding, what time the swallows and swifts are hawking around, and small birds warble in the reeds.

The Thames hereabouts, notwithstanding the prevailing officialism, is very

LECHLADE.—THE FIRST LOCK.

modest. At the period of my last visit I found that there had been no floods for eighteen months; and the river, as our boatman put it, had long wanted washing out. In its widest part in the neighbourhood of the first lock it is not more than twenty yards from bank to bank, and on each side there are thick margins of bullrush, sedge, and flag. The water is fairly deep, but cumbered with masses of weeds, out of which spring the tall rank stems of water-parsnip, while here and there small, compact, yellow water-lilies gleam like moidores on a silver plate. The river runs parallel with the lock channel under St. John's Bridge, a comparatively new and good-looking structure of one arch. It is about a mile from the town of Lechlade, and near it is the old-fashioned "Trout" Inn, still maintaining all the homely characteristics of the English countryside inn. A great sycamore overshadows it; there is an old-fashioned garden at the rear, and its little orchard, with a noble walnut-tree in the centre, offers a pathway to the pool. There is something in the semblance of a weir at St. John's Bridge, though it is of the most rudimentary kind, having fallen naturally into decay, and even into desuetude. Still, the small sluices are occasionally lifted, and serviceable streams are formed to keep the pool in motion, and prevent the patriarchal trout from giving notice to quit. At the end of St. John's Bridge tradition placed the priory of Black Canons, but of this there is no more substantial record than that of ink and paper. A few yards beyond the garden, on the left bank, the River Lech creeps into the Thames.

The regulation towpath exists from this point downwards, but for miles to come it is, like the boundaries of counties, a generally invisible line. The path is never trampled by horses, since barge traffic is unknown. For the towage of pleasure boats it is occasionally used, though these upper regions are rarely indeed penetrated by tourists. This is a pity; for there are a quietude and utter rurality about the river from Lechlade till within the precincts of Oxford that will be looked for in vain upon the busier haunts. Farther down there are glimpses—samples, so to speak—of what we have here in the bulk. We shall here find none of those notable Thames scenes that have been written about and painted from the olden times to the present day. Progressing from Lechlade downwards you feel altogether removed from the haunts of men. A patient angler sitting in his home-made boat under the overhanging boughs of a tree you will occasionally pass, and the presence of labourers toiling in the meadows informs you that this is not wholly a sleepy hollow. But river traffic, in the common meaning of the term, there is none. For a whole day you will probably not meet a boat; and there is no necessity to sigh for a lodge in some vast wilderness, some boundless contiguity of space, seeing that you have so excellent a substitute at hand. The solitude is, in truth, delightful. As you drop down between the banks you see drawn up in review order regiments of familiar friends—the dark glossy leaves of the water dock, bursting into seed in July; huge clumps of blue forget-me-nots that can only be plucked from a boat; ox-eyed daisies, well above high-water mark, gleaming as fixed stars in the floral firmament; the yellow-flowering great watercress, the purple loose-strife beginning to blossom, yellow iris, the white flowers of the common watercress, the pink persicaria, meadow-sweet, the comfreys, and sometimes a clump of arrowheads. From one to another flits the superb dragon-fly:—

> "One almost fancies that such happy things,
> With coloured hoods and richly-burnished wings,
> Are fairy folk, in splendid masquerade
> Disguised, as if of mortal folk afraid;
> Keeping their joyous pranks a mystery still,
> Lest glaring day should do their secrets ill."

A mile and a half of winding and uniformly narrow river brings us to rare old Buscot. The days of its weather-stained lock and weir are numbered too, but so long as they remain they will be, in conjunction, an object such as the artist loves, and a reminder to us all of other days when the world was not so jaded as now, when things were not so new, and when the ways of men were more primitive. There is a very fine tumbling bay on the farther side of the weir, and a sharp sweep of swiftly-running water coursing over a gravelly shallow, upon which the trout come out to feed at eventide, and the silvery dace and bleak poise in happy security during the long summer days. One is tempted naturally to land at the little village. The square, solid, countryfied-looking church tower, surrounded by old trees, and approached through a flower-garden, suggests, as your boat pauses at the lock, that it will be better to spend a quarter of an hour afoot than in the tedious process of passing through. Buscot is not a large or a pretentious place, but it is pleasant to look at, and deserves mention in passing, as giving name to the first weir of goodly size. A particularly pretty bit of the river, winding, tree-lined, and narrow, is followed by a long unromantic stretch.

Quaint, time-honoured Hart's Weir is so little a weir that the ordinary boat shoots the open half on the strength of a miniature rapid representing, at a summer level of the river, a fall of three or four inches only. The water there opens out into a wide bay that is purling rather than tumbling; and this is succeeded, in the ordinary course of nature, by a silted-up shallow, densely covered in their season with the white, yellow-eyed blossoms of the water crowfoot. Kelmscott on the north, and Eaton Hastings on the south, are the nearest villages. The banks of the Thames are now clearly defined by those trees which love to spread their branches to a river. The Thames does not, like the Loddon, encourage a monopoly of alder, but favours rather the familiar willow, the Lombardy poplar, and the plentiful hawthorn. Clumps of old elms, most picturesque of English trees when standing on village green, or as a rearguard to church or manor-house, vary the prospect. White and red wild roses are plentiful in the higher reaches of the river, and every turn of the narrow stream offers new combinations of wood and field:—

> "No tree in all the grove but has its charms,
> Though each its hue peculiar; paler some,
> And of a wannish grey; the willow such,
> And poplar, that with silver lines his leaf,
> And ash far-stretching his umbrageous arm;
> Of deeper green the elm; and deeper still,
> Lord of the woods, the long surviving oak."

After Hart's Weir the Thames settles down to an interval of insignificance. There is an indescribably soothing influence exercised by a river in the soft mood which characterises the Thames throughout. How pleasant it is to be simply moving with the current, which does so much and is heard so little. Loud, even, by comparison with the murmur of the waters, is the sighing of the breezes amongst the shock heads of the willows, and the silver shiver of the poplars. Under the spell of this influence the prosaic features of a reach like that between Hart's Weir and Radcot serve as a foil for the more lovely objects. Besides, occasional descents from the higher platforms of admiration, to which special points of interest are apt to summon you, give time for reflection and observation. Thus you will not be long in discovering that in these veritably upper reaches of the Thames what of animal life is still left may be seen without let or hindrance. The birds here are in no danger from the cockney fowler's gun. Amongst the water-fowl the most frequent appearance is that of the common moor-hen, which breeds as freely as ever, and still maintains its character as amongst the tamest of our wild birds. The coot is less often seen, but the heron will be often disturbed from its busy occupation on the shallows. Even in much-frequented reaches of the Thames the heron may still be descried at a distance, shy, watchful, and wary. On both of the days occupied by my voyage from Lechlade to Oxford I saw herons. They may have travelled twenty miles from the heronry for their nightly or early-morning forage, but you rarely can approach them within gunshot. The bird is most artful and shy at all times; but I have always fancied that the herons of the Thames Valley are the most wideawake of all. They hear the thump, thump of the rowlocks half a mile off, rise from their depredations, and wing their way slowly into the centre of a field; or perchance you may see one doing sentry on the upper boughs of a tall tree.

Between Hart's Weir and Radcot Bridge I descry three "herns" in one meadow that had been so disturbed by our gliding boat. As they stand motionless and lank in the fields, on a fence, or in the tree-tops, only a practised eye can identify them. In summer-time, though rarely, you put up a couple of wild ducks from the main river. The boating man, as may be supposed, meets with less bird life than the pedestrian, who, stealthily walking on the grass, will often obtain a passing flash of a kingfisher, or witness the alarmed flight of rarer birds. My July voyage brings me into constant companionship with troops of the wanton lapwing, in glorious plumage and full of noisy life; rooks, as a matter of course, busy, self-satisfied, and radiant in their blue-black vesture; swallows, swifts, sand martins, and reed warblers. The common sandpiper is about upon the shallows where the streams run swiftly, and the elegant water wagtails abound. At intervals throughout the day, near shrubby undergrowths and open meadows, the music of skylark, thrush, or blackbird charms the ear, though the eye seeks in vain the whereabouts of the performer. Four-footed creatures are few. The merry vole is an exception, and in some of the woods the cautious searcher may find squirrels in active play. The otter, seldom seen by the human eye in broad daylight, is plentiful enough in the earlier stages of the Thames, and of them, as of other wild creatures, it may be generally said that they are not so harried and wantonly destroyed as in the middle and lower parts of the river.

Radcot Bridge, of which we catch sight three miles from Hart's Weir, is understood to be one of the oldest bridges on the Thames, and its appearance is quite in character with this theory; moreover, it is an interesting piece of stonework, apart from its age, its three Gothic arches being curiously ribbed underneath. There is a very steep ascent to the crown, and over the centre arch is still preserved the socket in which, on the crest of bridges, the sacred cross was wont to be uplifted. There are, in point of fact, two bridges at Radcot, but the "real original" is the antique three-arched affair to the right, as we drop down. The river is here divided, a short cut to facilitate navigation and deepen the channel forming a new departure. The old stream wanders round, when the weeds will allow it, under the ribbed arches, leaving the channel of the new cut, like a newly-come tradesman who has a contempt for the old-fogeyish methods of the ancient inhabitants, to transact its business merrily, with promptness and despatch. For a couple of miles or so the

RADCOT BRIDGE.

Thames has now all the essential characteristics of a trout stream, with eddying pools and golden shallows, over which the water ripples at a moderate depth and at sparkling pace. In the hands of patient fish culturists and preservers this portion of the river might be made, no doubt, a trout stream; but *salmo fario* is as yet the least abundant of Thames fishes. What are called the coarse fish, or summer spawners, are on the contrary abundant, and most plentiful of all, under the willowy banks of the meadows, are the chub, which, for want of better game, afford passable sport to the fly-fisher, who, from the towpath, ought to be able to command any portion of the Thames at this stage of its course. In the wide deep pools marking the sites of old weirs, of which little trace but the piers remain, there should be, and as a matter of fact is, excellent angling for perch and pike.

Because the Thames has been so much praised, and so much the subject of picture and poem, it does not follow that it is all pleasing to the eye. After leaving

Radcot Bridge, and with the exception of these pools, once foaming and noisy with the action of the descending water, some indifferently furnished reaches have to be passed—reaches that are almost canal-like in the straightness of their course and in the uninteresting character of the low-lying land on either side. The country immediately bordering on the river is sparsely populated, and the world must revolve somewhat slowly for those who live there. Some indication of this may be gathered from the fact that at Rushy Lock, where there is a fine weir and pool, we had the pleasure of being our own lock keeper, opening the heavy gates, letting in the water, and releasing ourselves. The labour accomplished, a small urchin of six years of age was sharp enough to put in an appearance in time to take the toll; it was evident, however, that traffic was so unexpected that he alone had been left in charge, Indeed, during one whole day's progress we met but two boats.

Tadpole Bridge, a substantial structure of one span, between four and five miles below Radcot, carries the road from Bampton, where Phillips, the author of an almost forgotten work, "The Splendid Shilling," was born. The singular spire of Bampton Church is seen from Radcot Bridge, the view from which also includes Faringdon Hill, and some effective wooded heights around. The few tourists who make a pilgrimage from the source to the mouth of the Thames, or those who visit these stations for a sojourn of greater or lesser duration, turn aside from the river and visit both Faringdon and Bampton. At Faringdon any traces of the houses which withstood the hard knocks of the Cromwellian period are gone; nothing but the site remains, and that only as a vague tradition, of the castle built by the supporters of Queen Matilda, and pulled down by the supporters of King Stephen. Sir Edward Unton, who was Queen Elizabeth's ambassador at the French Court, is buried in the church.

Bampton, on the Oxfordshire side, is half town and half village, and it has an indirect connection with the Thames, although it is some distance from its banks, because the steeple, to which I have already referred, is a striking mark upon the landscape. Not without reason has the character of singularity been applied to it. From a square tower rises an octagon steeple, with belfry windows; pinnacles at each corner form basements of statues, and these are supported by slabs resting at right angles against the steeple. Skelton says that the church contains examples of almost every period of architecture, from the Conquest to the reign of George III. It has a fine Norman porch, an inner arch that is much admired, brasses, and a series of sculptures, probably work of the fifteenth century.

There is not much material for description in the next few miles. The river seems occasionally to lessen rather than increase in size, and right and left you look in vain for anything worthy of inquiry or admiration, save the comfortable old farmhouses and homesteads, environed by the usual clump of characteristic elms; and, at farther distance from the river, here and there a country mansion, secure in the privacy of its trim park, suggesting always the happy language of the poetess:—

"The stately homes of England, how beautiful they stand,
Amid their tall ancestral trees o'er all the pleasant land."

It is not till you have passed below Tadpole Bridge that the first beds of white lilies challenge attention. The lilies hitherto have been of the small yellow description; but now, in sheltered bays, thick beds of the gorgeous white variety shine gloriously from between the large glossy leaves. They are, fortunately, out of the line of every-day Thames traffic, and are so spared to develop to maturity in waters to which the steam-launch has not yet penetrated. About two miles below Tadpole one's attention cannot fail to be arrested by the high, skeleton-like, weather-worn bridge called Tenfoot Weir. This is another site of a weir long fallen into disuse. The wooden bridge consists of a central arch, or compartment, of staging set twenty feet high, with steep flights of steps on either side, the central division marking the outline of the old weir. A thatched cottage and thickly clustering willows in the bend which is here formed by the course of the river present an extremely picturesque variety to the monotonous character of the previous mile of the Thames. Another object of interest will be found a little lower down, at Duxford Ferry. Alongside a clump of willows lies a "sheer hulk," representing one of the long, narrow canal boats used when barges regularly plied where there is no water to float them now. Close to the blackened, slimy timbers of the wreck a promising family of calves cluster, as if pondering in bovine fancy over the former glories of the defunct craft and the industry it typified. A comfortable group of farm-buildings, thatched and tiled, nestles at the head of the ferry, which is not furnished with the usual horse-boat, for the simple reason that it may be crossed without any such assistance. A child might walk across on the hard gravelly shallow at ordinary times without being more than knee deep. It is, however, more than twenty yards wide, and the stream concentrates immediately afterwards to a width of not more than twenty feet; but it remains shallow.

There are two or three fords in the course of the next few miles, all of the same character. A rather notable one is that at Shifford. The legend runs that in the locality Alfred the Great held one of his earliest Parliaments, and there and then gathered "many thanes, many bishops, and many learned men, proud earls, and awful knights." This was to a great extent Canute's country. A mile or two on the Berkshire side of the river, near Bucklands, is kept the Pusey horn, given to the family by that king. The inscription upon it is, "I, King Knoude, give William Pewse this horne to holde by thy londe." There are some doubts, however, as to whether these letters are not of later date than the time of Canute. At Longworth, another village, there are the remains of an ancient encampment, Cherbury Camp, and here, it is said, a palace of Canute's once stood.

The River Windrush, a more considerable tributary than any previously received by the Thames, flows into the parent river from the north at Newbridge. The point of debouchment might, by reason of the weeds and rushes in the water and overhanging bushes of the banks, be easily overlooked by a casual observer, and the Windrush, in this peculiarity, closely resembles other feeders of the Thames, in looking its meanest where it offers its volume to the parent river. The Windrush is one of the

Cotswold brood, and at Bourton-on-the-Water it becomes a valuable trout stream. Great Barrington, whose freestone quarry furnished the stone for Christopher Wren's restoration of Westminster Abbey, is opposite the village of Windrush; the river afterwards enters Oxfordshire, and by the peculiar quality of its waters gives to the town of Witney a special pre-eminence in the whiteness of the blankets produced by its fulling mills. The river is thirty-five miles long from source to inlet to the Thames.

The oldest, and in truth the oldest looking, stone bridge on the Thames is called Newbridge, and this we approach below the place where Alfred held his Parliament. The bridge is an excellent sample of old English masonry. It has been New-

THE FERRY, BABLOCK HYTHE.

bridge for at least 600 years now, yet its groined arches and projecting piers seem as strong to-day as ever they have been. A public-house accommodates the traveller on either side of the bridge, one of them replacing a mill that perished for lack of customers. Strange to say, the river seems immediately to change its character when we have passed through these ancient arches. Not only is the presence of a couple of working barges, with gaily-painted posts of primary colours and vivid figure-subjects painted upon the panelling of the deck cabin aft, evidence that another era in the commercial character of the river is beginning, but the Thames, almost without warning, becomes wider and deeper, and altogether more like the Thames as we know it at the popular stations above the City boundaries, though of course it is still the Thames in miniature. The barges come in these days no farther than this station, and their business is mostly one not unconnected with coal. These boats, moored near the old bridge, seem to remind us that although heretofore we might have cherished the fancy that the Thames was almost an idyllic trout stream, lending grace to a rural district, it must henceforth be considered as being a recognised water highway with a mission that becomes more and more important as the distance to London Bridge is lessened. It is quite a remarkable change, and in a few moments your estimate of the river changes also. It is a thing now of laws and regulations. The very foliage on the banks seems to be of a more permanent character. Hitherto the Thames has been struggling with an indefinite career before it, winding through the meadows, streaming over the shallows, not quite certain whether it was to have a respectable position or not. But after Newbridge it has set up a substantial establishment, wherefore — Isis though it still may be and is called by the good Oxford people—it is to all intents and purposes Old Father Thames. We have seen the Seven Springs rill in its infancy and the Thames in its boyhood and lusty youth; here, however, it enters upon its early manhood.

CUMNOR CHURCHYARD.

Opposite Harrowden Hill, and to the west of Newbridge, Standlake Common may be explored by whosoever would benefit by its attractions, which, truth to tell, are very scanty. Snipe undoubtedly enjoy its boggy virtues during the winter; but the common is a marshy tract at best, and those who pass on to the village for the sake of its church of Early English architecture, and the farmhouse said to be built by John o' Gaunt and Joan his wife, do not care to linger there.

We shall pass two weir-pools, long disused, between Newbridge and Bablock Hythe, namely, Langley, or Ridge's Weir, and Ark, or Noah's Ark Weir. These and previous weirs referred to are of the very simplest kind, and, except in the two instances mentioned, perform their service independently of a lock. The object of this simple form of weir is to dam the river to the required height for such purposes as mill heads or navigation. The business is accomplished by the working of flood gates or paddles in grooves, and between rymers, to the sill at the bottom. In winter there may be a swift stream through the weirs, but, the weir paddles being withdrawn, there is very little fall. Shooting the weir stream—one of the

STANTON HARCOURT CHURCH.

adventurous feats of the upper navigation—is an amusement unknown below Oxford, and at times it is not without its risks.

Although Bablock Hythe by road is not much more than five miles from Oxford, the circuitous voyage by Thames is twelve miles. Bablock Hythe is a well-known station on the Upper Thames, albeit it does not boast the rank of hamlet or village, and has for the accommodation of man and beast only one of the small old-fashioned inns of the humblest sort, where the rooms are low, the beams big and solid, the floors flagged, and the apartments fitted up with all manner of three-corner cupboards and antique settles. The great ferry-boat, however, gives it a decided position of importance, and it is known to Thames tourists principally as the starting-point for visiting either Cumnor or Stanton Harcourt. Most people probably go to Cumnor from Oxford, the distance being only about three miles; but many are glad to make it an excuse for halting on the somewhat monotonous ascent of the river. The reader needs scarcely to be reminded that Cumnor Place has been made immortal by the pages of Sir Walter Scott, and that the sorrows of Amy Robsart have been wept over by the English-speaking race in all parts of the world. There is an inn at Cumnor still called after that hostelry over which Giles Gosling firmly ruled, and in the church there is a monument sacred to the virtues of Tony Fire-the-fagot and his family, who are thus handed down to posterity in a far different character from that suggested by Sir Walter as pertaining to the tool of the villain Varney. Cumnor is on the Berkshire side, and on the Oxford side is Stanton Harcourt, visited for the sake of the remains of its ancient mansion, and its fine church.

ABOVE OXFORD.

Visitors, probably, would not make the journey exclusively in the interests of either one or the other, nor of the two large upright stones called the Devil's Quoits, which one historian conjectures were erected to commemorate the battle fought in 614 between the Saxons and the Britons.

The real attraction of Stanton Harcourt is historical, and historical in several degrees. It was one of the vast estates which fell as loot to the half-brother of William the Conqueror, and was evidently a considerable possession. For more than 600 years the manor continued in the Harcourt family. Little is left of the grand mansion in which the lords of Stanton Harcourt dwelt. The Harcourt family gave it up as a place of residence towards the close of the seventeenth century, and it fell forthwith to decay. With the exception of the porter's lodge, the arms on each side of the gate, showing that it was erected by Sir Simon Harcourt, who died in 1547, and some upper rooms in the small remaining part of the house adjoining the kitchen, are all that remain. But there is a more recent historical interest attaching to Stanton Harcourt; in a habitable suite of rooms in the deserted mansion Pope passed the greater part of two summers, and to this day the principal apartment bears the name of Pope's study. The little man required quiet and retirement during his translation of the Fifth Book of Homer, and upon one of the panes of glass he wrote, in the year 1718, "Alexander Pope finished here the fifth volume of Homer." The Harcourts, however, removed this pane to Nuneham Courtney, where it is preserved—a piece of red stained glass, six inches by two. The old Stanton Harcourt kitchen, converted to modern uses, was always a curiosity, and Dr. Plott, the Oxford historian, says of it, "It is so strangely unusual that, by way of riddle, one may truly call it either a kitchen within a chimney or a kitchen without one, for below it is nothing but a large square, and octangular above, ascending like a tower, the fires being made against the walls, and the smoke climbing up them without any tunnels or disturbance to the cooks, which, being stopped by a

KYNSHAM WEIR.

large conical roof at the top, goes out at loopholes on every side, according as how the wind sets, the loopholes at the side next the wind being shut by folding doors, and the adverse side open."

The visitor at Stanton Harcourt should certainly not neglect an inspection of the beautiful church, said to be the finest in the country. It is cruciform in shape, and has a massive tower. The nave is Norman, of about the twelfth century, and according to "a custom established there time immemorial" the men entered through a large, and the women through a small, doorway. A wooden roof to the nave is understood to have been added in the fourteenth century, while the chancel, transepts, and tower arches are of the thirteenth. The oaken rood screen is reputed to be the oldest wooden partition of the kind in the country. The Harcourt aisle or chapel, erected about the same time as the mansion, is an example of the enriched Perpendicular style of Henry VII., and it is still the burial-place of the ancient Harcourt family. In the chapel, as in the body of the church, are several interesting monuments, and one of them is famous. In Gough's sepulchral monuments, where it is engraved, the following description is given:—

CROSS AT EYNSHAM.

"This monument of Sir Robert Harcourt of that place, Knight of the Garter, ancestor of the Earl of Harcourt; and Margaret his wife, daughter of Sir John Byron, of Clayton, Lancashire, Knight, ancestor of Lord Byron. He was Sheriff of Lancashire and Warwickshire, 1445, elected Knight of the Garter 1463, commissioned with Richard Neville, Earl of Warwick, and others, to treat of a peace between Edward IV. and Louis XI. of France, 1467, and was slain on the part of the House of York, by the Staffords, of the Lancastrian party, November 14th, 1472. His figure represents him in his hair, gorget of mail, plated armour, strapped at the elbows and wrists, large hilted sword at left side, dagger at right, his belt charged with oak leaves, hands bare, a kind of ruffle turned back at his wrists, shoes of scaled armour, order of Garter on left leg, and over all the mantle of the Garter, with a rich cape and cordon; his head reclines on a helmet, with his crest, a swan; at his feet a lion. His lady, habited in a veil head-dress falling back, has a mantle and surcoat and cordon, and a kind of short apron, long sleeves fastened in a singular manner at the waist, and the Order of the Garter round her left arm; her feet are partly wrapped up in her mantle."

The Thames takes a northerly course from Bablock Hythe, and winds and doubles in such contortions that in one part a strip of not more than twelve yards of meadow separates two reaches of considerable length.

A high, wide wooden bridge, bearing the name of Skinner's Weir, now crosses our course, and soon we come to Pinkhill Lock, so called from a farm of that name in the neighbourhood. The weir is a new one, a great contrast in its severe and formal cut to the weather-worn structures to which we have been accustomed. The lock-house is quite a dainty cottage, and the garden one of the prettiest to be found along the Thames. The lock garden is generally a winsome little preserve, with its kitchen garden, flower-beds, sometimes a beehive, its stack of fagots, and a general air of rusticity; but the lock-keeper, or probably his wife, at Pinkhill Weir has devoted special care and attention to a flower-bed running the whole length of the lock, which I found to be bordered by a blaze of summer flowers, prominent amongst which were white and blue cornflowers. From the lock bridge a commanding view is obtained of the hilly country to the right, and the woods and copses around its base, and straggling to the top.

The telegraph wires along Eynsham Road detract considerably from the rural flavour of the surroundings, and Eynsham Bridge itself does not look so old as it really is. It is a very conspicuous, and, indeed, handsome structure, with eight arches and a liberal amount of balustrading in the central divisions of the parapets. Eynsham, Ensham, Eynesham, or Emsham, has a history which goes beyond the Conquest, and it is by right, therefore, that the bridge is named after the village, though its real name, as decided by the Ordnance Map, is Swinford Bridge. Early in the eleventh century an abbey was founded here by the then Earl of Cornwall, and Ethelred, the reigning king, signed the privilege of liberty with the sign of the holy cross. At the Dissolution the abbey and its site passed into the ownership of the Stanley family, but no ruins have been preserved. Ensham, or Eynsham Cross, stands in the market-place of the village, opposite the church. The bridge, as we now see it, was built about sixty years ago. The village is pleasantly situated on rising ground.

A little below the bridge the picturesque materials of the weir are stored when not in use, and the rymers are piled in a stack close to the spot where they sometimes even now do effective service. Your boat passes through, however, generally without let or hindrance. A little farther on the Evenlode enters the Thames. Like its predecessors, it seems a poor insignificant stream as it delivers its waters through a reedy mouth to the Thames; but it has itself received the River Glyme, which passes through Woodstock and Blenheim Park, and feeds the large lake, now choked with weeds. The Evenlode is the last of the Cotswold offerings thus embodied in verse by Drayton :—

> "Clear Colne and lively Leech have down from Cotswold's plain,
> At Lechlade linking hands, come likewise to support
> The mother of great Thames. When, seeing the resort,

> From Cotswold Windrush scrowers; and with herself doth cast
> The train to overtake; and therefore hies her fast
> Through the Oxfordian fields; when (as the last of all
> Those floods that into Thames out of our Cotswold fall,
> And farthest unto the north) bright Elulode forth doth beare."

Woodstock is not more than four miles from Eynsham, but it is generally reached from Oxford. The river winds now round the foot of Witham Hill, and we are on close terms with the outskirts of the immense wood through which one could walk for eight miles before losing its shade. The portion that comes to within a few yards of the Thames consists of oak-trees, with an occasional ash, and as we halt to sit a while under the umbrageous canopy we receive as a salute the cooing of doves, agreeable contrast to the reception, a few miles higher up, conveyed in the harsh squawk of a couple of herons. Longfellow might have sat amongst these identical brackens when he wrote:—

> "But when sultry suns are high,
> Underneath the oak I lie,
> As it shades the water's edge,
> And I mark my line, away
> In the wheeling eddy play,
> Tangling with the river sedge."

The Thames describes a sharp horseshoe curve round the base of the hill. From the bank a fine view across the flat is obtained of Cassington Church spire, and of the last mill on the River Evenlode, making for the Thames midway between the bridge and Hagley Pool. The paucity of pleasure-boats on the river between Lechlade and Bablock Hythe may be attributed to the great weediness of the river, rendering it sometimes almost impassable; also to the prevalence of shallows, and the absence of anything particular to see, and the all-important consideration that there are few hotels to stop at. There is not a riverside house of call between the little cottage inn at Bablock Hythe and Godstow. An occasional steam-launch finds its way from Oxford up the Canal and into the Thames, by way of the Wolvercott Paper Mill; but this unpleasant type of vessel is very rarely seen so far up, since the forests of aquatic undergrowth are the reverse of favourable for the working of the screw.

What the steam-launcher, however, loses is gained by the angler. This mild sportsman I found to be very much in evidence below Bablock Hythe. Here at any rate he was able to pursue his pastime in peace; and the frequency with which he appeared on the bank from Eynsham downwards gives me an opportunity of interjecting a few timely remarks upon the Thames as a resort of fishermen. The professional fisherman, as we know him at Richmond, Maidenhead, or Marlow, with his punt, Windsor chair, and ground-bait, is unknown in the upper reaches of the river; but the fish are there. Although anglers have multiplied a hundredfold within the last half-century, the angling in the River Thames at the present moment is better than it has been at any time during the present generation. It is not to be hoped, with any reasonable confidence, that the efforts now being made by the Thames Angling Preservation Society to convert the Thames once more into a salmon river will be successful; and any one who makes personal acquaintance with the source of the Thames, and marks the character

OXFORD, FROM GODSTOW.

of the contributory streams, will be prudent in entertaining a doubt as to whether there are now breeding-grounds suitable, even if fish could be induced once more to run up through the filth of the Pool from the sea. The alleged scarcity of Thames trout is very often put down to the excessive disturbance caused by steam-launches, and the traffic by pleasure-boats upon all the reaches of the Thames, from Teddington Lock to Oxford. It is somewhat strange, therefore, that the higher you ascend the Thames the fewer become the Thames trout. There are a few large fish in most of the deep wide pieces that were once weir-pools, or that still may be so, between Lechlade and Oxford; but the water is too sluggish to encourage them much, and trout, with the exception of truants from Lech, Coln, or Windrush, are, therefore, few and far between. Pike, on the other hand, are more numerous, if not of so large an average size as those caught lower down. The Thames, from the start, abounds in chub, bleak, barbel, gudgeon, roach, dace, and perch; bream, carp, and tench are partial in their haunts. But the river above Oxford is not so accessible as the great body of modern anglers would require, and hence it comes to pass that these remote waters are little visited except by the local disciples of Isaac Walton. The weeds are, after a fashion, annually cut by the Thames Conservancy where their growth would be a serious hindrance, but otherwise they are not kept down, save by the uncertain operations of winter frosts and floods. The right of fishing is generally, above Oxford, claimed by the riparian proprietors, or their tenants.

Soon after putting Bablock Hythe in our wake, the flat country, varied by only occasional uplands, which had been the rule since leaving Lechlade, is exchanged for a bolder type of scenery, as, for example, the fine wooded eminence rising before us. This is Witham, of which we shall see a good deal, now from one point, and now from another, as we near the City of Learning. It requires no guide or guide-book to inform us that from the summit a widespread view is obtained of the valley of the Thames.

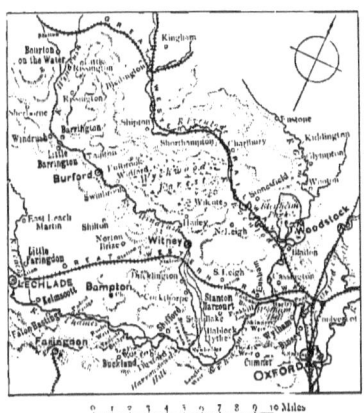

*THE THAMES FROM LECHLADE TO OXFORD.

Hitherto we have looked in vain for the typical eyot. With the exception of one small islet below Hart's Number Two, or Langley Weir, there has been nothing in the shape of an island until we arrive at Hagley Pool, where the first solitary island appears. The picture from here is exceptionally interesting. A rustic bridge spans a backwater trending towards Witham Mill, and in the direction of Oxford. The thickset woods stand out in prominent relief, and another farmhouse of the higher class, surrounded by ricks, appears to the left. Hagley Pool, which is merely a lake-like widening of the water at the bend, is covered with the yellow waterlilies. Three miles from Eynsham we are at Godstow Bridge. The spire of Cassington Church, a conspicuous landmark on the left hand throughout, is a pleasanter object by far than the tall chimneys on the right, which are not redeemed by the rows of poplars that would fain hide them. It is unfortunate, but true, that the first glimpses we get of the spires of Oxford are in conjunction with the tall red-brick chimney and not elegant University paper-mills. While following the bend at the broad part of the river the public buildings of beautiful Oxford open one by one into view, but again disappear temporarily at the next bend, at the head of which stands King's Weir. This serves as much the purposes of a lock as a weir, its gates opening when necessary to admit the passage of larger craft than those which can be conveyed over the rollers supplied for pleasure-boats. The river from the pool for some distance is almost choked with weeds, very narrow, and of hardly sufficient depth at low water to admit the passage of an ordinary pleasure-boat.

Godstow at once suggests the story, often told and always interesting, of fair Rosamond. The lady gives a flavour also to Woodstock, some eight miles distant.

The wrongs and the rights of Mistress Rosamond will never in this world be accurately known, but that she was poisoned by jealous Queen Eleanor at Woodstock, and that she was the mistress of Queen Eleanor's husband, Henry II., are facts which no one dare deny. According to Lord Lyttleton, Henry II. met the frail daughter of Walter, Lord Clifford at Godstow, in 1149, on his return from Carlisle, the lady being at the time, in accordance with the custom of the age, placed amongst the nuns to be educated. The nunnery is still known by the ivy-clad walls which remain on its site. It was a nunnery of the Benedictines, consecrated in the presence of King Stephen and his Queen in the year 1138. The nunnery was dispossessed, and has crumbled to ruins, but the brave river passes by even as in the olden times before Henry VIII., the spoiler, gave the house to his physician, Dr. George Owen. There was another nunnery at the foot of Witham Hill, but that was an older establishment, which existed as early as 690, on the spot where the Earls of Abingdon have their seat, partly built, it is understood, by the stones of Godstow, even as the modern buildings at Stanton Harcourt are supposed to have been erected from the stones with which the original mansion was constructed. The ruins of Godstow Nunnery, such as they are, catch one's eye first from the river. It may be that the pathetic romance touching the silken thread and the bowl of poison is not, as many hold, founded upon fact; but we cannot be equally sceptical with regard to Rosamond's connection with Godstow. She retired to the nunnery to pass the remainder of her days, after the marriage of the king, in seclusion. She died, and was buried in the choir, opposite the high altar, and Henry raised a grand monument to her memory. The nuns forgot the frailty of the lady, remembering rather the manner in which she had enriched the establishment, and the tokens of favour they had received from the king on her account; and we read that her remains were treated with much honour by the sisters, who hung a pall of silk over her tomb, and set it about with lighted tapers. This chronic honour was put an end to by Hugh, Bishop of Lincoln, who, going to the nunnery and requesting to know why one particular tomb should be so much honoured, was informed that it was the tomb of Rosamond, sometime leman to Henry II. In order that the nuns might not be led astray by having her example constantly set before them, and that other women might beware, poor Rosamond's bones were cast out of the church; but they were brought back again by the nuns, and wrapped in perfumed leather.

The farther arch of the old bridge at Godstow has been removed to admit of various improvements being carried out in one branch of the stream, which here divides, and in order to widen the structure; but the two arches of the ascent from the right-hand side remain as they were, and the well-known "Trout" Inn at Godstow retains all its characteristics of creepers, flowers, tiled roof, and pleasant waterside seats. A full view of Oxford, set back beyond the farthest confines of Port Meadow is obtained, while the smell of the roses in the pretty garden of the time-honoured "Trout" Inn still lingers about us. The village of Wolvercott lies to the left, and at the other end of the mill-stream, the entrance to which was noticed just above the King's Weir. Close by the ivy-covered gable of the nunnery a new weir is being erected, and it may be added that in the excavations incidental to the work four old stone coffins were discovered in the summer of 1885.

Passing by the village of Binsey, where in 730 there was a chapel constructed with dark room for the most stubborn sort of sisters, and where the saints caused St. Margaret's Well to be opened, in order that people coming there to ease their burdened souls might be rid of their diseases, one feels that the first stage of a voyage down the Thames is pleasantly terminated by the noble array of pinnacles, towers, and spires across Port Meadow, presented as a free common to the city by William the Conqueror, and so to this day preserved. The towers and spires have an imposing effect, with Shotover Hill behind. The most prominent objects are St. Philip's and St. James's Church, the Roman Catholic Church, the Observatory, the Radcliff, the Sheldonian, St. Mary's, All Saints', Tom Tower and the Cathedral, and, nestling down among the trees, the square grey tower of Oxford Castle. To the right is the "Perch" Inn at Binsey, and as you pass this Binsey Common opens out in the same direction, and there are once more the wooded slopes of the Witham Hills, which we have had in view for the last eight miles. The River Thames round Port Meadow is more disgracefully weedy and neglected than any other portion of its course. Beyond Binsey is Medley Manor House, at one time an oratory attached to Godstow, a place where any of the devotees, in case they were detained from the city or on their journey to Abingdon, could rest for the night, without going on to the nunnery. The flocks of geese in hundreds, just now giving the signal of rain, on the edge of Port Meadow, opposite Binsey Common, may still lead us to think that we are in the rural parts described on previous pages; but down yonder, on the other side of the cut leading to Medley Weir, are a fleet of ugly house-boats. There is also a semicircular iron bridge across the cut, and we are brought face to face with the fact that the next mile and a half of river will be essentially townified and crowded.

The division of the river at Medley leaves the business of practical navigation to the straight cut, and the original Thames, once flowing by the site of Bewley Abbey, will probably be soon choked out of existence. In succession now follow in a prosy catalogue Medley Weir, the Four Streams, the Railway, the Canal, Seven Bridges Road, and Osney Lock and Mill. A hoary gateway and fragment of wall, with its Perpendicular window absorbed in the mill fabric, are all that remain of Osney Abbey, the powerful and magnificent, whose abbots were peers of Parliament. One hastens under the railway bridge, and looks aside from the gasworks, knowing that beyond Folly Bridge a new phase of Thames life will begin for the intelligent voyager.

W. Senior.

THE BARGES.

CHAPTER II.

OXFORD TO ABINGDON.

Oxford, from the Upper River; the New Town—The Courses of the River, from Medley Weir to Folly Bridge—The Houses of the Regulars and Friars—The University and Parish Churches—The Halls and Colleges of the Seculars, from the Thirteenth Century to the Reformation—Jacobean Oxford—Classic Oxford—Convenient Oxford—The Architectural Revival—The Undergraduate Revival—The River below Folly Bridge, and the Invention of Rowing—The Navigation Shape of the River—Floods—The Barges—Iffley—Littlemore—Kennington—Radley—Sandford—Nuneham.

THE traveller down stream, who looks for Oxford across the flats of Port Meadow, is aware of a large town, dusky red in colour, skirted by a canal and a railway, and dominated by the slim brick bell-tower of a church, one of the pangs of the architectural renaissance. There, beyond the dingy quarter called Jericho, one may stray through many streets of villas in the Middle Victorian taste, by flower-beds gay with the geranium and calceolaria. Little is wanting that would be found in St. John's Wood or West Kensington. For this is, in late after-growth, that town of Oxford that meant to be like London, and was like London, before the University came to interfere. It had its Norman castle, its Gild Merchant, its charter, as good as those of London. It was a place where Parliaments met. It had a palace of the kings, and a rich Jewry, and a great mind to trade. But the University sprang up and choked these things. London never had a real University, but only colleges for students of Common Law, and so flourished. In Oxford the town went under, and the University was everything. The "nations" came, and after long war reduced the natives to servitude. But the wheel has turned. The down-trodden race is quickly hiding the University with its new towns of houses and churches, and the very University has lost the monastic rule that allowed its members to camp as an alien garrison in the place. Now they are surely being wrought into the fabric of the town.

For the University there exist two rivers; one, The River, below Folly Bridge, the other, The Upper River, above Medley Weir. Between the two there is not one stream, but many. The river goes out of itself and returns into itself again. And in this division it suffers various fortunes. It goes far afield and grows forget-me-nots. It turns mill-wheels, and is a servant of breweries. It is locked and sluiced for the passage of barges. It is constrained and laid away in low and discouraged quarters, where it keeps company with people out of repair, with philanthropic enterprises, with aimless smells, with exhausted dust, with retired hansom cabs. It is beguiled into obscure cuts for bathing. It is imprisoned under streets. And when it comes to itself again it is not allowed to have its name, but is called by the vain sound of Isis.

The two main branches of the stream enclose a space rather over a mile in length, and roughly of the shape of a slim ewer, with a handle broken off near the top, that is at Medley Weir. There is a minor junction of the streams by a cut across the narrowed neck of the ewer opposite Worcester Garden. The upper of the two islands thus formed is given up to meadows and the two railway lines. The lower island, Osney, holds the two railway stations, and the continuation southwards of the Great Western. South of the stations, and at right angles to the railway, the Seven Bridges Road runs out towards Botley. South of this again St. Thomas's Church lies east, and St. Mary's Cemetery and Osney Mill west of the railway havoc. The rest is meadow and garden land, scored with the streets of old and new settlements, and cracked by lesser dykes and courses of the stream.

The eastern branch, after defining the upper rim of the ewer, turns sharp southwards, and, keeping company with the canal, skirts red Oxford and Worcester Gardens. It is here that its interest begins. A little way above the first, or Hythe Bridge, a fresh division takes place, and a narrow irregular strip of low island is formed, running under three bridges to the Castle Mill, and below that occupied by breweries for some hundred yards. Now it is only on the upper stretch of this island down to the Castle Mill that any attempt is made by the town to come to public and pleasant terms with its river. The attempt is a shy one. The treatment is on a humbler scale than that of the River Witham at Lincoln. The Fishers Row of low houses—some new, some old, and one or two remarkable—straggles along a narrow quay, arched over by the bridges. In the doubled stream, where it fronts the houses, fleets of old punts lie moored to their poles among the choking weeds; not the varnished toys of the Cherwell, but the craft native to these shallow standing waters, as the gondola to the lagoons of Venice. At the back of the houses, their gardens abutting upon it in all variety of confusion and decay, moves a furtive and even feebler stream. There is a wealth of matter here for the artist to rescue from its odours; grey walls that have seen better days and other uses, bricks roughcast and timber, willow leaves and fluttering clothes, the most old and various dirt. All this is only to be won by glimpses from the bridges, or from the hospitality of back pigstyes and the like; and it is only just to add that the tenants of this picturesque quarter — people, pigs, and ducks — show to the curious visitor an unvarying courtesy. The best bit was till lately to be seen from Pacey's Bridge, the second in order down stream. Just there a house is bracketed out over the water, with windows disposed in graceful bays. But the jealousy that keeps the stream secret has shut away that last easy view, on the one side with a shop astride the water, on the other with a more wilful screen. Hythe Bridge is a poor new thing; Pacey's Bridge is defaced with a new top. The next bridge brings us to the Castle and the Castle Mill, the very heart of the old town; the Castle older than the University, the Mill of older foundation than the Castle. Then follow breweries, not without charm, but reticent about the river. Just below the Swan Brewery the streams come together again at a point marked by a summer-house; but it is only for a fresh separation. From a garden in Chapel Place may be seen the point of division; but one branch is now built over. Its name is the Trill Mill Stream, and it runs behind Paradise Square, and round by way of Rose Place, across St. Aldate's. Then it comes to light again behind the houses, and skirts Christ Church Meadows, to join the river near Folly Bridge. The other branch takes a stealthy course round the low quarter between Paradise Square and the gasworks. They are least ashamed of it in Abbey Place. From that point onward it shows at the end of poor little streets, with meadows and willows beyond. From one of these — Blackfriars Road — a bridge crosses to the bathing-cut, which rounds the base of our ewer, and leads into the navigation stream. At the tail of an island formed by the cut the navigation stream itself comes in, and the united water bends round the gasworks, and so to Folly Bridge, past some broken little gardens and backs of houses in Thames Street. Folly Bridge is as poor as the other Thames bridges in Oxford. It replaces the old Norman *Grand Pont* with its forty arches, and Friar Bacon's Study over the further end. A top storey added to the "Study" was the "Folly." There is another now almost in the same spot, built by a money-lender.

Of the navigation stream in its course from Medley Weir there is less to say. At the neck of the ewer, at the point called Four Streams, it goes so far as to form a regular cross. One of the arms is the cut already mentioned running towards Worcester. The opposite arm is known as the Old Navigation Stream, and runs out in a great loop under the Binsey Road and the Seven Bridges Road at New

OXFORD, FROM HEADINGTON HILL.

Botley, and back to the present navigation stream at the base of the ewer. A smaller concentric loop leaves the stream at the first bridge beyond the station, throws off a branch to join the outer loop at the Binsey Road bridge, and returns at Osney Mill. Here, just by the mill, there is a lock on the navigation stream. The island formed by the mill-stream and the lock runs down a hundred yards or so, and on the face of the island, made by the loop above, there is a meaner repetition of the Fishers Row. It may clear the maze a little to think of the two mills and islands, and quays balancing on opposite sides of the ewer.

But this is not all. We have still to account for a stream that left the Thames at Hagley Pool, above Godstow. From that point it describes a yet wider loop, passing first by Witham, then under the Seven Bridges Road at Botley, and on by the two Hinkseys. At Clasper's Boathouse under the Long Bridges it is reinforced by a fresh offset from the main stream, and does not return again till just above Rose Island by Kennington. The old men on the river have been heard to say that this branch from Clasper's to Kennington used to be the main stream for barges, and it is quite possible, for the Long Bridges and new towpath only date from the end of last century. The Hinksey Stream is not navigable throughout, because of two mills on its lower reaches. The low Cumnor Heights behind make a limit to the wandering and division of the water; but the whole flat between this boundary on the west and that of the Oxford Canal on the east is an amphibious country, now lake, now labyrinth.

It will have been observed how little the obscure region of the river we have traversed has to do with the University of our time. One is invited to think how the river of Oxford has come to be treated so; why the colleges shun it and give it over to railways and slums. And again, if one regards the college quarter with any attention, one is forced to ask by what steps the plan of building and habit of life we know as a college came to be as it is out of the old Benedictine conventual schools. What were the links of building between St. Frideswide's and Merton, and what has become of them?

The answer to the first question, and partly to the second, is that a more magnificent and more richly significant Oxford than the present once occupied the isle of Osney and the river quarter now so degraded;[*] but all that proper fortune of the river, all that beauty and history, has been incredibly blotted out, leaving only its first and last links in St. Frideswide's and Worcester, together with a few names and inconsiderable fragments. The buildings of Oxford are a story whose mutilated preface is followed by a great gap where the opening chapters should be. A line here and there marks the interval, and when the tale is taken up again it is abruptly and in a changed temper. Quickly it runs to a fluent mannerism that makes a great bulk of the text. Then it proceeds in a classical version till the time when our own century began to spell and imitate the archaic forms.

---

[*] On this subject *see* Goldie: "A Bygone Oxford;" and Fletcher: "The Blackfriars in Oxford."

The town before the University is better represented than the following period by its castle and parish churches. The ancient St. Frideswide's remains as Christ Church. But greater churches than St. Frideswide's, one of them, that of the Franciscans, twice as long, have been taken clean away, and not a stone remains to stand for the Dominican and Franciscan houses that moulded the early University. We must give a little space to this, and to another missing chapter, and then briefly read the rest of the story.

When the two great mendicant orders arrived early in the thirteenth century, there was already, besides the old foundation of St. Frideswide, at that time a house of Austin canons, the great monastic foundation of Osney, dating from early Norman times. In its church, over the tomb of the foundress Edith, English wife of the second lord of the Castle, was painted a tree full of chattering pies, whose voice assailed her in her walks. Her confessor knew them for souls in Purgatory, and the canons were installed to pray for them. By the time the friars came to Oxford the chattering souls were perhaps as much thought of as are the souls of those killed in the French wars of the fifteenth century by the Fellows of All Souls College now. At the Dissolution the great Abbey Church had a chance of safety. For a short time it was the cathedral church of the new diocese of Oxford. But that fortune passed to St. Frideswide's, and no one translated Osney Abbey into a college. All that is left of it now is an archway and part of a barn among the buildings of the Mill.

The two great orders of friars settled finally near one another on opposite sides of the Trill Mill Stream. The Dominicans were first in the field, and for a time encamped near the schools and the Jewry, with designs on both. They built a hospital for converted Jews, which afterwards was used as the Town Hall, and it is on record that at one time they had two Jews in the *Domus Conversorum*, but one of them, an acolyte, afterwards suffered a relapse. Soon, however, the friars migrated to the damp riverside, as a place more favourable to rheumatism and ague, just as in London they went from Holborn to Blackfriars. It was the happiest fact that the mendicant orders coming to towns in their young and ascetic days settled in outcast and uninviting quarters, and covered them, as they grew in riches, with pleasant gardens and splendid buildings. But the Black Friars of Oxford, like those of London, are only remembered now by names of streets, and of that bend of the river by the gasworks still known as Preachers' Pool.

The Grey Friars are even more completely gone. The Paradise given them by a pious lady has given its name to a square, but the groves and buildings of the Minorites have not even left a name to the streets that have replaced them. There must be many of the friars still below ground in their coffins; in the courtyards behind the dismal streets there are glimpses of provoking walls, but with no speaking stones; and in the wall that divides the garden of Trinity from Parks Street lie many old stones incognito, brought from the quarry of the Friars Preachers and the Friars Minor.

The White Friars, or Carmelites, had no better fate. Edward II., flying from Bannockburn with his Carmelite confessor, vowed a house to Our Lady of the White Friars if he should cross the Border in safety. To redeem the vow he gave over his palace of Beaumont to the Carmelites. Beaumont Street runs through the site. Some of the stones are in Laud's new quad at St. John's.

Of another great house a small witness remains. Above Hythe Bridge, the way by Fishers Row is continued from the tip of the eyot across a little bridge, and thence runs for a space alongside an ancient wall. This was a boundary wall of Rewley Abbey, the great house of the order of Citeaux. Almost covered by the wood stores of a wheelwright is a doorway with carved spandrils, and a label ending in sculptured heads. The wheelwrights, whose sheds lean against the old wall, show, a wooden peak, the last vestige of a "summer-house" lately pulled down. Only the other day they came upon a well in the garden behind. The London and North-Western Station occupies the site of the chief buildings. Before it was put up the remains were considerable. Rewley and Osney Abbeys between them accounted for most of the Osney island.

But the most speaking memorial of this lost University is that side of Worcester that still remains of old Gloucester Hall. Benedictine novices, from the many houses of the order scattered over England, were numerous among the early students of Oxford. The Benedictines were rich, and there were few University endowments. Gloucester Hall was the house founded in 1283 for those of the novices who came from Gloucester Abbey. Then other Benedictine abbeys had houses built alongside the first for their own students, till twenty-five abbeys were represented. Others in the same way sent their men to Buckingham College, Cambridge. Over the doorways of the halls still standing at Worcester may be seen the escutcheons of their several abbeys; the griffin of Malmesbury, the cross of Norwich.

The house of the fourth great mendicant order, the Austin Friars, has disappeared as completely as the other three. On its site, in the reign of James I., Wadham College was built, but the phrase "doing Austins" long survived as a memory of the University exercises that took place in the Austin Schools.

The friars of the Order of the Redemption of Captives have left as little sign. Their property is part of the garden of New College.

But some of the houses of the Regular Orders, besides Gloucester Hall, remain in a translation. The College of the Novices from the Convent of Durham is the old part of Trinity. St. Mary's College of the Regular Canons has left a gateway opposite New Inn Hall, and the latest-founded of the religious houses, that of St. Bernard for the Cistercians, still shows in the street with some changes as the front of St. John's.

We have to think, then, of the Oxford of the thirteenth and fourteenth centuries as chiefly made up of the schools of the Regular and Mendicant Orders, afterwards suppressed. The Colleges of the third class, the Secular Clergy, were only beginning; the prevailing influence was that of the Dominican and Franciscan teachers, particularly the latter, with their Roger Bacon and Duns Scotus and Occam, to set off against the Aquinas and Albertus of the Parisian Dominicans. At the time of the Dissolution the numbers and spirit of the various religious houses had run very low. However, they made a push to extend their buildings, hoping so

NEW COLLEGE, FROM THE GARDENS.

to ensure their wealth against the Royal Commission, just as the Colleges did in our own time. But Henry VIII. seems, in most cases, to have made for the lead from the roofs of the buildings as his own share of the spoil, and the later additions with their flat lead roofing would just meet his taste. The walls he sold or gave to others, who used them chiefly as quarries. A letter quoted by Mr. Fletcher, written to Thomas Cromwell by one of his agents, gives a good notion of the look of things at the time, and of the spirit of the reformers:—" The Black Fryers hathe in ther backsyde lykwise dyvers Ilonds well woddyd and conteyneth in lengith a great ground. There quere wasse lately new byldede and covered with ledde. It ys lykewisse a bigge Howse, and all coveryd with slatt, saving the queere. They have prety store of plate and juellys, and specially there ys a gudd chales of golde sett with stonys, and ys better than a C marks : and there ys also a gudd crosse with other things conteynyd in the bill. Ther ornaments be olde and of small valor. They have a very fayer Cundytt and ronnyth fresshelye. Ther be butt X. Fryers, being Prests besid the Anker, which is a well-disposyd man, and have l. marks yerly of the King's cofers."

Now we turn to the reaction against all this; to the quarter of the University that remains, the quarter of the old town parishes in which the Halls and Colleges of the secular clergy grew.

And first we feel the Gothic pattern of the streets. We have left the water behind, but the streams had to do in determining the flow of the "streamlike" streets. They kept the forms they were pressed into by the castle and city walls; St. Giles's bursting out wide from the point where the old north gate cramped it by St. Michael's; Broad Street broad because just outside the circuit; the rest winding and twisting with the happiest effects for the jostling buildings.

Then when we look closer at this large mass of Gothic work, of great establishments squeezed into the old shapes, and elbowing scanty strips and corners of the displaced houses, the notable point about most of the work is, not how old, but how new it is. The

ST. MARY'S, FROM THE HIGH STREET.

Gothic is late, even belated. Little of it is earlier than the Perpendicular Period; much of it is more recent still, and of a kind to which purists grudge even the name of Gothic. It is true that in Oxford buildings, when made of local stone not cunningly laid, become shabby and theatrically aged in the shortest time. They look not venerable, but battered and burned, the stone hanging in rags, and leaving where it falls raw yellow patches. Mouldings and carvings drop away; pinnacles, battlements, and gables, and all outstanding features, thaw like blackened snow; walls are suddenly found wasted and thin, the rooms and towers depending on the merest crust of stone. The heads about the Sheldonian Theatre shed their beards of a rainy night. But all this is a very sham antiquity. Some of the later buildings suffer from it most, and some of the oldest look, and are, newest because of sedulous restoration. One has to search diligently for hints of the older work, and to entrap it as it looks out of its new body in some favourable light.

It is the churches, the parish churches and the towers of St. Frideswide's and St. Mary's, that seem most to promise age in a distant prospect, and to strike a recurrent note of antiquity as one goes about the town. The old Town Church of St. Martin at Carfax, with its picturesque altered gables and clock and "penniless bench," is much wanted; but it was pulled down inconsiderately and rebuilt in haste. But St. Peter's, St. Michael's, St. Mary Magdalen's, and St. Giles's, are rich in beauty and interest. The lovely spire of St. Mary's, panelled with pomegranates for Queen Eleanor, stands almost alone of the old University Church. There is, indeed, on the north of the chancel, and set at a divergent angle, a yet older building—the two-storeyed Ancient House of Congregation. Its two storeys simulate on the outer side the appearance of one to conform with the new church, but the groined roof remains of the lower room, now half-buried and given to lumber. This and the old church were the real centre of the University. In the five chapels the regents of the five faculties assembled for the Act at which disputations were held and degrees given. Not only the Schools and Theatre and Convocation House, but the University Library, too, lived off from these buildings. The first books were kept in chests in the "soler," or upper room, and there, too, those other chests were stored that were the earliest form of University endowment. In them the money left by benefactors was kept, and lent out to poor students, who in return pawned books and daggers and other articles of value.

The colleges began as a counterpoise to the schools of the regular and mendicant orders, more particularly the latter. The friars, learned and powerful, naturally drew to them great numbers of the poor unattached scholars. Statutes ineffectually made eighteen years the lowest age of consent. The University had a hard fight to keep even its degrees in its own hands. This third great body of scholars, unattached to monks or friars, consisted of the ordinary or secular clergy, men qualifying not merely for the work of parish priests, but for what are now the lay professions of lawyer and doctor. They had a bad time of it while the friars were still popular. They had few endowments, and were forced to labour for a living, or to beg their way. It was common for poor scholars to serve as scouts. They lived either in private lodgings or in the numerous private Halls, Inns, or Hostels that covered the sites of the present colleges. These are the second obliterated chapter among university buildings. They were simply lodging-houses, rented from the owner by a Master of Arts, who was styled Principal. By an early statute, that marks the encroachment of the University on the town, the owner of the hall was bound to let it to the first applicant who deposited the needful caution with the Vice-Chancellor. The Principal was paid by the inmates for board and tuition. The first colleges were such halls, furnished with an endowment for poor scholars, and with a set of statutes to regulate its administration. At first the scholars went to service at the nearest parish church; but gradually, as funds allowed, chapel and hall and library were built, and the familiar front with its gate-tower screened the old and new buildings. The full-grown College, as it had taken shape before the times of the Reformation and rich lay undergraduates, was a society incorporated for the benefit of poor scholars of the secular order. Its buildings replaced the single Hall or group of Halls that had been converted from private to corporate use, or else the old tenements were recast in the new mould. That new mould followed with modifications the plan of the monastic houses.

Some of these Halls still remain. But the form of university life they represented, and to a great extent the buildings themselves, have gone as completely as the Oxford of the Religious. The colleges swallowed most of them. New College accounts for ten, Merton for eight. From old prints one can gain a notion of the splendid jumble of gables and chimneys of all degrees of dignity that enriched the streets; and one is tempted to regret that some of the colleges gave up the picturesque grouping and domestic style of the clustered halls for the more monotonous and pretentious manner of their latter shape. As Henry suppressed the religious houses, so Laud suppressed the private halls, leaving five only as academic halls. Of these, one—Magdalen Hall - has left its beautiful bell-tower to Magdalen College, and its second site to Hertford. Of the rest, three are now absorbed in colleges.

The great date in college history is 1264, when Walter de Merton gave statutes to the college he had founded. University Hall, afterwards University College, had already been founded from a legacy administered by the University. But in Merton the idea of a great college was first clearly struck out, and its statutes were an exemplar for all succeeding societies both at Oxford and Cambridge. Merton, however, was not built in one heat. The old quad and parts of the chapel are early work, but the tower and other parts are later. The chapel is so large, because it is not only chapel to the college, but church to the parish of St. John, a great part of which the college absorbs. The library is one of the most beautiful as well as one of the most ancient rooms in Oxford.

Balliol and Exeter, Oriel and Queen's, are also early colleges, but they do not stand for so much historically. They group with Merton, and have all changed their first bodily shape. The next great moment in the college history, the beginning of a new group, comes about a hundred years after Merton. This was the foundation of the College of St. Mary Winton—called New from a sense of the importance of the event—by William Wykeham, Bishop of Winchester. Several things are important about this great creation. To begin with, the foundation was of a new magnificence. It provided for seventy scholars, a term at that time synonymous with fellows. There was a stronger accent about it of opposition to the regular clergy. Its lands were bought from impoverished monastic bodies. It was made self-sufficient by its nursery and counterpart in architecture, the College of St. Mary at Winchester. It was saved from the jurisdiction of the University by the power granted it of giving degrees, and from the jurisdiction of the Bishop of Lincoln, in whose diocese Oxford then lay, by the appointment of the Bishop of Winchester

MAGDALEN TOWER, FROM THE CHERWELL.

as its Visitor. But above all, not only was the plan of its institution a great educational achievement, but the building itself was by the same author, the work of a man of genius, coherent and complete. The quadrangle has been altered out of all knowledge by the addition of the third storey with battlements, and the re-shaping of the windows, but even so it shares with the added wings of the garden front a wonderful dignity and purity. The original towers are so dominant everywhere that one reads their spirit into the encumbered translation.

The stroke told. Henry VI. echoed the idea in Eton and King's College, Cambridge. In Oxford, the chapel and cloisters of Chichele's All Souls' were imitated from New College. But a richer reverberation followed. Waynflete's Magdalen is another New more magnificent, and later by a hundred years. It is not more beautiful. Feature by feature is reproduced, with just that luxury of virtue and grace that one would expect from Mary the Magdalen walking in the footsteps of

Mary the Virgin. The chapels are planned alike, and in either college set back to back with the hall. Magdalen has a cloister quad, a more spacious one; a higher and richer tower, wider and more lordly grounds. But one can turn from the baffling and haunting charm of Magdalen Tower to be satisfied with the simple fighting tower of New College; and it is dangerous to go from the blackened walls and gaunt arches, the austerely divided daylight of the cloistered walk at New College, to the coarser forms and less single purpose of the other. The older cloister is still the walk of a recluse, overlooked only by the tower and gable of the chapel, and interrupted by rare and funeral writing on the walls. The other is built in an easier temper. Staircases open upon it below, and many windows occupy it above. It is the covered thoroughfare of the College.

STONE PULPIT, MAGDALEN.

It should be remembered, however, that Magdalen cloisters have suffered much. They have been pulled down almost throughout and rebuilt. An upper storey has gone from the north side, ugly Westmorland slate has replaced the grey Stonesfield kind, and windows have been made bigger and more regular. Historically it is to be noted that Magdalen superseded a collegiate building of another kind, the old hospital or almshouse of St. John the Baptist. The stone pulpit in St. John's Quad marks this. On the saint's day a sermon used to be preached from the pulpit, and the quad was strewn with rushes and hung with boughs to represent a wilderness. At last a Principal caught his death of cold by going out into the wilderness, so they gave it up, and had the sermon in chapel instead. There was some grudging show made of keeping up the almshouse. A low vault under a chapel was given over to the poor. A report was drawn up for the year 1596 giving the following cheerful particulars:—" In sommer the resort is greater, in winter very smale, bycause of the coldnes and onwholesomenes of the vault; which is in verie deed so moyst and dampish that we have the last yeare removed the beddes into another house not far of, for that everie winter they are subject to rottennes." However, they were going to repair the floor " as well for the safetie of our beddes as for the health and ease of the poore." *

The building of the old quad of Lincoln went on by stages during this same fifteenth century, and Corpus followed early in the next. Neither of these has been

---

* Macray : "Notes from the Muniments of St. Mary Magdalen College, Oxford."

rebuilt, but both have been defaced so as to lose almost all interest; but they stand for points in history. Lincoln was a college of priests, to make head against Lollardry; Corpus stood for Greek.

"TOM" GATEWAY.

Then follows the notable foundation of Wolsey's Cardinal College, afterwards Christchurch. All Souls' had been founded with the spoil of "alien" priories—cells, that is, of foreign monasteries in England. Magdalen had taken the place of a religious society; but the final step was taken when English religious houses were suppressed to form one great educational foundation. St. Frideswide's was preserved to be its chapel. The huge ungainly quad was planned out and partly built. After the suppression of religious houses stone was cheap, so the building went on even after Wolsey's fall. The Tom Tower was added much later. It is one of Wren's essays in Gothic, masterful and striking in general design, but unfeeling in detail. The fan-vaulted roof of the hall staircase is a lovely piece of later work, but the staircase itself is badly managed. The cathedral is a rather disconcerted building; but there is plenty in it to study and enjoy. The story of the saint may be read in a window by Mr. Burne Jones. Other four windows by the same artist were executed by Mr. Morris, with the result that in colour as well as in design they rank with the best of old workmanship, and can be compared with nothing new, except those from the same hands in other places.

If New and Magdalen stand for the enriching sunset of Gothic in Oxford, the great group of buildings that follows the Reformation stands for a strange and prolonged after-glow of the art. It is this period that more than any other belongs to Oxford, gives it a peculiar character. Nowhere else is it so largely represented. The Renaissance, coming all this way, was too weak and distressed to create forms of architecture quite its own; but it passed as a principle of change into the veins of the old style, and broke out here and there in the strangest features. The main ideas of the Gothic structure held their own—the sloping roof, the traceried window; but a languor and a fever seized upon the mouldings and details of the old work. At any moment the sedate lines of the Perpendicular tracery might run wild into twirls of trivial scroll-work, or one whole side of a building speak a sleepy Gothic and another stammer the queerest Greek. But the whole seldom fails to please, because it is ordered throughout by the most sure and delicate sense of proportion. It is the work of men whose hand is well in, whose ideas are running few and thin, but are dealt out and recombined with the utmost freedom and familiarity. One is often blankly disappointed by the flatness, the poverty, the childishness of the decoration; but however meagre and thoughtless and alien the elements of the design may be, there never fails an artistic sense in the way they are set out, so that the most incongruous lendings of various styles meet and are subdued to perfect comfort in one another's company. Perhaps the salt that saves the whole is the sense of humour that pervades it, just as it does the rich enjoyed sentences of the contemporary literature. The buildings do not expect to be taken quite seriously; the figures on the tombs are very much at play with death. Sometimes, indeed, the windows of grave buildings like the chapel of Wadham stiffen out into the older and more decorous manner; but it would be hard to match for rollicking irresponsibility the porch that Laud added to St. Mary's Church.

Colour, too, was near the heart of the builders. They revelled in gilding, in paint, in marbles and alabaster. And in the weighty matters of architecture that go beyond the mere building, in the recognition of its neighbourhood, of its place as a mass in the streets or a kindly growth among fields and trees, they were very much at home. The presence of such buildings is one of comfort, of fun, of flexible tradition and generous possibilities. The style begins at Oxford under Elizabeth, and continues under Charles; but it centres under James, and hence is conveniently called Jacobean. Not only university and college buildings belong to it, but most of the beautiful domestic work of the streets, like Archbishop King's house off St. Aldate's, and the house off the High, used as a police-office.

It was in a building of the Jacobean time that the University idea first found adequate expression, gathered out of the scattered lodgings in which it had been housed. Already, by 1480, a noble room had been built for the Divinity school, with the library of Duke Humphrey above it. Sir Thomas Bodley's first act was to give this library a new roof and fittings, and to add to it at right angles the building that forms above an extension of the library, below the Proscholium or ambulatory of the schools. It was the day after his funeral, in the year 1613, that the first stone was laid of his magnificent plan for completing the quadrangle, of which the Proscholium forms one side. This quadrangle is a plan or map of the University's theory of knowledge. As one enters under the gateway tower the scholastic sciences announce themselves in gold letters above the various doors. The faculties—the faculty of Arts with its subdivision into the Trivium and Quadrivium, the faculties of Canon and Civil Law and of Medicine lead up to the fifth and crowning faculty, the science of sciences, Divinity, lodged behind the richly-panelled front of the Proscholium. Before this, the faculty of Arts had been housed in the thirty-two schools that gave their name to Schools Street. In these the Regents, that is the young M.A.'s, the ruling and teaching body of the University, gave lectures and sat, at stated times, to determine in the disputations that preceded, as examinations do now, the B.A. degree. The public *viva voce* in the schools is the remnant of this formal exhibition of logical skill. The disputant went round to solicit the presence of his friends, and statutes were passed to restrain the system of touting for an audience as well as to limit the regular supper that followed. At Cambridge it was the duty of the Bedells to go round to the various colleges and halls where the questionists were, and " call or give warninge in the middest of the courte with theas words : ' Alons, alons, goe, Mrs., goe, goe,' " and any tendency to a real *viva voce* was rudely checked by the same officer. " If the Father shall uppon his Chyldren's aunswer replie and make an Argument, then the Bedel shall knocke him out "—which seems to have meant that he hammered loudly on the door.* The Act, or public contest of degrees, still took place in St. Mary's, till the Sheldonian Theatre completed the new group of the schools in 1669. The new Convocation House, with the Selden Library above, had already been added in 1640 at the further end of the Divinity school. About the same time as the new schools Wadham College had been built. Complete at the outset, it is remarkable among Oxford buildings for its singleness and symmetry of design, and its skill of building or fortune of stone; it is one of the most ancient of the colleges in the sense that it is authentic.

The rebuilt University College and Oriel and the new Jesus may be grouped together. They have in common the beautiful treatment of the upper windows as a series of little gables in place of the tiresome screen of battlements. The front of Jesus is a modern disguise, the clever but unsympathetic work of Mr. Buckler. It replaces the old Elizabethan front with its gateway in the fashion of the beautiful one of St. Alban's Hall. The Jesus gate, however, had been obscured by a heavy rusticated screen. Brasenose gained in the Jacobean Period its exquisite dormer windows; Lincoln its homely second quad and lovely chapel. Another fine example is the hall and chapel of St. Mary Hall. In Merton four of the five orders of the Schools Tower were reproduced. The chief author of all this work

---

* Wordsworth: " Scholæ Academicæ."

was a Thomas Holt of York. Among his followers were the brothers Bentley, and Acroide, Oxford builders. A greater name is associated with the new quad of St. John's. In this Inigo Jones was mastered by the genius of the place, and constrained to build the wonderful garden front. Inside the quad he had his own way in the colonnades, but he was more in character still when he designed the Danvers Gateway of the new Physic Garden, and plotted its wall and walks. Here, at last, in a quiet corner of the place, where science was beginning in a gentle way to stir, the English Gothic tradition of building was fairly broken, and the key struck of the manner that in the end of the seventeenth and first half of the eighteenth century gave Oxford its sturdy and picturesque English Classic. Soon after, the troubled times of the Civil War, the rather farcical, but disastrous siege of Oxford, "leaving no face of a University," and the subsequent spoiling of the colleges by the Puritans, must have served very effectually to snap the chain of building tradition, and make a blank for the new ideas.

THE DOME OF THE RADCLIFFE, FROM BRASENOSE.

When the strange holiday time was over, and the University was in a frame for building again, the period of Wren and of his school began. The Chapel of Brasenose, built under the Commonwealth, marks the point when the relations of the mixed styles were becoming too strained. The Sheldonian Theatre announces the rupture. It is in Wren's happiest manner. There is no building where the audience is more artfully disposed so as itself to be a great part of the architecture. This was followed by various buildings of the school of Wren. He revised Bathurst's design for Trinity Chapel, though he clearly thought it a bad job, and he is said to have had a hand in Hawksmoor's work at Queen's and All Souls. Certainly the robust screen and gateway of the Queen's front are not unworthy of him. Aldrich's All Saints' Church and Peckwater Quad at Christ Church belong to the early eighteenth century. The last great building in this manner is Gibbs's Radcliffe Library. It gives the University a comfortable centre as only a dome can, and counts for quite half in any distant view of Oxford buildings.

The rest of the eighteenth century has nothing very notable to show. Hawksmoor, in his nightmare buildings at All Souls', had proved how dead Gothic was. A good deal of Classic went up, the work of academic amateurs, dabbling in Vitruvius and Palladio. But one holds one's breath till the period is well over. Dilettanti among the dons travelled to Italy and came back terribly ashamed of their barbarous Oxford. It is a matter for thankful wonder that all the old buildings were not replaced by Palladian colleges. The clean sweep of old Queen's and of the mass of buildings that made way for the Radcliffe, must have tempted many a common room. Hawksmoor actually prepared a design for a brand-new Classic Brasenose, with four domes and a High Street front. Magdalen had the narrowest escape. A Mr. Holdsworth, a Fellow of the College, "an amiable man and a good scholar," returned from a sojourn in Rome full of enlightenment. His first scheme was to pull down the whole building, tower and all; but he had to give up the tower chapel and hall, and content himself with the destruction of the cloister quad. However, he began his scheme with the New Buildings on unoccupied ground, and somehow it was not carried further; so the great new quad, with its three colonnades, remained on paper.

But in 1771 the University set itself to a wider change. The rough unpaved streets in which the buildings were rooted like trees, the island markets that blocked the traffic, the narrow rambling bridges, and above all, the North Gate or Bocardo by St. Michael's, and the East Gate above Magdalen, hurt the best feelings of dons dreaming of vistas and piazzas. The place, besides, was no doubt very dark and dirty. An Act of Parliament was obtained for cleaning, lighting, and paving the town, removing gates and other obstructions, building markets, and repairing or rebuilding Magdalen Bridge. So Oxford became convenient and lost half its pictorial effect. The old bridge over the Cherwell at Magdalen was everything that a good bridge can be without being convenient. It had a chequered course of six hundred feet over land and water, leaping the water in a series of arches of different height and width just as was necessary; occupied by houses and shops where it crossed the land; and throwing out, at irregular intervals, angular bays of varying width and projection. But in some places it was as narrow as thirteen feet, some of the arches were ruinous, and the city and county were responsible for the repair of different parts, which they both appear to have left alone. So it had to come down. The new bridge, as well as the market and other changes, was the work of an engineer named Gwyn. His bridge kept something of the old picturesqueness, though in a formal way. It had the same places to go over; a circular bay in the centre stood for the old angles; and the lines at either end swept out in graceful curves. But people were very angry because it was so narrow and high. The roadway was afterwards heightened to reduce the pitch of the bridge, the parapet was lowered, and in our own time the width has been doubled for the convenience of trams. Old St. Clement's Church, too, has gone from the road on the further side, and has been rebuilt in another place and manner in full view of every one who crosses the bridge.

Many of the old houses had been shorn away in the process of widening the streets; but some people were not satisfied. The old Gothic buildings had begun to command a certain zeal without knowledge; but people disliked the Gothic pattern of the streets, and the irregular patches of domestic buildings. They wanted to have things cleaned up, and made regular; to have "views;" to see the great buildings in solitary distance with no interference from house-roofs or trees, a thing that very few buildings in the world can stand. An interesting evidence of this state of mind is a little book by a certain rector of Lincoln, Dr. Tatham. It was he who defaced the old quad of his college with stupid battlements and other changes. He is remembered by Mr. Cox as an old gentleman, who lived out of Oxford, but might be seen landing his pigs in the market-place on Saturdays, and who, in defence of the faith and the Three Witnesses, in the University pulpit, wished all the "Jarmans" at the bottom of the "Jarman" Ocean. The book is called "Oxonia Explicata et Ornata: Proposals for Disengaging and Beautifying the University and City of Oxford." The buildings of Oxford, he thought, were "too crowded and engaged. Our forefathers seem to have consulted petty convenience and monastic reclusiveness, while they neglected that uniformity of Design which is indispensable to elegance, and that grandeur of Approach which adds half the delight. If the Colleges and Public Edifices of this place were drawn apart from each other, and dispersed through the extent of a thousand acres, so that each might enjoy the situation a man of genius would approve, we might boast," &c. He prefixes to the book a little design of his own for a martyr's memorial, "a triumphant monument to be placed across Broad Street, the whole so airy as very little to obstruct the view of the buildings." Of this design one may say that it would have been much more interesting than the Eleanor Cross of Sir Gilbert Scott.

This cross of Scott's was one of the first new works at Oxford of the Gothic revival. Wyatt and others had already worked at what they called restoration, and Pugin's gateway, lately removed, had been set up at Magdalen the year before. Oxford has suffered its full share of buildings that were the costly grammatical exercises of men learning a dead tongue. In architecture such exercises are more expensive and obtrusive than in any other art, and it will be long perhaps before people will have the courage and sacrifice to pull them down again. Some are merely learned and lifeless, like Buckler's Magdalen School and Jesus, and Scott's Chapel of Exeter. Others are hopeless, sullen blocks, like Scott's extensions of Exeter and New College, and Butterfield's new buildings at Merton. Others are fancifully bad, like the conscientious ugliness of the Museum, and the *recherché* ugliness of Balliol, and the mixture of both in the Meadow Buildings of Christ Church. Butterfield in Balliol Chapel and Keble College shows a great power of geometrical invention in form and colour, an invention for the most part greatly astray. It is refreshing among all this to come upon the strong, though wayward artistic temperament of Burges in the decoration of the hall and chapel of Worcester, or even the respectable classic Taylorian buildings of Cockerell, unpleasant in colour and jarring on the spirit of the place as they are.

Very different work has been done of late years. There is less about it of defiant expression of undesirable artistic personality, or pedantic exhibition of a style—more recognition of the power of the place, more actual artistic instinct. Even the much abused Indian Institute of Champneys, in spite of the heavy frivolity of its details and interior, is, in the disposition of the wall and window space, the invention of its tower, and the way the whole building takes its place in the picture, a piece of architecture, which such things as Balliol are not. The space of blank wall on its corner tower is worth more than all the geometrical troubles that fret the face of Keble. At Magdalen, again, the genius of the old build-

THE 'VARSITY BARGE.

ings has been lovingly reproduced by Mr. Bodley in the new. His tower is not good, and it was carrying faithfulness too far to reproduce the stupid gargoyles and grotesques of the original; but much of the rich decoration in wood and stone is refined in design and workmanship. One can praise, too, the extension of St. John's. It may be said of all the new Oxford buildings that they are apt to be heavy within, owing a good deal to the fear of fire that makes all the staircases of stone.

But there is an architect who has taken up the tale where Thomas Holt dropped it, and who has carried it farther, with results almost as important for the appearance of Oxford. A like moment in the history of the place seemed to have come round again. The new ways of the University needed new schools — for examination this time instead of disputation — and a great extension of college building coincided. Never, perhaps, since Wren had the churches of London to rebuild has one architect had such a fling in so important a place as Mr. T. G. Jackson in the Oxford of our time. And there can be no question that, whatever the faults of restlessness and overcrowding that sometimes mar his designs, Mr. Jackson has been worthy of his opportunity.

The last great addition to Oxford is its undergraduates. It is not very long since the colleges, in respect of undergraduates, were normally as All Souls' is now, with its four Bible Clerks; but they were not as All Souls' is in respect of its Fellows. The long deserts of theological and political war had left them for the most part mere club-houses, whose members existed to drink port together of an evening, and abuse one another in little pamphlets during the day. The Common Room was the great invention of the late seventeenth century, and the eighteenth was spent in bringing it to perfection. Then came the Fellows of Oriel, the Examination Statutes, the genius of the Masters of Balliol, the Commission, the new

A "BUMP" AT THE BARGES.

Statutes, the Unattached. It is an exciting show for the visitor, this incongruous surface of new and old, this great bustle and pulse of the machine among the frail and crumbling walls. Each morning dilatory tides of men in cap and gown set about the streets under the jangling bells; each afternoon, in punctual haste, a steady stream of the same men, in flannel, makes for the river, and flows back for the monastic evensong and refectory dinner. There was a time when the dinner-hour was ten o'clock in the morning, and it was thought that so late an hour was a sign of the decay of learning.

Meanwhile the streets grow emptier, and the visitor, in the abstraction of the growing darkness, will gather hints of the antiquity about him. He will see the society grimace of the buildings relax. Their features will relapse into startling meanings, and the presence of other centuries will strike in upon his senses. If he is an American, like Mr. James's *Passionate Pilgrim*, he will feel about it all the pang of a forfeited possession. It is part of himself that was lost and is found, a history forgotten long before he was born. Now he remembers it.

Nowhere is midnight so late as in Oxford. It is announced from so many towers at so many moments by bells of the most various tone and cadence; but by all, even to the most maundering and belated, with the same precise conviction, as if one could hear all the lecturers saying the same thing in their own words— *It is midnight here, now.* And faint and loud another and another awakes and insists, *It is midnight here, now.* Through the middle clamour the chime of St. Mary's drops down three pathetic steps and climbs up through the same intervals. The University is older by another hour.

The great deed of the new undergraduates was the discovery of the river. In the early years of the century it was still only a place for fishing in; occasionally a heavy tub was rowed down to Nuneham. Bell-ringing had gone out as an exercise; cricket was the game of one exclusive club; the nearest approach to a healthy rivalry between the colleges was a competition between New College and All Souls' in making negus. New College won by putting in no water. It was not till 1837 that the old boats had their sides cut down. About ten years later outriggers came in, and after another ten, keelless boats. Another ten brought sliding seats from America, and so the skiff and the four and the eight reached their perfect economy of

IFFLEY MILL.

construction, and the quality of beauty they share with their counterpart, the bicycle, on land. Both bicycle and skiff are extensions of the human machine within such limits that they remain as it were mere developed limbs working at every moment as parts of one balancing frame, projections of the person.

In 1839 the University Boat Club was started, and the great Oxford school of rowing shot up to overshadow the older faculties. Before this time college racing had begun on the admirable bumping system, that not only makes the race a prolonged spectacle for those who stand still to see, but allows of so much spirit of body in those who run by their college boat. At first the boats started out of Iffley Lock. The stroke of each boat, as its turn came, ran down the thwarts pushing out, and the next boat followed as soon as he cleared the lock.

IFFLEY CHURCH.

The river between Oxford and Abingdon in its present shape is a sort of free canal, locked at Iffley and Sandford, and again just above Abingdon. There used to be a lock at Nuneham, but it was taken down. More recently, too, the Folly Bridge Lock was carried away, and has not been replaced. The history of the river as a work of art is a long and interesting one. Obstructions to its natural course, and to its navigation, began with mills and mill-weirs. In early charters there are provisions for removing "gorees, mills, wears, stanks, stakes, and kiddles." Commissions and Acts of Parliament tinkered at the navigation, but till the end of the eighteenth century no progress was made beyond the old mill-weir, with sluices in it to let the boats through. This arrangement was called a "flash" lock. The flashing emptied the reach of the river above the lock, but all the water was needed to move the heavy barge sticking on the "gulls" below. Navigation was of course terribly slow. A bargeman had sometimes to send on ten miles ahead to get a flash when going up stream, and sometimes lay for a month till enough water had accumulated. When he got it there was none for any one else. Leonardo da Vinci, who thought of everything, had invented the pound-lock long ago; and other people

LITTLEMORE CHURCH AND KENNINGTON ISLAND.

before and after him had hit on what seems a very obvious contrivance. But that ingenious people, the Chinese, are said still to hoist their barges over weirs, and it was not till the time of the great period of canal-making that began with the Bridgewater Canal in 1760 that the pound-lock made any way in this country. The canal from Birmingham reached Oxford in 1790, and shortly afterwards the locks and towpath were put into their present shape. Then followed jealousies between rivers and canals, till the railways came and made inland navigation of less importance. Nowadays there is little barge traffic through Oxford (Folly Bridge was always a difficult point), but another question, that of floods, presses as much as ever. Quite recently the engineer to the Thames Commission brought out a scheme for doing away with Iffley Lock and Weir, and dredging a deep and narrow channel between Iffley and Folly Bridge. The Vice-Chancellor and the Dean of Christchurch, both by virtue of office Commissioners, were in favour of this scheme with a view to Oxford health; but it has been proved to the satisfaction of the Thames Conservancy Board, whose officers have examined the place, that so sweeping a change is not needed. The effect of the change would be to give the river banks instead of brims, and it has been argued that it would kill the elms of Christ Church and the fritillaries in the water meadows about Iffley. Most serious of all would be the loss of Iffley Mill. We may hope that Oxford health need not be bought so dearly.

Meantime Iffley Lock ends the shorter course for rowing practice, Sandford the longer; while Nuneham and Abingdon, to keep to an Oxford point of view, mark the longer picnic courses. It is all Frideswide ground; the saint was rowed to the outhouse near Abingdon by an angel, when she was warned in a dream and fled. The meadows of her convent now are lined by the college barges. These are an interesting study in development. The first of them were old procession barges of the London City Companies. One of them, the Oriel barge, still remains, with its delicate form, and long sharp prow, in which the rowers sat. The bronze figures by the door of the saloon are untouched, the oval windows, the tarnished gilding within. But the spirit of utility rebelled and the model changed. The long prow was chopped off close, the semblance of the high stern went, and there was left merely

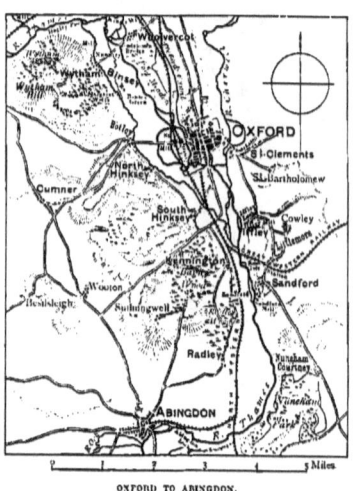

OXFORD TO ABINGDON.

a square floating dressing-room with railings round its roof, and seats for the spectators of races. Then the sense of beauty mutinied, perhaps alleging the use of the toy for picnic excursions, and the prow and stern were restored. The University barge is a monument of the Gothic revival. Several architects have tried their hand in designs for these craft, and new ones are from time to time constructed. It is the oddest little street, this row of motley Noah's Arks; and when the high poles shake out their amazing flags, and the men come down in fearless college colours, and a vast and diverse millinery decks every foot of standing room the roofs can give, there would seem to be some touch of an Arabian Night about a very English day, were it not that the vigorous people wear many more colours than Arabia would allow.

A little hill at Iffley lifts up the rusty-grey rectory and church. The church is, for its size, made in an absurd number of styles, beginning with Late Norman. The heavy arches inside are carved round with sunflowers, looking like an ancient imitation of modern work. Outside, there is the strangest confusion of carvings; a centaur strayed from Phigaleia, and other pagan images among the Christian symbols. The Gods in Exile have visited Iffley too. On the south side a great yew has been building all through the Transition and the Perpendicular and the Tractarian times, and the people who Decorated, and the people who Late Middle Pointed, and the rest make the ground quite uneven round its roots. An undated villager, who styles himself Archdeacon of Iffley, and has a venerable humour, comes among the graves for company.

Behind the hill, and a little beyond Iffley, lies Littlemore. Here is the little church that Newman built, and came to from St. Mary's for the last two years of the Via Media. Near it is the range of low buildings that people called a monastery, where Mark Pattison and others came to be with Newman, and where, on October 8th, 1845, Newman was received into the "One Fold of Christ" by Father Dominic the Passionist, the good father making holy puns upon the name of the place. Now The College, as the building is called in the village, is given as almshouses to the poor. The largest room is a public library. In the kitchen lives an old woman who served the Newmans in her youth. Her husband, an old toy peasant, with smock-frock and silver hair, and a fine rheumatism that I am sure his country gladly supports, sits by. She stands up and remembers Newman. He lived there with his pupils "before he became a Pope. The Pope of Rome, that's the real Pope, over-persuaded him, and he went away and never came back again. She did hear that the Church of England gave him some punishment for leaving, but didn't rightly know. And the clerk's wife had been to see him, and found him in a bare room with no carpet on the bricks, like any poor person, and had said that it was to be humble and like his Master that he did it all."

Meanwhile on the river we have to pass some ornate sewage works, and the wanton embankment of a railway, that here crosses to Littlemore. Below lies the Rose Isle, with its "Swan Inn," and on the right the heights come nearer with the little village of Kennington. A beautiful tree-planted road runs along the top to Radley, with its school in the old park of the Bowyers' house, and against the tall trees is a little grey church and thatched cottages, where women come out and sit with their sewing machines on the summer evenings. From this the road goes on through a corn country to Abingdon.

Next on the river comes Sandford Mill, with a leaning chimney, that has all the interest and all the beauty of the leaning tower of Pisa. Sandford Church lies away from the river, nearer the Nuneham Road. The porch proclaims, "Condidit me domina Eliz. Isham. Anno gratiæ 1652," and adds—

> "Thanks to thy charitie, religiose dame,
> Which found me old, and made me new againe."

It is proper, at the same time, to speak strongly of the taste which found the church Norman, and made it something very new indeed. But it is worth while going in to see the curious carving in the chancel of the Assumption of the Virgin.

Another mile, and the heights of Nuneham close in on the left with woods sweeping down to the very edge of the water. Presently one comes upon a little

A PICNIC TO NUNEHAM.

island, connected with the Nuneham side by an intensely rustic bridge. By the landing-place is a cottage with exaggerated thatch. Here they make tea. They make most not for the University picnics that the summer term brings to these hospitable woods, but when the great revolt of the town sets in with the long vacation. The river is as populous as ever then with dashing young fellows in flannel, and enchanting young ladies dressed in the depth of fashion. Great and many barges are towed down to Nuneham, and there merry people dance round Carfax, and float up again to Salter's in the heavy purple dusk, trolling snatches of songs. Carfax reminds us what a place of shifting Nuneham is. To begin with, the family was removed hither from Stanton Harcourt in the last century; then they moved the church and the village to new places; then the river was moved into a new cut, and the town of Oxford presented Lord Harcourt with Otho Nicholson's conduit. It was a work that gave the town the final accent of the Jacobean style; but it was in the way of cabs. May one hope that perhaps it came to Nuneham, like other pilgrims from Oxford, only for a season, and that it is waiting in that hospitable retreat till its home is worthy of it again?

The conduit did good service to Nuneham at the time. It took the place of a projected "Gothic castle." Gothic castles and abbeys, well ruined, were in vogue,

THE BRIDGE AND COTTAGE, NUNEHAM.

to cap a rising ground or to conceal a dairy. It was the time of "landscape gardening." People corrected their land as much as possible after the ideas of Claude. It is for his pictures one must look in grounds like those of Nuneham. The trim Elizabethan garden, with its pleasaunce and mount, and bowling-green and wilderness, its fountains and clipped trees, give place to a carefully-arranged disorder. Foregrounds were picturesquely grouped, middle distances were plotted, and sunk fences, palings painted green, grottoes with stalactites and stalagmites and other devices went to make up what was called Nature. The disciple of Rousseau felt that he had indeed returned when he could sit upon a jag of extremely difficult geology fresh from the contractor's hand, and drop the tear of sensibility into the cascade that his own fingers had turned on. The man who had the chief hand in the laying out of Nuneham was Lancelot Brown, called "Capability;" but Lord Harcourt kept, besides, two tame poets, Mason, the author of "The English

Garden," and Whitehead, the Laureate, to help him to be elaborately natural in his gardening, and to write verses on the seats. The two had a genuine contempt for one another. Mason had the last word. He wrote the verses on Whitehead's memorial urn. He said—

> ". . . let the sons of fire
> The genius of that modest bard despise,
> Who bade discretion regulate his lyre,
> Studious to please, but scorning to surprise;
> Enough for him, if those who shared his love
> Through life, who virtue more than verse revere,
> Here pensive pause, while circling round the grove,
> And drop the heart-paid tribute of a tear."

Mason could do most things badly. His patron says of him :—"In the church there is a barrel-organ, upon which is set Mr. Mason's music for the responses to the Commandments, and his Sunday hymns. The adjoining flower-garden was formed by him, and he suggested the alterations on the north terrace. So that in a very small space we have specimens of his genius in music, painting" (the altar-piece of the church was his work), "and poetry; of his taste in improving the beauties of Nature; and what is most soothing to those who loved him, a proof that he applied his talents to the noblest purpose, that of celebrating the praises of Him from whom he received them."

All this has only a little to do with the look of Nuneham Reach from the river. One may discern upon it perhaps the seal of Claude, of this and the other poet. It is so with all planted landscape. But apart from that, Nuneham, in its architecture of hill and wood and water, has the trick of the great places in Thames scenery. It is an early feat that promises Richmond.

It is best to see Nuneham Reach from the railway bridge. From any other point it is necessary to see the bridge itself. Soon after its great achievement the river is claimed again by a distant spire. Above Nuneham the towers of Oxford linger, holding the landscape; for the rest of the way it belongs to St. Helen's at Abingdon.

D. S. MacColl.

DISTANT VIEW OF ABINGDON.

CHAPTER III.

ABINGDON TO STREATLEY.

Abingdon—The Abbey—St. Nicholas' Church—The Market Cross—The Ancient Stone Cross—St. Helen's Church—Christ's Hospital—Culham—First View of Wittenham Clump—Clifton Hampden—The "Barley Mow"—A River-side Solitude—Day's Lock—Union of the Thames and the Isis—Dorchester—The Abbey Church—Sinodun Hill—Shillingford Bridge—Bensington—The Church—Crowmarsh Giffard—Wallingford—Mongewell—Newton Murren—Moulsford—The "Beetle and Wedge"—Cleeve Lock—Streatley.

UNLESS they be absolutely black and squalid, all old country towns have a charm of their own. They may possess historical or personal associations of the supremest interest, like Lichfield; they may hold a central place in some dire story of battle and siege, like Colchester; their renown may be architectural, as at Salisbury; or that of a vice-metropolis, as at York. Others there are, and they are in the majority, which have for all attraction quaint streets of gabled houses, and rural environs gay with birds and flowers, and ancient timbered parks watered by quivering streams. The town of Abingdon unites most of these attributes, save that it has seen little of war, and that it is unassociated with any commanding personality. It is handsomer and more shapely than most of the riverside towns in the Thames Valley; and although it is little more than a big village in the centre of a moderately prosperous agricultural district, it is entitled to take upon itself some of the airs and graces due to the possession of distinctions for which many larger places sigh in vain, since it has been a municipal and parliamentary borough since the days of Mary Tudor. A town seated upon a river is nearly always seen to best advantage from the water, and the view of Abingdon immediately after the bridge is shot is very pretty and reposeful. The bridge itself, although not remarkably graceful, is yet exceedingly picturesque. Of great antiquity, it is greyish-brown of hue, and profusely mossed from water-line to coping. Several of the arches are dry, and serve only to carry on the road above, with its irregular rows of oddly-gabled cottages. To the left all is level meadow, backed by belts of woodland; to the right lies the town, the tall, handsome spire of St. Helen's Church, with its flying buttresses, rising high above the red-tiled roofs of the waterside buildings. Abingdon is a land of chestnut-trees. Along the waterside, on the eyots, in the quiet gardens of the old red-brick houses, there are chestnuts. To the stranger chestnuts and grey-stone villas are, indeed, the two most notable characteristics of this pretty little town. In the late spring and early summer the place seems to be surrounded with the peculiarly lovely blossom of a tree which, whatever the season may be, is always pleasing to look upon. The chestnut in England has in modern times been treated with less courtesy than it deserves. It is a better tree in many ways than the elm, which is usually placed only a step or two below the oak. It may not be so graceful, but it is beautiful notwithstanding, and far less treacherous. In such esteem, indeed, was it held by our ancestors that many of their beautiful half-timbered houses, which a careless posterity supposes to have been invariably built of oak, were largely constructed of chestnut, while many an old house is full of admirably carved and polished chestnut furniture.

All that there is of interest in Abingdon centres round the bridge—the two ancient churches, the ruins of the abbey, and the market-cross. So many rich and flourishing towns

ABINGDON, FROM THE RIVER.

grew up, in the far monastic days, around the great abbeys that it is a not unfair presumption that before the Dissolution Abingdon enjoyed comparatively greater importance and prosperity than it does now. It is still a flourishing place, and although its streets are quiet they present no signs of decay. It is true that it did not become a borough until after the Dissolution; but since the charter was granted by Queen Mary, it may have been intended as some solatium for what the townspeople had lost. That they really did lose much is clear. Abingdon was a mitred abbey, and very ancient, having, all legend says, been founded in the seventh century. At the Conquest the abbot held great landed possessions in his trust, and the house was no doubt rich in the portable wealth for which the monasteries were renowned—vessels of gold and silver, censers encrusted with gems, jewelled crosses, and vestments embroidered with cloth of gold. As the abbey grew in riches and independence the monks seem to have taken very little trouble to keep on good terms with the townspeople or with the country-side. Quarrels were constantly brewing, provoked, no doubt, by each side alternately; but the town was stronger than the abbot and his chapter and all the brethren, and about the time of Edward III.'s accession the men of Abingdon and Oxford united to read the monks a lesson they were not likely to forget. A great riot occurred, in which the Mayor of Oxford and the more muscular students of the University lent their aid, with the result that a large portion of the abbey buildings was burned. The town was gradually becoming independent of the large revenues disbursed by the abbot, for it conducted a very remunerative commerce in cloth, and, indeed, an old chronicler tells us, "stood by clothing." Nevertheless, when, in 1538, the abbey went the way of all the other monasteries, Abingdon necessarily received a heavy blow. The remains of the monastic buildings, although not extensive, are picturesque and exceedingly interesting. The abbey precincts probably sloped to the water's edge, since the gateway, which is still in fair preservation, is close to the river, near the market-place. It has been shorn of much of its ornamentation, and now possesses no very remarkable features, either of architecture or of decoration; but it has been carefully conserved, and remains whole and sound. The most attractive portion of the abbey buildings still existing is used as a brewery, and this, like the gateway, has been religiously shielded from other injury than Time inflicts. This portion consists of the abbot's apartments and the crypt beneath. The abbatial parlours have been converted into lofts, while the crypt has returned to what may not improbably have been its original uses—the storage of great casks of the ale for which Abingdon is well famed in its own neighbourhood. The crypt is entered beside a backwater, where grow more of the abounding chestnuts; but to reach the lofts, where once the abbots of Abingdon transacted such secular affairs as the regulation of accounts and the inditing of business letters, one has to ascend a short flight of time-worn steps. The doorways have pointed arches, and the windows likewise, in the main, preserve their ancient appearance. In one of the lofts are the remains of a handsome fireplace, which has been assigned to so remote a period as the reign of Henry III. The gigantic chimney served by this fireplace presents a remarkable and picturesque appearance as seen from the road. To those of an antiquarian turn of mind these monastic remains are very interesting; and they deserve to be better known.

At the corner of the Market-place, adjoining the Abbey Gateway, is the church of St. Nicholas, which, although far less interesting than St. Helen's, nearer the river, yet contains much that is worth seeing and describing. Architecturally it is not remarkable, save for a Norman doorway and an unusual little turret which surmounts the tower, and forms the roof of a minstrels' gallery of great antiquity. Here is the tomb of John Blacknall and his wife, who left many bequests to the town, one of which is still enjoyed by forty-seven poor persons, who receive each a loaf of bread at their benefactors' tomb every Sunday. The monument to this united pair is of great height, and records that, by a rare coincidence, they both died on the same day—the 21st of August, 1625. The epitaph insists upon this touching unity even in death in the undignified language common to inscriptions of the kind :—

> "Here death's stroke even did not part this pair;
> But by this stroke they more united were.
> And what they left behind you plainly see —
> One only daughter and their charity.
> What though the first by death's command did leave us,
> The second, we are sure, will ne'er deceive us."

Among the ancient treasures of the church are a carved font, an ancient lantern in the porch, and the remains of a painted window, with an illegible inscription. Opposite this church, at the

ABINGDON BRIDGE.

side of the Market-place, is the Market Cross, designed by Inigo Jones, erected in 1667, and far too extensively restored in 1853. It is really, like so many similar buildings, a covered market, with space for a considerable number of persons to congregate. The fine timber roof, which has happily not been interfered with, is supported upon stone pillars. This building occupies the site of one of our most famous stone crosses, which the town owed — as it, no doubt, owed much else — to one of the religious foundations. One of the fraternities connected with the Church of St. Helen was called the Brethren of the Holy Rood, and of this godly community no less a personage than Thomas Chaucer, son of the father of English poetry, was a governor. The Brethren of the Holy Rood erected this cross at their own expense, and it has always been believed that Thomas Chaucer had some hand in designing it. Leland, the antiquary, did not overstate the matter when he described it as a "right goodly cross of stone, with fair degrees and imagerie." It had a decorated base, and two tiers of canopies containing statuettes, while upon the top was a carved tabernacle. The treaty with the Scots in 1641 was celebrated by the singing of the 106th Psalm at the foot of the cross by a gathering of two thousand people. Three years later it was demolished by Waller's army, as being a "superstitious edifice." So much admired were the graceful proportions of Abingdon Cross that it was taken as the model for that which Sir William Hollis erected at Coventry. If Chaucer's son really had any part in designing it, we do not know; but it is at least pleasing to fancy that he had. The existing market cross is a not unpleasing piece of work; but many a masterpiece of Inigo might be spared could we but have restored to us the graceful sculptured rood built by the Confraternity of the Holy Cross.

The Church of St. Helen, with its precincts, is by far the most interesting part of Old Abingdon. St. Helen's is an exceedingly handsome, well-proportioned church, such as one rarely finds in so small a town. There has been some internal restoration, and the tower, from which springs the slender arrow-like spire, was renovated at a very large expense in 1885; but, at least in the interior, little violence appears to have been done, judging from the undisturbed condition of the tombs and mural monuments. The church is of unusual size, and its generous proportions speak well for the pious large-heartedness of the founders. The timbered roofs are admirable, carved boldly and simply, and still quite sound. In the chancel the roof is more elaborately carved, and the timbers of the north aisle retain faint blurred traces of once brilliant religious paintings. The church possesses the unusual number of five aisles, named respectively the Jesus aisle, Our Lady's aisle, St. Helen's aisle, St. Catherine's aisle (in which most of the Abingdon worthies are buried), and the aisle of the Brotherhood of the Holy Cross. There are said to be only two or three other five-aisled churches in the kingdom. There are two or three good old tombs to bygone Abingdon worthies, and upon one of them the inveterate punning propensities of our ancestors, where inscriptions of any kind were concerned, is oddly exemplified. This is the tomb of "Richard Curtaine, gent., a principall member of this Corpá," whose epitaph reads:—

> "Our Curtaine in this lower press
> Rests folded up in natur's dress."

The real Abingdon shrine, however, is the resting-place, against the chancel, of John Roysse, who founded the Grammar School. This "pious ancestor" died in 1571, yet there is usually a wreath of flowers lying upon his breast. It is an altar-tomb, with a reclining full-length effigy and a partially-defaced inscription. Good Master Roysse was one of the many charitable benefactors who seem to have flourished in the genial soil of Abingdon. The Grammar School was founded in his lifetime; but in his will he left at least two other charities. He was clearly not the stamp of man who has a mind to be forgotten after death. The upper stone of his tomb, he ordained, was to be the "great stone" in the summer arbour of his garden in London, and the twelve old widows who were to receive each a loaf, "good, sweet, and seasonable," kneeling round his stone every Sunday, were to say, upon receiving their doles, "The blessed Trinity upon John Roysse's soul have mercy." This once picturesque ceremony, shorn of its olden formalities, has, since 1872, been performed in the hall of Christ's Hospital. To play upon numbers and words was one of the conceits of the time, and so it was ordered that, since the Grammar School was established at once in the 63rd year of its founder's age, and in the 63rd year of the century, the foundation should educate 63 boys "in sæcula sæculorum." A small room shut away from the church, and called the Exchequer Chamber, is used as the muniment room of the famous "Hospital of Christ," concerning which much hereafter. Yet another interesting tomb—interesting because it exhibits the monumental sculpture of a century ago in all the fulness of its bathos. Mrs. Elizabeth Hawkins, whom it commemorates, died in 1780, and ordered that a sum of £400 should be expended upon a fitting memorial. The money was duly laid out, the lucky recipient thereof being one Mr. Hickey; and now, after a hundred years, the only people who can look with satisfaction upon the transaction must be they to whom Hickey bequeathed his money. The sorrowing stone cherub in the foreground looks very much as though he had just undergone nursery correction. More attractive is a good and very curious bit of wood carving affixed to the front of the organ. The date is unknown, but its antiquity is probably not great. It obviously represents King David, who, with a gilded crown upon his head, plays upon a dazzlingly gilded harp. Near the door of the vestry hangs the elaborate genealogical tree of one W. Lee, who was five times Mayor of Abingdon, and lived to see 197 descendants. It is dated 1637. In the vestry is a copy of Foxe's "Book of Martyrs," together with a number of Bibles and books of homilies, all having still attached to them the ancient chains by which they were formerly secured. The church registers go back to an unusually early period.

Upon the south-western side of St. Helen's churchyard are the picturesque arcaded buildings which form the more ancient portion of the "Hospital of Christ"—a long low range of half-timbering extending to the river front, where it is joined by a more modern wing of stone. Age has blackened the timber and barge-boardings, and in the sunshine the contrast between the chequer-work of the old and the grey stone of the newer buildings is exceedingly charming. The porch is quite romantic, and such as one rarely sees save in a water-colour. It is supported upon stout oaken piers, the heads roughly but effectively carved; at the partially-open sides is more carved work. It has a steep roof with wide projecting eaves and a diamond-paned lattice in the gable. Almost immediately behind rises, from an irregular red-tiled roof, the graceful carved cupola which lights the entrance hall. When the wide door of stout oak, over which one could climb at need, stands open, the passer-by gets a delightful glimpse of a panelled hall, into which the sun streams not too glaringly through the cupola and the little lattices, imprinting quaint arabesques upon the floor and the black polished wainscot, and making life lovely for the six-and-thirty aged men and women maintained by the charity first instituted by the pious Brethren of the Holy Rood. Over the porch are some curious paintings illustrative mainly of works of mercy. One of them is a view of old Abingdon Cross; and another a portrait of Edward VI., in whose reign the hospital was refounded. These buildings are strongly remindful of the better-known Leycester Hospital at Warwick, but they are not nearly so lofty. This antique porch is on sunny days the favourite spot "where want and age sit smiling at the gate." The interior of the hall is very quaint, and contains some sturdy furniture, suitable for the support of a giant in complete plate-armour. A large table of oak with carved legs was presented, as an inscription upon the frame of the picture hanging over it records, by "Franncis Little, one of ye governors of this hospital" in 1607. In the Exchequer Chamber in the church is preserved a manuscript history of the hospital, written by good Master Little, with the title, "A Monument of Christian Munificence." Here is another portrait of Edward VI., and a curious picture of the building of Abingdon Bridge. It is of great age, dedicated, it would seem, to "Jefforye Barbur and John Howchion." When the brethren of St. Cross first became a corporate body is now doubtful. Francis Little in his history says that the foundation existed in 1388, and there is reason for supposing that it had come into being long before. At the Dissolution the confraternity was abolished, but it was revived by Edward VI., and endowed with three-fourths of the old foundation. The Charter enjoins upon the Governors that they are to keep in repair the four bridges over the Thames and the Ock, to provide food and lodging for fourteen poor persons, and to devote the surplus of their funds to other charitable uses. These funds have grown to such a bulk that thirty-six poor people are now maintained, while Abingdon Grammar School has been rebuilt, and a public park given to the town out of the surplus. Surely, roses should blossom upon the graves of those who founded and re-founded "the Hospital of Christ at Abingdon." Near to this dreamy old-world churchyard is Ock Street, one of the longest and finest streets to be found in an English country-town. It is broad as a boulevard, and is literally crowded with old Jacobean and Georgian houses, some of them so large as to be fairly called mansions.

The Berkshire shore is lined with pleasant houses for half a mile or so below Abingdon Bridge. The towing-path is here upon the Oxford bank, and skirts rich meadows picturesquely studded with large shade-trees. Away to the left lie heavy masses of woodland, such as engirdle the whole of the Thames Valley; on the facing bank are the straggling environs of Abingdon, having, when seen from this point, somewhat of the foreign aspect so often worn by these little waterside towns. But in less than a mile we are amid scenes that are very English. The meadows at first are flat, which, the rather than a blemish, I esteem to be a beauty. The perfection of sylvan and pastoral river scenery, as distinguished from the bold and rocky loveliness of some of our wilder English streams, demands flattish banks, the better to throw into relief the undulating fields and shimmering woodlands which so often close in a homely scene having for relief merely some grey church tower almost hidden among the lofty elms, and the mellowed ruddiness of a farmhouse gable. A little below Abingdon the tiny Ock enters the stream, and so ends its independent existence. Any time from eight to ten in the morning—for, oddly enough, boating-men are rarely up with the lark—camping-out parties may be seen engaged in the serious business of breakfasting, or in the lighter but less exhilarating task of washing-up the cups and saucers, and generally "making tidy" before the day's leisurely pull. As a rule, however, the river is deserted during the whole of the forenoon, even in the height of the season, as, indeed, the towing-path always is, whether it be late or early—at least, upon this portion of the stream. The river banks, from the bridge at Abingdon to Culham Lock, are very charming in summer, to those

CULHAM CHURCH.

who are content with ordinary scenery, and do not expect a famous view on entering every reach. Nearing Culham, the river bends very sharply to the right, and just at the curve a white wooden bridge crosses a beautiful little back-water, brilliantly pied with water-lilies, and thickly bordered with graceful aquatic grasses. Then come fields of standing corn, the sturdy ears sheltering the frail crimson poppies —wheat and tares intermingled. From some hidden spot in the centre of the field comes the loud, harsh cry of the corn-crake, that bird so often heard and so seldom seen. Sometimes the crop is the drooping oats or the "bold and bearded barley;" but whatever be the grain, there is the fat, solemn rook, who reluctantly wheels away from his farinaceous banquet, to hide for a few minutes in the long row of elms in the adjoining field. Close to Culham the stream divides, a broad rushy channel flowing past Sutton Courtney, with its venerable Edwardian Manor House and the well-known weirs, while a straight, narrow, and not very picturesque cut, makes direct for the Lock. In passing there is a very pretty glimpse of Culham Church, which stands out effectively from a background of trees, and looks in the distance the ideal of an old parish church. A nearer view reveals that most of the building is very modern, and that even the square tower dates only from the days of William and Mary. Culham is a pretty and interesting little place, and still happily preserves its village green.

A few yards below Culham Lock the river assumes its old proportions, the water from the deep millpool at Sutton, where there are fishes indeed, now forming the old main channel with the cut as a mere contributary. Hereabouts there are usually one or two camping parties, the proximity of a lock-keeper's cottage being a convenience which none can appreciate so well as a tired oarsman. Although the immediate banks continue flat, the country around grows more rugged, the meadows and cornfields become billowy, and sloping gently up long miles ahead, although apparently no farther distant than the next parish, is seen Wittenham Clump, with its smooth grassy sides and little grove of trees atop. Hence away to Shillingford it is rarely out of sight, for the river winds so sinuously through the valley over which the Clump watches that between Clifton Hampden and Day's Lock it describes a perfect semicircle. The Clump forms a majestic background to many a stretch of varied timber and parti-coloured fields. Something like a mile below Appleford Bridge commences another unlovely necessary cut—a kind of graduated penance in preparation for the severer *supplice* of Clifton Lock. He who elects to see the river-land from the towing-path has a decided advantage over the oarsman, where these cuts and locks are concerned. This particular cut is more tolerable than some of those which the exigencies of navigation have rendered necessary. The Berkshire shore has a fringe of plantations and mossy creepered banks, which compensate somewhat for the nakedness of the Oxford bank at this point. At the end of this straight channel is Clifton Lock. The keeper's cottage is in summer a lovely picture, for it stands in a little garden ablaze with brilliant flowers of the old-fashioned stock viewed with disfavour by the scientific gardener; while the cottage walls are covered with creepers yellow and russet. Just clear of the Lock the main stream re-enters the channel, and a bend in the river's course reveals the heights of Clifton Hampden and the beauteous vale beneath. The long, red-brick bridge of six pointed arches, which has only of late years superseded the ferry, is in itself a picturesque object. The surrounding country is flat, and so is most of the village; but the bold hill which rises with a sharp slope from the water is crowned by the church and the vicarage. From the summit to the edge of the stream the bluff is densely timbered, and thick belts of woodland line the Oxford bank for some distance below the bridge. The delightful little village relies upon Nature for all its charms, for it has no history. Nor can it be said that the church is very interesting, save as a favourable example of Sir Gilbert Scott's early skill as a restorer. Sir Gilbert's work here was done in 1844, when he was comparatively a young man. The old work is really ancient, for Clifton Church was originally a chapelry served from Dorchester Abbey. The reredos is in mosaic; but the most remarkable thing in the church is an altar-tomb to the late Mr. G. H. Gibbs, at whose cost the building was restored. The recumbent marble figure is a portrait. The churchyard is kept with unusual neatness, and numbers of the graves are covered with flowers.

Its altitude is such that it affords delightful views up the river towards Abingdon, and down towards Day's Lock and Sinodun Hill. The serpentine course of the river is very striking as seen from this height; and even here, with the naked eye, Wittenham and Sinodun seem to bar the stream.

At Clifton Hampden, in the season, there is usually a house-boat or two moored among the masses of water-lilies which profusely strew the stream near the bridge, and a more charming spot, away from such "fashionable" places as Goring, Henley, or Maidenhead, could hardly be selected as the anchorage of these leviathans of the upper Thames. The neighbourhood abounds in rural walks, and in subjects both for the pencil of the artist and the pen of the man of letters. One of the most charming "bits" at Clifton has neither been sketched nor described quite so often as it deserves to have been. The "Barley Mow" is assuredly the oddest

CLIFTON HAMPDEN CHURCH.

and quaintest of inns on the river. It lies on the Berkshire bank, in a little roadside corner all to itself. What its age may be it would be difficult to tell; but its high, overhanging roof is thatched and its walls are half-timbered. The diminutive casements, about the size of the door of a rabbit-hutch, admit just enough of light to heighten the interior effect. Broad masses of light are out of place within such venerable walls. The brick-floored kitchen—or maybe it is the parlour—is delightfully snug; the walls panelled darkly all round; the honest raftered ceiling so low as to do away with the necessity ever to stand upon the naked wooden settles to reach things; the fireplace extending across one whole side of the room, the oddest imaginable cross between an old-fashioned ingle-nook open grate and a modern kitchen range; the chimney-piece garnished with many a brightly-burnished pot and pan. No demure Phyllis makes her appearance; but the cider—we are in a great cider country—is nectar. At the back of the inn is just such a queer little garden as Dickens loved to write about. All the flowers were our great-grandmother's, and, indeed, modern daintinesses would sadly mar the antiquated aspect of this typical roadside inn of a day that is long past.

At Clifton Bridge the towing-path crosses to the Berkshire shore, and for the next two miles the scenery is, perhaps, the prettiest, with the exception of Clifton itself, between Abingdon and Wallingford. The Oxford bank is clothed luxuriantly with trees, out of which now and again peeps, half unperceived, the canvas shelter of a camping party. These occasional encampments are almost the only sign of life, so far as the banks of the river are concerned. Between Clifton and Day's Lock the country is remarkably solitary. The waterside meadows are nearly all empty; but here and there a herd of cattle browses leisurely, or, if it be high noon, shelters itself from the heat and the tormenting flies under the lee of the thick hedgerows. Pedestrians are never seen. That it is good to row upon a beautiful river, but undesirable to walk by the side of it, appears to be the popular idea; but despite the physical exhilaration and the æsthetic delight of the rhythmical swing of oars, the river can be seen best from the towing-path, and if the love of walking-tours had not very largely died out we might expect to see the banks of the upper Thames as much frequented as its waters. It is often possible to pass between Clifton and Day's Lock without meeting either man or boat, which seems a little odd, since that reach is in high favour during the season. To the walker upon the towing-path this silence and vacancy become oppressive, and the sudden splash of a water-rat striking out from among the rushes is quite startling. The Berkshire shore is flattish here; but there are swelling uplands beyond, and the Wrekin-shaped Sinodun Hill looms quite close upon the left. Presently there stands out from among the trees on the Oxford bank an old church with a very long nave and tall tower, with an unusual high-pitched red roof, topped by a vane. That is the famous Abbey Church of Dorchester, the solitary remnant of the ancient grandeur of the olden capital of Wessex. A little farther is Day's Lock, with the ferry between Little Wittenham and Dorchester, where, even in a season of drought, the water is unusually full and brimming, the result, perhaps, of the wedding near by of the little Thame with the more classic and magnificent Thames, or Isis, as the poets have preferred to call it. This conceit owes its origin almost entirely to such comparatively modern poets as Warton and Drayton, though Spenser, in the "Faërie Queen," seems to have originated the legend in somewhat of a backhanded way :—

> "The lovely bridegroom came,
> The noble Thamis, with all his goodly traine,
> But before him there went, as best became,
> His auncient parents, namely, th' auncient Thame;
> But much more aged was his wife than he,
> The Ouze, whom men doe Isis rightly name.
> Full weak and crooked creature seemed shee,
> And almost blind through Eld, that scarce her way could see."

Nearly opposite Dorchester there is an eyot adorned by a remarkably fine chestnut, while between Clifton and Day's Lock are others which bear little save the humble, useful osier. At Day's the towing-path crosses into Oxfordshire. Dorchester, which makes a very picturesque appearance from the river, since it stands upon a greater elevation than the country through which we have passed, is about half a mile from the Lock. The field-path, which runs for some distance through a most unpoetical turnip patch, skirts the famous Dyke Hills, the Roman fortifications upon

ABINGDON TO STREATLEY.

which sheep most peacefully browse. The fortified camp of which these earthworks formed part is supposed to have been guarded on one side by the Thames, on the other by the Thame, and must, consequently, have been of enormous strength. Dorchester, which fell from its splendour and ceased to be a capital more than a thousand years ago, is a quaint little village, in which the antiquarian

DORCHESTER, FROM LITTLE WITTENHAM.

voyager can spend some hours of crowded interest. Its three or four old streets are full of strange twists and oddly-gabled houses, and the number of old-fashioned inns is remarkable, it being remembered that the population of the place but slightly exceeds a thousand. There was surely never a more complete fall from a high estate than that suffered by Dorchester. Not only was it the capital of Wessex, but it was the seat of the great bishopric eventually removed to Lincoln; and the Venerable Bede records that Dorcinea was full of richly-garnished churches. Twelve centuries and a half ago Cynegils, King of Wessex, was baptised there, as of right in his capital, by the sainted Birinus. The bishopric, after being removed to Sidnacester, was restored to Dorchester, and it was not until after the conquest that Lincoln was finally selected as the home of the Bishop-stool. The Abbey Church is the glory of the place, since it is not only exceedingly fine in itself, but is the sole survival of the dim ages in which Dorchester was a cathedral city, and the capital of one of the Heptarchical kingdoms. The Church of Dorchester Abbey was undoubtedly built upon the site of the Saxon Cathedral, of which some fragments, such as the north wall of the nave, and an arch or two, probably formed part. As it stands now, the church is a patchwork of styles, from the Norman to the Tudor. It is of great size, the length from east to west being 183 feet, and the area over 10,000 square feet. Dorchester churchyard has sometimes been considered handsome; but it is too ragged to be fairly so described. Near the south door is an ancient churchyard cross, the shaft of which is very much dilapidated, but the head has been well restored. The porch to the south door is Tudor work in stone, with a good timbered roof. The interior of the church is not unremindful, at a general glance, of St. Albans Abbey, since the nave is entirely blocked by the tower. Restoration was commenced by Sir Gilbert Scott; but there is so much to be done, and the cost of doing it is so considerable, that the work will probably not be finished for years to come. At the bottom of the north aisle is a large collection of sculptured stones, which, no doubt, before the Dissolution, formed part of the monastic buildings. They were mainly obtained from an old house in the village, which would seem to have been largely built with materials taken from the Abbey, and it is intended to build them into the fabric as opportunity offers. The western end of the building is somewhat gloomy, a defect which might without difficulty be removed by the uncovering of the handsome west window, which has long been bricked up. Dorchester has one of the very few leaden fonts of Norman workmanship which now remain to us: there is another at Long Wittenham, on the opposite side of the river. Round the bowl are cut, in high relief, the figures of the eleven apostles, Judas being, of course, inadmissible. What, had not the tower intervened, would have been the western end of the nave, forms an antechurch, which is used for the minor services. A pillar in this chapel has some quaint carvings near the base. One of the most ancient portions of the church seems to be the Lady Chapel, at the eastern extremity of the south aisle adjoining the chancel. The altar here was erected in memory of Bishop Wilberforce, of Winchester. There are four altar-tombs in the Lady Chapel, the survivors of probably a much larger number. Two are to ladies; the others represent Crusaders. The feet of each rest upon a lioncel. Close to these tombs is the brass of Richard Bewforest, to whose piety posterity owes the preservation of this Abbey Church. In 1554 Master Bewforest purchased the church from the hands of the despoilers, paying therefor £140, which, although a goodly sum for his day, was assuredly not extravagant. Here, too, is an unornamented brass to an undistinguished person, named Thomas Day, with the following odd epitaph, dated 1693:—

> "Sweet Death he came in Hast
> & said his glass is run;
> Thou art ye man i say,
> See what thy God has done."

In the chancel there have been many fine and elaborately ornamented brasses, but only a few remain in their integrity. One of the most perfect thus records another Bewforest:—" Here lyeth Sir Richard Bewfforeste: I pray thee give his sowl good rest." This Richard was not a knight, but an ecclesiastic, as the brass, upon which he is represented with cope and crozier, proves; and the prefix was given him according to an ancient custom, of which we have an example in the person of Sir Oliver Martext, the priest in *As You Like It*. On the north side of the chancel is the wonderful "Jesse Window," which has been so often

SINODUN HILL AND DAY'S LOCK.

described that it has become one of the best known of our ecclesiastical antiquities. The ornamentation of the window takes the form of a pictorial pedigree in stone, the tree having its root in the body of Jesse, each progenitor of the line of David being represented by a small stone figure; but the effigies of Christ and His mother have disappeared. Upon the glass of the window are somewhat rude representations of the chief members of the line of Jesse. This very remarkable window is in good preservation, notwithstanding that it is now at least five centuries old. A word must be said of the fine timbered roofs of the Abbey. That of the nave, supported upon most graceful clustered columns, is really magnificent, while the groined roof of the Lady Chapel possesses a lightness and grace which such work often lacks. There are still many brasses, together with an enormous number of flat stones in the church, but the majority of the brasses and incised stones have been damaged, apparently with wilful intent. Here and there an elaborate matrix sadly suggests the treasures we have lost. Against the lych-gate at the western end of the churchyard is one of the largest and most luxuriant chestnuts to be noticed even in a neighbourhood full of large chestnut trees. The gate and the tree, with the great grey church for background, fashion themselves into a lovely picture. Beyond the church and the quaint old houses there is nothing of interest in Dorchester save the building now occupied as a national school, which was formerly the grammar school. The interior, full of great timber beams and joists, is very picturesque. It is believed to have been the refectory of the Abbey, and an antiquity of some seven centuries is assigned to it.

Opposite Dorchester is Sinodun Hill, which has been growing gradually nearer for several miles during our leisurely progress down-stream. If it be good climbing weather—that is to say, not too hot—Sinodun should not be passed heedlessly by. The climb is a stiffish one, but once the shelter gained of the little clump of trees atop, there is ample compensation for an exercise such as Englishmen are not usually afraid of. From this eminence the country lies displayed as though upon a map. The shining river twists and curvets like a snake in agony; upon its timbered banks repose tiny villages, distinguishable in the mass of foliage only by the vanes upon the steeples and the thin quivering lines of smoke which melt into nothingness just above the tree-tops; roads and railways look straight and uncompromising indeed beside the sinuous stream. The country is multi-coloured—the fields green and brown and yellow, with here and there a great square of black woodland. The sun seems to shine upon some and to leave others in shadow, while over all there move flecks of trembling light. The view in the direction we are travelling is closed by swelling downs destitute of all colour but the dim grey of distance.

Down below us, near the weir, industrious anglers are barbelling or spinning for jack, for hence almost to Shillingford are fine fishing grounds. Here the river bends somewhat towards Dorchester, and it is long ere we pass out of sight of the Abbey. Upon the Berkshire shore are uplands, broad, swelling, and cultivated to the utmost rood. These rolling uplands never look better than in haymaking or harvest time, when the cocks and sheaves are yellowing in the sunlight. The regular, almost square, boundaries of the fields suggest a green and yellow chessboard, and at seedtime the mathematical furrows are as straight as though cut by a machine. The nicety of vision, and the accuracy of touch with which a ploughman cuts a furrow are astonishing in one who usually has instinct and eye alone to guide him. After all there is something intellectual in the following of the plough, and the peculiar qualities required of the ploughman are such that it is not altogether surprising that both science and letters have drawn notable recruits from the furrowed field. Almost until we reach the next ferry, a couple of miles below Day's Lock, Dorchester still straggles along parallel to the river, and the last glimpse of its red roofs from a bend in the stream is exceedingly picturesque. The towing-path ceases abruptly at the ferryman's quaint little cottage, and the *venue* for the pedestrian changes for a time into Berkshire. The stream just here is very charming to the lover of rivers, for although both shores continue flat they are dotted

SHILLINGFORD BRIDGE.

with clumps of woodland, and the water's edge is gaily caparisoned with verdure. The towing-path for a short distance grows almost wild for so highly civilised a country as that through which the Thames flows, and the pedestrian wades to the knees through rank brambly grass. A few more minutes and we reach Shillingford Bridge, with its four grey arches. At the Berkshire end of the bridge is that pretty rural inn the "Swan," a favourite abiding place of boating parties which include ladies. The little lawn is dotted with gay costumes of coolest tints and softest texture, for a lazy afternoon hour or two is not ungrateful upon the banks of Thames in the dog-days. On the Oxford bank is a cluster of tiny cottages, each in an ample garden full of those brilliant old-fashioned flowers which the cottager loves so well. The diminutive latticed windows are garnished, too, with geraniums and fuchsias; honeysuckle climbs to the not very lofty gables, and the little trellis-work porches are aglow with the cool foliage and delicate tints of clematis. The road is thickly bordered with elm and beech, and beyond, shining brilliantly in the afternoon sun, are long red ranges of barns and cow-sheds, darkly-roofed and golden-walled ricks of last year's hay, side by side with the brand-new thatch of the yellow stack that has just left the thatcher's hand. From the bridge itself there is a pleasant view up and down the river over what our grandfathers would have called a "fine champaign country," flat and pastoral on the Oxford shore, but swelling into bold wooded undulations on the opposite bank—such a stretch of varied scenery as most becomingly wears the sober darkling tints of autumn. When the wind swirls the brown sapless leaves into the turbid river, and the bare stubbles echo to the crack of the breechloader, Nature hereabouts has that distinct autumnal charm which is never more delightful than in a sylvan and pastoral landscape.

From Shillingford to Bensington the towing-path is again in Oxfordshire. The river banks become more frequented, and the complete angler abounds; for most renowned baskets are constantly obtained from this pretty stretch of water. The eyots are luxuriant with osiers, and in the osier harvest punt after punt lies heavily laden with the lithe, flexible sticks which the men cut and tie into bundles with astonishing deftness and rapidity. Many of these little osier-covered islands are surrounded with white and yellow water-lilies, which seem to have an affection for such a situation. The square tower of Bensington Church has a venerable appearance; but the really ancient church has been restored into newness. Consequently, nothing remains of any great interest; but, most happily, the reforming zeal of the re-builders stopped short of interfering with the handsome chancel-arch. On the south wall of the nave is an inscription which, from its very oddity, deserves to be recorded:—

> M. S.
> To the pious memory
> of Ralph Quelch and Jane his wife
> who slept } together in 1 { bed by ye space of 40 yeares.
> now sleepe }   { grave till Ct. shall awaken them.
> He } fell asleep Ano. Dni. { 1629 } being aged { 63 } yeares.
> She }   { 1619 }   { 59 }
> For ye fruit of their { labours } they left { ye new inn twice built at their own charge.
>   { bodies }   { one only son and two daughters.
> Their son being liberally bred in ye University of Oxon thought himself bound to erect this small monument
> of { their } piety towards { God
>   { his   }   { them
> Ano. Dni. 16....

Epitaphs in this form are by no means uncommon; but it would be difficult to find one of quainter conception. Even the surname of the worthy proprietors of the "New Inn" has a Dickens-like grotesqueness. Bensington is interesting to lovers of English literature as having belonged to the Chaucers, from whom it descended to the De la Poles. Bensington Lock is below the village, and oarsmen pulling up to Oxford have learned to beware of the dangerous cross-current at the weir. Near the lock the tow-path crosses again to the Berkshire shore. Hence away to Wallingford the country becomes much more picturesque. The Oxford bank is most profusely wooded; groves of willows and alders edge the stream; while farther ashore glades of elm and chestnut perfume the air. Overshadowed by trees, whose branches intertwine, is a pretty red-brick boat-house, into which as we pass disappears a gaily-freighted boat, seeming to pass from brilliant sunshine and rippling river into the dark recesses of some dusky cavern. Then the woodland opens out, the scenery becomes park-like, and through the clumps of oak which stud the foreground we get glimpses of Howberry Park, a more than usually handsome Elizabethan house, the successor of a hardly more picturesque Jacobean building destroyed a century ago by the flames which await every country-house, be it soon or be it late. Howberry Park, once the seat of the Blackstones, lies in the parish of Crowmarsh Giffard, almost opposite the town of Wallingford. The vestry-door of Crowmarsh Church is riddled with bullets—reminders, it is said, of the last siege of Wallingford, at which time this door hung in the west entrance to the church. The first view of Wallingford is not very prepossessing. Against the bridge rises the tall and unutterably inelegant spire of St. Peter's Church, the hideous product of a mind unhappily diverted from law to ecclesiology.

Wallingford possesses interesting memories, although its visible antiquities are not numerous. The town was of consequence in Roman times, and a line of splendidly-preserved earthworks, thrown up by Latin-tongued warriors, is to be seen in a field near the railway station. The Castle of Wallingford underwent sieges innumerable, since its comparative nearness to London rendered its possession of importance to each side in the dynastic wars of the Middle Ages. It was held for the Empress Maud; it resisted stoutly in the behalf of that clever scoundrel, John Lackland; it was garrisoned for Charles I., but was compelled to surrender, and the Parliament made short work of its keeps and battlements. The fortress was not entirely destroyed, and the mutilated remains are carefully preserved in the gardens of the present Wallingford Castle. In the museum at the Castle there is an interesting collection of antiquities relating to the town and the fortress. The importance or the piety of the town must have been far greater previous to the Cromwellian civil wars than either is now, since there were then fourteen churches, whereas there are now but three. Beyond one or two tablets to local benefactors, there is nothing interesting in St. Mary's Church on the Market Place. St. Peter's is

ST. PETER'S CHURCH AND WALLINGFORD BRIDGE.

the burial-place of Sir William Blackstone, "one of the judges of His Majesty's Superior Courts at Westminster," and Recorder of Wallingford, who built the flint tower, with its uncomfortable spire—both conspicuous monuments of the architectural decadence—and died in 1780. In the Council Chamber of the Town Hall there is a modern portrait of the judge in robes and bag-wig. It is charitable to suppose that his lordship's legal acumen was superior to his architectural taste. The most interesting tomb in the churchyard is that of Edward Stennett, the friend of Bunyan, who may have died any time between 1705 and 1795, since the third figure of the date has become obliterated. Among the portraits in the Town Hall is one of Archbishop Laud ascribed to Holbein. The date of 1635 upon the painting indicates that the author of the ascription was daring even beyond the usual audacity of such persons. The presence of Laud's portrait is explained by the double fact of his being a Berkshire man and a benefactor to the town. In common with most of the towns in the Thames Valley, Wallingford contains many good red-brick houses, chiefly of Georgian date.

The river, after leaving Wallingford, widens a little, and there is a continuation of the park-like meadows. A short distance down stream is Wallingford Lock, which is a lock only in name. Here the towing-path deserts the Oxford for the Berkshire shore, and the long and lovely reach which ends at Moulsford Bridge begins. This spot marks the commencement of the stretch of meadow, hills, and woodland, which makes the delight of Goring and Pangbourne. The Oxfordshire bank is not merely studded, but is thickly overhung, with trees and undergrowth, beneath whose shade many a boat is moored for those aquatic flirtations which are among the most enchanting of summer diversions. Directly one gets clear of Wallingford the wooded

MOULSFORD FERRY.

heights about Streatley come in view, with a glowing "scarf of sunshine athwart their breast." On the Oxford bank, halfway to North Stoke, more or less, is Mongewell House, a delicate bit of white in a setting of green lawns and venerable trees. Once Mongewell was an episcopal retirement, to which the Bishops of Durham resorted for relief from the fatigues of administration. It was admirably suited to such a purpose, since it is a silent and contemplative spot—the more peaceful, perhaps, from the contiguity of the little Church of Newton Murren, a marvel of the miniature, with a tiny chancel, and a belfry no bigger than a dove-cote. Any monotony there may be from this spot to the ferry at North Stoke is relieved by the Streatley Hills, looming ever larger as the boat swings down the reach, and by the fine clumps of timber which line the river bank on each side. Many a sweet rural picture is passed on the oarsman's highway between Newton Murren and Moulsford Bridge, and in such a country all seasons of the year, and all times of the day, have their charm. The early-morning hours upon the riverside provide unending delight to the real lover of nature. Everything is fresh, crisp, and blithe, for the life of the fields and hedgerows is busy and bustling long before the earliest man's breakfast-time. The ideal climate, cool but not cold, exhilarating, buoyant, redolent of the delight of life, would be a perpetual summer morning, such as it is from five until nine. Every sight gratifies the eye. Then the dew is still heavy upon the hedgerows and the tall aquatic grasses, and where there is a bit of furzy country, there is a tear in every golden flower of gorse. The atmosphere is clearer and more elastic than later in the day. The far-distant rush of trains, the only reminder that there is a world beyond the horizon, and that its daily fret has begun, which at noon is a mere rumble, in this crisp air is sharp and almost shrill. The ring of the scythe under the whetstone many fields away sounds but a few yards off, and the metallic clang of the stable clock at some country house, hidden behind the belts of woodland, half-an-hour's walk as the crow flies, is distinct as the raspy cry of the corn-crake in the yellowing wheat near by. It is hard to say at what season of the day this stretch down to Moulsford Bridge is most charming. To my taste it is the early morning; but poets and lovers would probably prefer sunset, not to say moonlight.

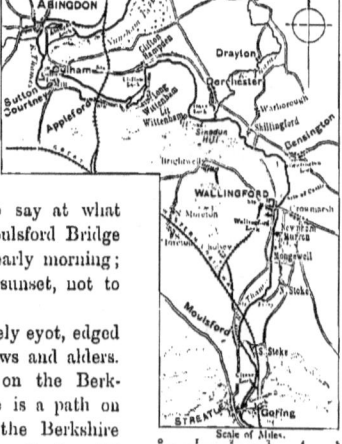

ABINGDON TO STREATLEY.

Against Moulsford Bridge there is a lovely eyot, edged with flags and rushes, and bushy with willows and alders. In time of drought the furthermost arch on the Berkshire shore is not uncommonly dry. There is a path on each side of the river just here; that on the Berkshire bank is the more enticing, for it is quite romantically wild and undulating; but the towing-path proper crossed into Oxfordshire at Stoke Ferry a little further up. It is well worth risking trespass and climbing to the railway bridge for the sake of the fine view up and down the river. Looking back the way we have come, the country is rich, pastoral, and full of trees; ahead the prospect, while equally sylvan, is far more varied. The river winds but little, and the long reach past Moulsford Ferry is in sight for some distance, but the banks are more park-like, and the land begins to swell towards the background of hills that closes in the view, the outposts of the range of downs which beautifies the river beyond Streatley. The brimming, almost straight, reach of water immediately below the bridge is one of the most interesting spots on the river to the muscular generation, since upon it are rowed the trial eights of the Oxford University Boat Club. Close to the bridge the perch-fisher is usually in great force, for around the eyot the perch dwells in numbers. It is but a short distance hence to the Ferry, where the water is remarkably deep and limpid. Opposite thereto is the oddly-named "Beetle and Wedge" Inn, a quaint, three-gabled old place, overgrown with ivy and shaded by clumps of luxuriant elms. "The Beetle" is a grateful halting-place, and its brick-floored parlour a cool retreat from the glare of the outer world. There is usually a garrulous villager or two, in the long-descended smock-frock beloved of the older generation of peasants even in these changeful days, who will pause in the discussion of their mugs of brown home-brewed to greet the stranger with the old-fashioned courtesy which still happily clings to their class. The "Beetle and Wedge" is an odd old place, and although not nearly so original as the "Barley Mow" at Clifton, it has the low roofs and capacious fireplaces which add so much to the comfort of an ancient hostel. It is really astonishing how large a number of our old wayside inns have survived the crushing blow dealt them by the abolition of the stage-coach. There they stand still, with their venerable gables, handsome red roofs, and ample chimneys, eloquently suggestive of warmth and good cheer for tired travellers. In a comfortable old-fashioned inn the crusty loaf, the hunch of well-seasoned Cheshire, and the tankard with "a good head to it," like David Copperfield's birthday treat, have a zest and flavour which are always lacking elsewhere; the result, no doubt, of their being usually eaten during the exhilaration following upon physical exercise. These ancient Thames-side inns possess a charm peculiar to themselves, due largely to their lovely surroundings and to the river flowing beneath their windows.

From the "Beetle and Wedge" to Streatley and Goring Bridge, the goal of our pilgrimage in this chapter, the towing-path keeps to the Berkshire bank. As we near Cleeve Lock the scenery becomes yet more sylvan. The river is densely lined with trees, the more especially on the Oxford shore, and the stream winds just enough for picturesqueness. Groups of splendid beeches dot the country, and the water is enlivened by many a boatful of flannelled rowers and pink-vested sirens. Ladies appear to have recognised, with intuitive taste, that pink and white are two of the most effective colours for river wear, and the Thames, in all the fashionable reaches, owes much of its vivacity to the brilliant hues of its attendant water-nymphs. However solitary the river may be in some parts, as between Clifton and Dorchester, for instance, there is enough of life and movement within hail of Goring. The neighbourhood of Cleeve Lock is a favourite haunt for house-boats and campers, since there is nothing prettier on that side of Abingdon until such famous spots as Henley and Maidenhead are reached. The house-boats which take up their moorings hereabouts are usually of the larger and more elaborate pattern. The little muslined windows are gaily decked with flowers, there is a miniature flower-garden upon the flat roof, and where the roof overhangs are suspended Chinese lanterns, gorgeous with many a brilliant stripe and spot. A graceful white-robed figure, in a coquettish pink sash, seated in the stern, is not the least attractive object in the landscape. The roar of the Streatley weirs below is plainly heard, and many are the lovely glimpses of the brimming, rushy river between the lock and the bridge. Overhead rise, close at hand, the broad, rolling hills, upon which the sun casts shade and shine in successive flecks. The clouds, alternately deep blue and flaky-white, seem to cast their moving reflections upon the crest of the hills, for the gilded sunshine melts with delicate gradations into soft, shimmering shadow. Half a mile or so below Cleeve Lock the stream divides, the cut to the left going to Goring Lock and the main channel to Streatley. From the point of divergence to Streatley and Goring Bridge is but a brief pull, and few pilgrims of the Thames will desire to push on without halting for a while at this pretty village. Near the bridge is a mill, fed from the river, looking very picturesque with its steep gables and high-set dormer windows. The weirs here are favourite sketching grounds, and almost daily in summer and early autumn easels are pitched in the wise represented in the final illustration to this chapter. These weirs are exemplars of the picturesque. Roughly built up with stone and stakes, they are overgrown with furzy vegetation, to which the water, as it pours foaming down the cascade, forms a charming contrast. There are few prettier glimpses of Thames scenery than are to be had from the long white toll-bridge which connects Goring with Streatley. Looking down are the thick woodlands about Cleeve Lock, with the rich, timbered meadows on the Berkshire

STREATLEY MILL.

bank. Upward, towards Goring and Pangbourne, the course of the river is seemingly stemmed by the downs, which are covered with herbage and timbered to the water's edge. The weirs, with their tumbling waters, and the little eyots, cumbered with tall osiers, add to the picturesque diversity of the scene. The twin villages themselves are embosomed in foliage, which in the wane of summer takes many changing tints.

Although it is not a very distinguished spot, historically speaking, Streatley has far-reaching memories. Ina, King of Wessex, is mentioned in the Cartulary of Abingdon Abbey as having given a piece of land there in 687. After the Conquest the manor was part of the rich booty secured by that bold brigand Geoffrey de Mandeville. The church, which nestles among some grand old trees at the foot of the village, near the waterside, is ancient but hardly picturesque. Its patron, oddly enough, is doubtful, but is believed to be either St. Mary or St. John the Baptist. The massive square tower is well preserved and dignified. There is some uncertainty as to the date of the church, but it appears to have been built by Pone, Bishop of Sarum, in the first or second decade of the thirteenth century. He it was who endowed it, and some of the architectural details are similar to those in the bishop's own famous cathedral. The oldest funeral inscription in the church is upon a brass, dated 1440, in memory of Elizabeth Osbarn. This brass, like one or two others, is very well preserved, and still bears the full-length figure of the lady. Large families appear to have been very common in the Thames Valley in the olden times, as numberless inscriptions in riverside churches testify; and it is not surprising to find here a brass, dated 1603, to a parent of eleven daughters and six sons. The village has a pleasant street on the brow of the hill, with some good old houses shaded by older trees. Streatley is a delightful place to halt for the night on a boating or walking excursion. Its material advantages are that it has capital accommodation for the tired walker and rower, and that the proximity of Goring Station makes it easy to bring up the heavy luggage, without which ladies are not happy, even on the river. Of its more æsthetic attractions I have already spoken. To the dweller in towns it is unspeakably delicious to be lulled to sleep and gently awakened by the musical plash of the weirs, while a stroll at dusk along the river bank is full of delights. In the gloaming the ruminating, sweet-breathed kine loom mistily as they lie sociably grouped under the lee of a protecting hedge. On the river twinkle through the gathering night the lamps of the house-boats, the Chinese lanterns, depending from the overhanging roofs, glowing through their fantastic filaments like great transparent fire-flies. And but for the rush of the weirs, the dip of a belated oar, and an occasional ring of laughter from the huge, blackly-outlined boats, the night is silent.

J. PENDEREL-BRODHURST.

THE THAMES AT STREATLEY.

CHAPTER IV.

STREATLEY TO HENLEY.

Streatley, the Artists' Mecca—Goring *versus* Streatley—Goring from the Toll-gate—Streatley Mill Weirs and Backwaters—Antiquity of Streatley and Goring—Goring Church—Common Wood—Basildon Ferry and Hart's Wood—A Thames Osier Farm—Whitchurch Lock—Pangbourne Hardwicke House and Mapledurham Caversham Bridge—Reading and its Abbey—A Divergence to the Kennet, with calls at Marlborough, Hungerford, and Newbury—The Charms of Sonning—"The Loddon slow, with verdant alders crowned"—St. Patrick's Stream Shiplake Weir—Wargrave and Bolney Court—Park Place—Marsh Lock—Remarks on Thames Angling—The Approach to Henley.

"THE village swarms with geniuses and their æsthetically dressed wives," was the touching lament written in "Our River" by Mr. Leslie, R.A., with regard to the Berkshire village of Streatley. The sentence is, in some senses, both a description of the place as you may see it at almost any time during the summer season, and an indication of the reason of its popularity amongst artists. No doubt it is the fashion in the sketching months for mere idlers at the palette to saunter and pose up and down the village street, in company with the strangely-dressed women-kind who some years ago provoked an outburst from a Royal Academician. But it is also, and has long been, the resort of genuine workers with the brush, who make Streatley their temporary home because, in the long white bridge, shady backwaters, lively weir, busy mills, woods, and hills, they find materials worthy of their ambition and their care. In the Thames Valley this portion of the river may be pronounced the Mecca of landscape painters. Streatley, however, is the fashion, because it is honestly deserving of such a distinction. Unlike many popular stations, it does not owe repute to one distinguishing attraction, but to many advantages which, in combination, raise the village to a high position upon the catalogue of places to be enjoyed, talked about, sketched in water colours, immortalised in oil, and haunted by the inoffensive people referred to in the first line of this chapter.

Streatley receives more assistance from Goring, however, than is generally acknowledged in set phrase. The Oxfordshire village on the left bank is, indeed, as by common consent, ignored in conversation, the word Streatley doing duty for both sides. The two communities are separated, not only by the river, which, after the straight length above, widens out into unusual breadth, but by the toll-bridge, which fixes a coin of the realm as an additional barrier between the few hundreds of persons who constitute the respective populations of Goring and Streatley. The villages possess certain characteristics in common. To each is allotted a mill. That of Goring is the more modern, and probably best furnished with appliances for contributing to the trade and commerce of the country, and its rapid little stream is marked "private" to warn off the ubiquitous angler who may look with longing eye upon the shoals of barbel which congregate in its deep strong current. The mill at Streatley is quite another affair—time-stained, decidedly picturesque in its antique pattern of architecture, and maintaining to this day a simple half door, suggestive of—

"The sleepy pool above the dam,
The pool beneath it never still,
The mealsacks on the whiten'd floor,
The dark round of the dripping wheel,
The very air about the door
Made misty with the floating meal."

The country behind Goring is stamped with strong characteristics by the receding hills, which soon develop into the historic Chiltern range; and at the rear of Streatley we mount, direct from the village, the grassy sides of the chalk downs of Berkshire, which, geologists maintain, were once a continuation of the Chilterns. The lock and weir are on the Goring side, but the distinction is to some extent nominal, since, standing upon the crown of the long wooden bridge, it will be seen that there are two weirs and backwaters, imparting a special animation to the character of the river. The eye wanders with delighted satisfaction from the merry streams to the reedy eyots, to the grand trees of the one side and the osiers and green meadows of the other, while weirs and backwaters play and plash throughout the livelong day. Above the lock, the Thames, after broadening for the express purpose, throws an arm each to Goring and Streatley, and the Goring weir, within the distance of half a mile, is the primary cause of several of those sequestered backwaters which add so many potent and diverse charms to the Thames. The stream issuing from the Streatley mill is too near, for proper effect, to the spectator who stands upon the bridge; and requires to be looked at from the meadow, to which the tow-path crosses at the bridge from Berkshire to Oxfordshire. While at Goring there are many private grounds adorning the bank, with a rich background of shrubbery, ornamental walks, gay flower-beds, and pleasant residences, at Streatley we have the inn, the boat-builders' and timber-yard, and the pretty cottage gardens of the waterside and of the straggling street extending therefrom up towards the foot of the downs. The toll-bridge invites excellent acquaintance with the river at close quarters, that down stream being exceptionally fine; but it is the high land sheltering either Streatley or Goring which commands rare birdseye views of the river and adjacent country.

Streatley is supposed to have derived its name from Icknield Street, a Roman road continued from the other side by a ford. The cartulary of the Abbey of Abingdon refers to a gift of land at "Stretlea" by the King of Wessex who ruled A.D. 687. Domesday Book deals with the manor, whose tithes at the time of the production of Magna Charta were under the assignment of Herbert Pone, Bishop of Sarum, who probably built and endowed the church, which has lately been restored, and whose square tower is always a distinctive object amidst the trees. The neighbourhood cen-

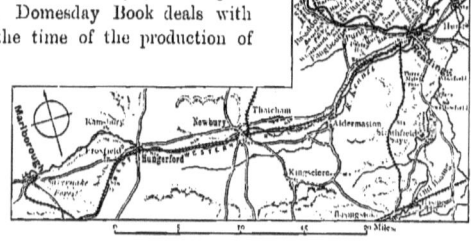

STREATLEY TO HENLEY.

turies ago obtained a reputation for health-giving qualities, and one of the memorial brasses in the church, dated 1603, incidentally bears testimony thereto by recording the virtues of an inhabitant who had six sons and eleven daughters. More than a hundred years since a medicinal spring at Goring was somewhat famous for its powers of healing, and Plot, the historian, mentions the water of "Spring Well" as celebrated for its curative properties in certain cutaneous disorders. The church at Goring, close to the river, is a historically interesting as well as picturesque structure. The grey square tower, with its round-headed windows divided into two lights by a central pillar, bespeaks its venerable age, and gives promise of the specimens of Norman and Early English architecture to be found in and around the edifice. Built in the reign of Henry II., dedicated to Thomas à Becket, and enlarged when King John tried to rule the country, it was connected with an Augustinian nunnery, of which traces still exist; and the remains of a priory have been built into a farmhouse some two miles from the village. The body of the church is singularly composite in its character. To its one lofty original Norman aisle without chancel, a north aisle, porches, and other appurtenances, have been at different times added.

A road ascending from Streatley skirts Common Wood, and at its highest point opens out a magnificent panorama of Thames Valley. The tow-path, however, as mentioned in a previous paragraph, now runs along the Oxfordshire bank, and the line of pedestrian traffic is therefore on that side. The short distance intervening between the banks lends undoubted enchantment to the shady recesses and warblings of the feathered songsters of Common Wood. When you have re-crossed by the wooden bridge towards the southern end of the hilly wood, the scene changes. Admirably situated at a bend of the stream stands the substantial, and the reverse of hidden, modern mansion termed the Grotto, surrounded by a clean-shaven lawn, which is intersected by gravelled walks, one of which follows the bank of the Thames, and is o'er-canopied with trees. From the sharp and picturesque curve of the river the bold round-headed hill of Streatley, the cosy village, the broad, divided river, and the Norman tower and delightful grounds of Goring, stand out a clear broad picture, which is almost suddenly lost round the bend studded by the eyots below the Grotto grounds.

The undulating chalk lands, rich in corn, roots, or pasture, as the exigencies of crop rotation may require, and dotted here and there by dark clumps of firs and larches, are absorbed opposite the Berkshire village of Basildon by Hart's Wood. The trees fringe the Thames closely, and densely clothe the wooded steeps. At all seasons these fine hanging plantations are fair to see; but there are special effects in spring and autumn, the intermixture of larches, at the former period, giving the wood a glow of dainty colour before other trees have put forth their leaves, and the abounding beeches, elms, oaks, and chestnuts, when mellow October comes round, making it equally conspicuous by the wondrous tints of decay.

The tow-path terminates for the time being abruptly opposite the snug village of Basildon; this may be gained by enlisting the services of the ferryman, who dwells in the solitary cottage under a line of full-headed pollards. The village and even its church are half-hidden in foliage, and there is an effective background formed by the plantations of Basildon Park. Passing peeps of the house are vouchsafed as we descend the river, steering by the Berkshire side of the group of islets in the middle. The parted stream marks the site of yet another Hart's Lock (Hart's Old Lock), of which no token remains. The chalk downs still appear on the Berkshire side, and the ridge that maintained Hart's Wood swerves in unison with the course of the river, and, well covered with wide-spreading oaks, shelters Coombe Lodge from the north and east. The osier beds of the Thames give employment to numbers of women and children, and maintain a distinct riverside trade. On a recent visit I was fortunate in witnessing the operations of an osier farm, thus described in my note-book:—The men, women, and children clustered on the farther shore, busily engaged in an occupation which is not at first apparent, come upon you with a surprise as you enter the next meadow. Out of the river formal growths of tall green sheaves seem to flourish within a ring fence. There is a rude building, half shed and half cottage, at the mouth of a gully, and in an open space between it and the Thames the above-mentioned people are working. Proceeding down the path the mystery gradually unfolds. We are facing an osier farm. The tall slender sheaves are bundles of withies that have been reaped from the islands and osier beds, and punted to this depôt. Here, in a square enclosure, they are planted *en masse* in the water, and the cut branches make the best they can of divorcement from the parent root, and preserve their vitality until they are required for use. The girls and boys are very handy at the operation of peeling. They take up a withy from the bundle last landed from the pound, draw it rapidly through a couple of pieces of iron fixed to

GORING, FROM THE TOLL-GATE.

a stand, and in a twinkling the bright green osier has become a snow-white wand. This humble colony of workers, about whom little is generally known, is one of many engaged in an out-of-the-way industry, hidden from the eyes of the world in some nook of the Thames. It is the first which meets our observation on the journey from Oxford. But even this simple form of industry has challenged the attention of the scientific. At the Inventions Exhibition of 1885 at South Kensington, an apparatus for willow-peeling was shown amongst the labour-saving machines.

Whitchurch Lock, two miles and a half below Basildon Ferry, is the halting-place for Pangbourne, the twin villages of Whitchurch and Pangbourne occupying similar positions, and enjoying the same type of communication as Goring and Streatley. St. Mary's Church, before its restoration, must have been a remarkably quaint building, and its singular wooden steeple attracts a considerable amount of attention even now. Amongst the curiosities in the interior, besides the memorial windows of stained-glass, is a monument to a sixteenth-century lord of the manor of Hardwicke, and his dame, represented kneeling at a *prie-dieu;* and a tablet with the following very original inscription:—

> "To Richard Lybbe, of Hardwick, Esq., and Anne Blagrave united in sacred wedlock 50 years are here againe made one by death she yielded to yt change Jan. 17, 1651, which he embraced Ivly 14, 1658.
>
> EPITAPH.
>
> "He, whose Renowne, for what completeth Man,
> Speakes lowder, better things, then Marble can :
> She, whose Religious Deeds makes Hardwick's Fame,
> Breathe as the Balme of Lybbe's Immortall Name,
> Are once more Ioyned within this Peacefull Bed ;
> Where Honour (not Arabian-Gummes) is spred,
> Then grudge not (Friends) who next succeed 'em must
> Y'are Happy, that shall mingle with such Dust."

The resemblance of the twin villages of Pangbourne and Whitchurch to the dual communities with whose concerns this chapter opened is sustained in several features. The reach immediately above Pangbourne, which is one of the very lovely stations of the Thames, is straight and uninteresting. The cut on the Whitchurch shore makes an abrupt curve to the lock, and the breadth of the river above the wooden toll-bridge, own cousin to that at Streatley, and the two islands side by side near the lock, produce a vivacious backwater, and a fine weir-pool, twenty-five feet deep, abounding in holes, eddies, and scours intimately known to London anglers, to whom Pangbourne is as much the object of worship as Streatley is the haven of desire to the artists. The wooden bridge, as at the last-named station, is the best coign of vantage from which to obtain adequate views of the three distinct streams, which gallop in joyous ebullitions of foam from the obstructions planted in the channel. A goodly current rushes from the very new-looking mill on the Whitchurch shore. The lower part of the church is concealed by trees, but clear above the rooks' nests in the swaying tops may ever be seen the wooden spire. The turbulent pool at Pangbourne weir may best be studied from the timber-yard on the Berkshire side, and there is a subsidiary weir which assists the larger body to create a homely and miniature delta before the scattered forces are collected in one uninterrupted volume of water at the bridge. The scenery at Pangbourne is not less charming than that of Streatley, and it is in both places of a character peculiar to the hilly country through which the Thames now flows. A wide-spreading prospect of the valley may be obtained from Shooter's Hill. Both Whitchurch and Pangbourne lay claim to a past history of some importance, but the old church, save the red-brick tower, which only dates from 1718, was replaced in 1865 by the present building; and this contains, amongst certain architectural qualities, an oaken pulpit, probably of the time of Elizabeth, carved in arabesques. The Pang bourne, which gives a name to the village, is a pretty trout stream joining the brimming river, straight from the village, at the tail of the noisy weir, and coursing with its overflow down the gravelly shallow.

The undulating chalk hills, prolific of agreeable changes in the scenery, continue without cessation for many miles below Pangbourne, but on the opposite side we have once more the flat meadows, neat farms, and humble cottages of agricultural Berkshire. The Thames, which had arrived at Pangbourne by a south-easterly course, moves for a short distance from west to east along a straight and deep-running

WHITCHURCH CHURCH AND MILL.

reach. The recurring woods on the left are a welcome foil to the level land on the right, and the distant landscapes are now very striking.

Under the hill on the Oxfordshire side, about a mile and a half below Pangbourne, Hardwicke House, a notable specimen of the Tudor manor-house, is a conspicuous feature. From the meadow on the opposite shore you have a perfect view of this most picturesque exterior. The colour of the brickwork has deepened, in the course of time, to the darkest of red; and its gables and clustering chimneys are clearly defined against the screen of noble elms which intervene between the house and the north wind, and cover the slopes behind it. The trim terrace is raised safely above the river; old yew, cedar, oak, and elm-trees cast long shadows upon the mossy turf, and indicate alleys and bowers such as those in which Charles I. spent some of the time passed by him at Hardwicke, "amusing himself with bowls" and other sports. Numbers of the trees upon the lawn, and some of the cool, quiet nooks of its shrubberies, are, no doubt, precisely what they were two hundred years ago.

Hardwicke House is, however, but an item in the catalogue of strong and varying attractions of the section of the Thames which began with Streatley, and

MAPLEDURHAM: THE CHURCH AND THE MILL.

which may be said to end at Mapledurham, something less than a mile farther down. Many lovers of the River Thames declare that, take it all in all, there is no sweeter spot from source to sea than this. In 1883 the hand of renovation was laid upon one of the overfalls, introducing of necessity an element of change; but the lock, weir, and lasher, the great bay of swirling water by them formed when there is no scarcity of supply, the backwaters, brook, and shallows have not been interfered with. As of yore, the whispering trees overhang the swift current, the lazy lilies wave in the tranquil backwater, and the rare old mill, first, perhaps, of its class upon the river, remains, like the face of a familiar friend, to greet the visitor, who, with each returning season, will assuredly, on the moment of arrival, bestow his earliest attention upon it. Mapledurham has, indeed, an almost unrivalled collection of good things to offer in the grounds of Purley on the west and those of the Elizabethan mansion on the east. Mapledurham House, largely concealed behind the foliage, is not at first so visible to the passer-by as Hardwicke; but it is too celebrated as a genuine example of Elizabethan architecture, and too well worthy of deliberate examination, to be neglected. The house was built in 1581 by Sir Michael Blount, who was Lieutenant of the Tower of London, and in the Blount family it has ever since remained. The name is a corruption of Mapulder-ham, and mapulder was the old English designation of the maple-tree. The glorious avenue of nearly a mile in length by which the front of the house is reached is, however, of handsome elms, but around the mansion are grouped poplars, oaks, beeches, and firs in picturesque profusion. From the right bank below the lock the gables, bays, oriels, roofs, and decorated chimneys, amidst such surroundings, constitute a striking picture. In the house are secret rooms and passages, supposed to have been used in the time of the Civil War by the Royalists for the concealment of priests or soldiers. By-and-by, in resuming your voyage down the river, Mapledurham House becomes the central object of another type of picture, composed of the delightful old mill, the curious church tower, the symmetrical trees, and the bright streams gathering from between the islands, and fresh from the mill race, and so continuing the sober volume of the Thames by Purley, and parallel with the railway. Mapledurham Church is near the manor-house, from whose grounds access is obtained to the churchyard by a pair of huge old-fashioned iron gates. It is a restored church, the south aisle of which is claimed by the Blount family as a private mortuary chapel. Purley is a small rustic Berkshire village, standing back half a mile from the river. The church, however, is nearer, and the ancient tower bears a scutcheon with the arms of the Bolingbroke family, and dated 1626.

A horse-ferry below Mapledurham conveys the pedestrian to the Oxford side, where, for less than half a mile, the tow-path continues. The ferryman is not always to be found, and the pedestrian, stopped by the iron railing, had better follow the footpath skirting the beautiful park at Purley. Backward glimpses may thus be indulged in of the mill, church, and manor-house, with a breadth of fertile meadow intervening; and, walking up the steep road towards Belleisle House, the temporary desertion of the river will be amply repaid by the extensive general view of the Thames Valley which has just been traversed. Purley Hall, built by South-Sea-Bubble Law, was the residence of Warren Hastings during his trial. For the boating-man, the river makes no exceptional demand upon his strength or imagination for several miles. The divergence by land, as above suggested, brings you presently to the "Roebuck," where a second ferry within the half mile assigns the tow-path once again to the Berkshire shore. The old-fashioned boating-tavern has not been demolished, but perched upon the hill above the Caversham Reach a more modern hotel tempts the oarsman to pause and refresh, and the holiday-maker to look out upon the remarkable map of river and landscape for which the situation is celebrated. The thatched roof, ancient kitchen, and tap of the original wayside inn are left standing—an eloquent contrast by the side of its successor.

The Thames between Purley and the eelbucks at Chasey Farm is studded with a variety of islands. They are at their best but small and tiny, bearing a few trees, or a crop of osiers, or amounting to nothing more important than a bed of rushes. Insignificant, however, though they may be, they preserve the character of the river, breaking as they do the monotony of the current, which, in the more level tract now watered by the Thames, shows an increasing tendency to the commonplace. The conclusion will be irresistibly forced upon us that we have at length, with reluctance, parted from the beautiful section which includes Streatley and Goring, Pangbourne and Whitchurch, Hardwicke and Mapledurham—scenic pearls of price lying within a convenient range of not more than seven miles.

Notice-boards upon a willowy eyot, and a fence athwart the stream, forbidding the passage of boats round the considerable backwater to the left, introduce us to a permanent line of eelbucks. Soon the bridge and church of Caversham appear afar; and, dimly, to the right, the chimneys and roofs of Reading. The Thames is again bordered on the north by hills, a continuation of the range which began at Hart's Wood. From Mapledurham Lock, however, the river, instead of running parallel with the hills, made a detour, and ran side by side with the railway, until, at the Chasey Farm eelbucks, it turned north again to meet them. "There is not," wrote Mary Mitford in her "Recollections of a Literary Life," "such another flower-bank in Oxfordshire as Caversham Warren," and this reference is to the breadth of country extending from the sedge-lined river to the tree-crowned chalk hills which have terminated their guardianship of the northern banks of the Thames. From the brow of the hills, upon which modern residences have of late years multiplied exceedingly, there are widespread prospects through which the silver Thames pursues the even tenor of its way, more beautiful from the distant stand-point than, for some miles above Caversham Bridge, it is when near at hand.

The bridge at Caversham is one of the plainest on the Thames, and this suburb of the county town is not in any way remarkable for its romantic adornments. The bridge was nevertheless of sufficient importance to draw from "Cawsam Hill" (the rustics to this day so pronounce the word Caversham) a furious onslaught from the troops of General Ruven and Prince Rupert, who "fell upon a loose regiment that lay there to keepe the bridge, and gave them a furious assault both with their ordnance and men—one bullet being taken up by our men which weighed twenty-four pounds at the least." Sir Samuel Luke's diary, in which this scrap of history is preserved, goes on to state that the "loose regiment" made the hill "soe hott for them that they were forced to retreat, leaving behind seven bodyes of as personable men as ever were seene." And, according to Leland, there stood in the time of Henry VIII., at the north end of Caversham Bridge, "a fair old chapel of stone, on the right hand, piled in the foundation because of the rage of the Thames." In consequence of the danger in which the meadows stood of floods, in the old pre-drainage days, when the river often played pranks unknown to modern times, the bridge was constructed of stone in its most critical part, but extended partly in wood by a number of arches over the pasturage. Before the days of the Cavaliers, as far back, indeed, as 1163, Caversham Bridge was the scene of a trial by battle, adjudged by His Majesty Henry II. Henry of Essex, the King's Standard-bearer, had charged Robert de Montford with cowardice and treachery. At a fight in Wales the Standard-bearer had thrown down his flag and fled, and his plea was that he believed at the time that the king was killed. The trial by sword is said to have been performed upon one of the islands near the bridge, with almost fatal results to the challenger, for though he recovered from what were at first supposed to be mortal wounds, he was obliged to retire to the abbey, where he exchanged the accoutrements of the soldier for the habit of a monk.

The Thames leaves Reading to the right, but according to some topographers the town derived its name from the Saxon "Rheadyne" ("rhea," a river), or from the British word redin (a fern), the plant, as stated by Leland, growing thereabouts in great plenty. Hall, however, makes light of these derivations, urging that the name simply meant that Reading was the seat and property of the Rædingas family. The Thames approaches close to the town below the pretty island, of about four acres in extent, which monopolises more than half the river, midway between Caversham Bridge and Lock; and is to the traveller by rail from London one of the earliest indications—with its line of willows on the farther bank, and the playing-fields intervening on the southern side—that the town is at hand. The facilities inherited by the inhabitants for bathing, boating, and angling are a boon appreciated to the full, and the Thames materially contributes to the reputation enjoyed by Reading as one of the most desirable country towns of England. The principal branch of the river below the swimming-baths sweeps to the left, but the navigable channel runs through the lock south of the small island. The divisions by islets and curvature of the course between the lock and Lower Caversham make the Thames a beautiful feature of the locality.

Full of historical memories (it is supposed that the Danes brought their war-ships up the Thames to the mouth of the Kennet), Reading is proudest, perhaps, of the abbey, of which so many interesting portions are well preserved in connection with the Forbury, the name given to the pleasure-grounds for the people, most creditably maintained by public subscription. There were four noted abbeys in the south of England—Glastonbury, Abingdon, St. Albans, and Reading, and Reading was not the least important. The wife of King Edgar founded the establishment as a nunnery, and Henry I. pulled it down to make room for two hundred Benedictine monks. It was given out that the hand of St. James the Apostle was deposited in the abbey, and the so-called relic "drew" a perennial inflow of support. Royal bones were laid in the abbey. Henry himself expressed a wish to be buried within its walls, and his body, accordingly, having been rudely embalmed at Rouen, was wrapped in bull-hides, and conveyed to Reading for ceremonial interment. At the Dissolution the royal tomb was destroyed and the king's bones ejected, with other *débris*, to make room for a stable. But the abbey during its existence was a power in the land. In it John of Gaunt married his Plantagenet wife, and there

FLOODED MEADOWS, FROM CAVERSHAM BRIDGE.

the marriage of Henry IV. to Lady Grey was proclaimed. The abbots of Reading were peers of Parliament, ranking only below their brethren of Glastonbury and St. Albans. They had the right of coinage; they gave to the abbey much wealth; and amongst the relics was one sent to Cromwell, and described by the commissioner who was sent down to inquire into the revenues as "the principell relik of idolytric within thys realme, an aungell with oon wyng that brought to Caversham the spere hedde that percyd our Saviour is syde upon .the crosse." The last abbot of Reading, defying the bulky Defender of the Faith, was hanged, drawn, and quartered, with a couple of monks, within sight of his own abbey gateway. What of the building was left after the energetic measures of bluff King Hal, was finally razed by Commonwealth victors. Portions, however, of the ancient chapel and chapter-house are left, and the old gateway stands, patched up with modern materials, in excellent preservation on the south side of the Forbury. It is understood that the abbey stones have been worked up into some of the public buildings of the town, and some of them were undoubtedly carted right and left, far and near, for miscellaneous use. The most interesting fragment is a Norman archway belonging to the abbey mill, and still spanning the mill race known as Holy Brook.

When the Plague raged in London, king, statesmen, and judges, with their courts, removed to Reading. Later, the royal troops held temporary possession of the town, and, after a ten days' siege by the Roundheads, the garrison displayed a flag of truce. Charles, and the looting Rupert, operating from Caversham Hill, tried in vain to retrieve the disaster, and when they were driven back, the garrison surrendered. In the reign of James II. the royal troops and those of the Prince of Orange, had a tussle in Reading market-place, one December Sunday morning, James's men, after a brief engagement, promptly leaving the enemy masters of the position. Archbishop Laud was a native of Reading; and John Bunyan, as related by Southey, was a frequent visitor to the town:—"The house in which the Anabaptists met for worship was in a lane then, and from the back door they had a bridge over the River Kennet, whereby, in case of alarm, they might escape. In a visit to that place Bunyan contracted the disease which brought him to the grave." Valpy was head-master of Reading Grammar School; and Judge Talfourd was one of the later worthies of the clean, thriving, Berkshire capital.

The River Kennet, referred to in the previous paragraph, runs through Reading. The great abbey was built upon it, yet within view of the broader Thames flowing through the level meads northwards. The Hallowed or Holy Brook, in which the Reading schoolboy of to-day angles for roach and dace, was a timely tributary turned to ecclesiastical uses, and employed to grind corn for the Benedictines, and minister generally to the refectory. The Kennet is, with the Loddon in the same general portion of the home counties, one of the most considerable tributaries in the great watershed of the Thames. Drayton, as usual, fastening upon some quality that accurately describes the character of his stream, says:—

> "At Reading once arrived, clear Kennet overtakes
> Her lord, the stately Thames; which that great flood again,
> With many signes of joy, doth kindly entertain.
> The Loddon next comes in, contributing her store,
> As still we see, the much runs over to the more."

The clear Kennet is, moreover, in other respects an exceedingly interesting river, and a stream, too, of some practical importance. It rises on the edge of the Wiltshire Downs, and for three or four miles runs in modest volume until it passes

THE THAMES AT READING, FROM THE OLD CLAPPERS.

through the old town of Marlborough, a steady-going Wiltshire borough, deriving its life not from manufacture, mining, pump-room, or esplanade, but from the land, as represented by the cattle, corn, malt, cheese, and woollen fabrics which are the subjects of barter and exchange at its periodical markets. In the palmiest days of coaching, four-and-thirty four-horse coaches used to stop at Marlborough on their journey between Bristol and London, the high road at that time running through what is now the centre avenue of the College grounds. The Vale of Kennet is here bounded by the Wiltshire Downs on the east, and Savernake Forest on the west. The forest is about a couple of miles from the town, and is the stateliest forest in the kingdom belonging to a private proprietor. It is sixteen miles in circumference, finely timbered, and possessing that too - often - lacking essential of a forest, harmonious alternation of hill and dale. There is a glorious avenue of beech-trees five miles long; and in the spring season the hawthorn-trees, of immense age, with heads that often compete in size and shape with the ordinary forest trees, and each standing bravely by itself, are a marvel of fragrant bloom. Amongst the groves of oak, beech, and chestnut, and undergrowth of bracken, fern, bush, and briar, there are hundreds of fallow deer; and a considerable head of red deer is still successfully maintained. The Kennet ornaments the Park of Ramsbury Manor, and touches Littlecote Park, a tragic reminiscence of which is given in the notes to Sir Walter Scott's poem of Rokeby.

So far, the Kennet has watered Wiltshire; but soon after leaving Chilton Lodge it enters Berkshire, meandering through a tract of marsh, and, dividing into two streams, runs through the decayed but once considerable town of Hungerford. Pope signalised the river in the line—

"The Kennet, swift, for silver eels renowned;"

and the successful attempt recently made by the Flyfishers' Club of Hungerford to introduce grayling into it reminds one of the super-excellent quality of the fish indigenous to its waters. The Kennet and Avon navigation makes the connection of this portion of Berkshire with the River Thames direct and valuable. The canal navigation, forming a waterway between the Thames and the West of England, is for the first nineteen miles, namely, from Reading to Newbury, the River Kennet itself; from Newbury to Bath, the canal proper is cut for a distance of fifty-six miles; and the Avon river completes the communication to Bristol. The numerous locks in the Vale of Kennet are connected with this system of navigation, which is practically associated with the concerns of the Great Western Railway. Hungerford, the town which has been here noticed as standing upon the Kennet, was described by Evelyn as a "town famous for its troutes," and it has well preserved its reputation. Amongst the inns of the town is one named after John o' Gaunt, who was a person of note in both Hampshire and Berkshire. His association with Reading has been already signified in the reference to the burials and funerals which took place in the abbey; and in Hungerford is a horn, highly honoured as a gift of John o' Gaunt to the town, and as a memento of the right of fishing enjoyed by the commoners, who still maintain the custom of fishing the Kennet three days per week. At Hungerford, in 1688, the negotiations which ended in the substitution of James II. by William of Orange were conducted.

The Vale of Kennet, from the Hungerford meadows to within a few miles of Reading, is a compact stretch of rural loveliness. We hear of the Vale of Avoca, the Vale of Llangollen, and the Vale of Health, but we do not find the valley through which the Kennet flows magnified in song, though of the smiling and peaceful order of valley landscape it has few competitors in England. Its green pastures lie by still waters, and its little hills seem to drop fatness. Between Reading and Marlborough the eye may, right or left, almost at any moment, rest upon limpid and often rippling water. Narrowed here to the dimensions and restless volume of a goodly lowland trout stream, it there journeys at an even pace, betraying anger and vexation only when subject to artificial restraint; as, for example, when it boils and swirls at a mill-tail, or races impetuously round into the repose of a backwater. The Kennet and Avon Canal is mixed up rather bewilderingly, to a run-and-read stranger, with the river. Pleasant brooks and brooklets thread the water-meads, garnished with forget-me-nots and cuckoo-pints; while in the moist hollows the marsh marigold blossoms in golden clusters. Ancient roofs of thatch-covered tenements, built in another generation, appear now and then; and long-established farmhouses and beautiful mansions vary the prospect on either side of the valley in whose typical English country scenery there is no break of continuity.

At the town of Newbury the Kennet becomes navigable, and so continues throughout the remainder of its course, which is concluded a little below the town of Reading, at the point where the Thames dips to the south as if to meet it, and almost touches the Great Western Railway line. Newbury is a very old town, as the description in Foxe's "Book of Martyrs," of the burning of Palmer, Askew, and Gwyn, in the middle of the sixteenth century, will show. In the fifteenth century Newbury was famous for its cloth weaving, and "Jack of Newbury," who may almost be said to be the patron saint of the town, was a wealthy cloth manufacturer. He kept a hundred looms at work, and on the invasion of the country by the Scots marched the entire force into the field, and received much compliment upon their martial bearing and superior garments. The two battles between Charles I. and his masterful parliamentarians are historical, and the canal near Newbury Lock passes the ground where the Roundheads camped prior to the first battle of Newbury. In the corn-fields and grass-lands of the rural outskirts of the town, occasional traces are unearthed of a battle in which six thousand men were killed, and a suitable monument, raised by public subscription, stands to commemorate where—

"On this field
Did Falkland fall, the blameless and the brave,"

and to record that Lord Carnarvon, Lord Sutherland, and other Cavaliers also, perished in the unfortunate cause of their unfortunate king.

The Thames from Reading to Sonning calls for no marked comment, and I must confess to a habit, when in these parts, of leaving the waterside at Caversham Bridge and travelling to Sonning along the high road that passes Lower Caversham, by farmhouses, corn-fields, and pastures, and one of the osier farms described on a previous page. A road at right angles conducts to the "French Horn" Inn, and to the bridges here spanning the Thames. Arriving at Sonning by river, however, you

SONNING-ON-THAMES.

glide underneath the woods of Holme Park, and so take into calculation the church and village from a point of view highly favourable to their scenic pretensions. No visitor can do justice to the exquisite beauties of this village without leaving the water and exploring the bridges, islands, and waterways which are so lavishly distributed between the widened banks. On the "French Horn" shore, the left branch sweeps round and streams abroad in a skittish shallow under a lightly-built bridge. At first it is difficult to decide whether this is a backwater or the main stream. Looking upwards, you notice that another channel yonder follows a row of pollards and orchard-trees on the "White Hart" side. There are separate streams, apparently, on either side of the bridge; and a shoulder-of-mutton-shaped eyot and other islets create a rapid current in another direction, overhung by a perpendicular bank. This is topographically confusing, but most agreeable in its endless motion and diversity. There are two divisions of the bridge; and beyond the first an independent backwater gallops down from the mill, past which, and its chestnut-trees, is the brick county bridge. The houses of the village, clad with creepers, and often embowered in fruit-trees, and the square tower of the church, as represented in the engraving, constitute one of the most familiar pictures of the Thames. A charming walk, immediately above and below the lock—locally termed the Thames Parade—extends along the skirts of the woods of Holme Park, the projecting boughs of which o'ercanopy the towing-path, and are reflected in the water. The eyot is connected with

SONNING WEIR.

the shores by the lock and weir, duly illustrated on another page from a favourite point of view. One of the choicest views at Sonning may be obtained by standing on the Parade, say a hundred yards above the lock, and peeping under the boughs of the trees towards Reading, which sometimes looks almost romantic in the dreamy obscurity of an enveloping haze.

Sonning, or Sunning, was not, in all probability, as some maintain, the seat of a bishopric, though it was a standing residence of the Bishops of Salisbury, who had a palace here through successive generations. Even in Leland's time it was "a fair olde house of stone, even by the Tamise ripe, longying to the Bishop of Saresbyri; and thereby a fair parke." The church, without which the charming landscape would lose one of its most harmonious features, contains curious monuments, a celebrated peal of bells, and rich carved work. It is peculiarly rich in memorial brasses, many full-length figures of the Barker family dating from the middle of the sixteenth century. Very different is the view down the river, when the back of the observer is turned upon the graceful trees drooping into the water, the masses of chestnuts and elms interspersed between the houses, and the divided stream and osier-bedded islets. The sinuous course is for a couple of miles between low banks; while in the somewhat distant background appear the towering woods, with which we shall become by-and-by more intimately acquainted. On the lower side of the bridge the river at once collects its scattered forces, and proceeds stately and slow until a chain of islets diversifies the course, and, with the assistance of sundry sharp twists in the left bank, gives increasing strength to the current, and braces itself for the press of business demanded by the mill and lock at Shiplake. The Rev. Jas. Grainger, author of the "Biographical History of England," was Vicar of Shiplake, and, in his dedication to Horace Walpole, remarks that he had the good fortune to retire early to "independence, obscurity, and content." The rev. gentleman, who considered Shiplake as synonymous with obscurity, died at the altar of his church while performing divine service, and is buried within its walls; and the tablet which marks his grave refers, as does the dedication, to the obscurity which at Shiplake accompanied the content. The church stands upon a very charming slope. The southern face of the tower is mantled over with ivy, and the sacred edifice does not lose in dignity by the near neighbourhood of farm buildings, rickyards, and orchards. From the porch there is a fine view of the valley of the river. The church, in which Lord Tennyson was married, was restored in quite recent times, but the stained-glass windows are so ancient that they are supposed to have been originally in the Abbey of St. Bertin at St. Omer.

The singular vagaries of the mouth of the Loddon introduce an unexpected variety above Shiplake. It was this tributary, mentioned after the Kennet by Drayton, in the lines previously quoted, which gave Pope a hint for his fable of Lodona, and he stamps the character of the Loddon in the line—

"The Loddon slow, with verdant alders crowned."

The Loddon is, nevertheless, scarcely a river on its own merits to inspire a poem, though it is in an especial degree the kind of stream which has attracted the consideration of pastoral poets. Almost any portion of the country watered by the sluggish Loddon might have yielded just such scenes as Gray describes in his immortal Elegy. The river rises in the North Hampshire downs, and flows by the site of that Basing house which is famous in the annals of Cromwellian warfare. Fuller, the church historian, resided in the mansion during the siege, and amidst the confusion of the battle is reported to have composed some portion of his "Worthies of England." Fragmentary ruins of the house are yet shown. Every visitor must bear witness to the debt owed by Strathfieldsaye Park to the Loddon, which divides it into two unequal parts. The quantity and quality of the water gave the late Duke of Wellington an opportunity, of which he perseveringly availed himself, of indulging privately in the pursuit of trout breeding, a project which was abandoned soon after his death. The Loddon in Berkshire passes by Swallowfield, where in his son's house Lord Clarendon wrote his "History of the Rebellion." Two centuries earlier than that the manor was the property of John, Duke of Bedford, Regent of France; and it has been in later times of more immediate interest to the admirers of Mary Russell Mitford, as being her home and burial-place. Her ever-delightful book, "Our Village," is composed of rural word photographs, taken when the lady lived at Three Mile Cross, and all the scenes are faithful pictures of Loddonside life. On returning from a recent visit to the Loddon, an old friend of Miss Mitford's, in Reading, gave me, as a memento of the authoress whom we both admired, a note in her handwriting, and after it had been some time in my possession I discovered that the small envelope in which it was enclosed was one which had been previously sent to, and turned by, the industrious old lady. The operation had been performed with wonderful neatness, and it was only by accident that I discovered inside, and in faded ink, the original address, to "Miss Mitford, Three Mile Cross, Reading, Berks." Arborfield succeeds Swallowfield, and the river here feeds the picturesque lake in Mr. Walter's park at Bearwood. The Loddon next touches Hurst, and flows in its lazy way to Twyford, so called from the two fords, which are represented in these days by bridges, crossing the two arms of the river. After a north-eastern course of some twenty-four miles, the Loddon here runs into the Thames. It should perhaps be stated, with reference to Pope's fable of Lodona, that it was not connected with the Loddon proper, but with one of the inconsiderable tributaries of a tributary that ripple through part of Windsor Forest. The poet was, nevertheless, quite accurate in his description of the Loddon as "slow," and "with verdant alders crowned." It is an altogether different river from the Kennet, which is bright, and abounding in gravelly shallows, after the fashion of the Hampshire chalk streams, and is a famous trout river. The Loddon, on the contrary, is deep, dark, sluggish, almost troutless, and thickly furnished with the alder, of which it has been written—

"The alder whose fat shadow nourisheth,
Each plant set near to him long flourisheth."

This attribute is only a poetical fancy, but the alder is essentially a tree whose roots are at home when planted by the river, and which is always contributing some evidence of its vigour—in the winter with its catkins hung out to freeze, in the spring with its queer little black cones, and in the summer and autumn by the glossy green leaves which are merciful to the defects of shape in its branches, and which sturdily hold on when the leaves of other trees have been snatched and scattered.

The water of the Thames flows into the Loddon through the private backwater known as St. Patrick's Stream, but the Loddon finally joins the Thames below Shiplake Lock, after indirectly opening into it by means of the mouths of St. Patrick's Stream. There is also, intersecting Burrow Marsh, a backwater, irreverently termed Burrow Ditch, and this joins with St. Patrick's Stream in swelling the volume of the Loddon. It should perhaps be explained that although under its normal conditions the Thames, through both branches, runs into the Loddon, in times of flood the position is reversed, and the Loddon pours its current into the Thames.

Shiplake has more than the ordinary share of backwaters and bye-streams, and on this account is a favourite resort of anglers. Independently of the virtually three outlets by which the waters of the Loddon escape, there are Phillimore Island and Shiplake Mill to be considered. By following the course of the loop formed by St. Patrick's Stream, the lock may be avoided; but the stream is a very strong one. Half-way round the bend the comfortable farmhouse of Burrow Marsh will be noticed, and the upper portion of the backwater is generally so choked with rushes as to be almost imperceptible. The weather-board mill and the weir are prettily set, and the islets abounding above the lock are links in a chain of choice Thames scenery. Near Shiplake Lock, as the illustration to that effect will signify, stands an island which is a favourite camping-out spot for boating-men who do not fear the risk of rheumatism, and who prefer a night on shore under canvas to the cramped and unsatisfactory repose attempted by those who decide to spend the night in their boats. For miles downwards from this point the Thames winds through scenery in which hill and woodland again take their welcome place. The views on water and from land may change in degree, but the general character is ever that of quiet beauty. The commanding situations upon the elevated ground overlooking the valley have long been built upon, and, on brow, slope, or level, mansions of varying styles succeed each other. Phillimore Island takes its name from the late learned owner of Shiplake House opposite. It is a dainty little bit of dry land in the midst of the water, covered with willows, poplars, aspens, and one or two chestnuts. Down stream Wargrave Hill with its imposing white house finishes the view for the time being.

It was at the "George and Dragon" hostelry, at Wargrave, about a quarter of a mile below Shiplake Weir, that Mr. Leslie and Mr. Hodgson, R.A., entered into a temporary partnership in the production of a humorous signboard. Wargrave was once a market-town, but it is now, happily for those who seek its quietude, a mere village far removed from the noise of the world. Sequestered backwaters between and at the rear of the islands, suggest a change for the visitor who is tired of the shaven lawns, pretty villas, and park-like grounds behind the public ferry and the sleepy village. The railway runs the other side of the river, crossing it below Shiplake Lock, and so passing by Bolney Court to Henley. A high road to the latter place runs past the "George and Dragon," and, under the towering woods, are the eyots opposite Bolney Court; while on the other side of the space, known as Wargrave Marsh, the Hennerton backwater, or Wargrave Stream, extends for over a mile, and is crossed by two modest foot-bridges. This backwater is well known for its aquatic offerings, and the artist has appropriately "happened" upon it at a characteristic moment, when a bevy of fair boaters have discovered that the lilies are in flower, and have ventured up to gather the æsthetic blossoms. In the secluded village of Ruscombe, between Shiplake and Wargrave, Penn, the founder of Pennsylvania, died, and was buried; and the notable objects of the neighbourhood may be concluded by mention of the monument, in Wargrave Church, to the memory of Thomas Day, who wrote "Sandford and Merton," and was thrown from his horse and killed on Bear Hill close by. The heights, of which there are no lack in the neighbourhood, give many picturesque and wide-spreading views of the river and the surrounding country.

SHIPLAKE: A CAMPING-OUT PARTY.

The islands in the Thames opposite the remarkably plain mansion of Bolney Court are a truly beautiful group, even if they have escaped the popularity accorded to less charming reaches of the river. Up stream a fine pine wood will be noticed; Hennerton House, to the right, stands on a lofty steep, embowered in trees; and below are the dark woods and white cliffs of Park Place.

Park Place now absorbs all the notice of the downward traveller. For miles above, the wooded heights have been visible, increasing in beauty as we approached nearer and nearer. They will now be close on the right hand, until progress is temporarily checked at Marsh Lock. The mansion was built originally by one of the Dukes of Hamilton. The father of George III., when Prince of Wales, lived

14

there; and George IV., before he came to the throne, and the first Earl of Malmesbury, there abode. The marvellous beauty of the situation, and the splendid success attending the efforts of those owners who understood how to compel Art to assist, by judiciously developing, Nature, have made Park Place what it is. The principal agent in this latter work was Marshal Conway, who, nevertheless, in many respects, carried his notions of improvement to excess. Towards the end of the last century much had been done to endow it with the attractions which made it so desirable a residence; but the Marshal, devoting all his time to additional embellishment, ran no little danger of pushing from the sublime to the ridiculous. The inhabitants of Jersey, to mark their appreciation of his governorship of the island, presented him on his departure with a Druid's Temple or Tomb, which had been found by workmen during his reign on the summit of a hill near St. Heliers. The relics were brought to Park Place and set up on the summit of one of the lesser eminences. Forty-five stones, averaging seven feet in height, four in breadth, and from one to three feet in thickness, were arranged in a circle sixty-five feet in circumference, and in the exact positions, so far as could be understood, which they occupied in the dim era of antiquity. The Marshal built also an artificial Roman amphitheatre, approached by a long underground passage leading to a valley planted with cypress; constructed a bridge from materials carted over from the remains of Reading Abbey; overhung a walk, at the end of which was a marble tomb, with weeping willows; and elsewhere excavated a cavern, and left other tokens of his eccentric restlessness. The mansion was rebuilt by its present owner, Mr. Noble, in the French-Italian style; but its principal merit is the incomparable situation (300 feet above the level of the Thames), and surroundings of nine hundred acres of superbly wooded hill and dale, velvet lawns and romantic glades, mossy dells and tangled thickets. The domain is entered by seven lodges, and east of the house a cedar is pointed out as having been planted by George III.

The latest considerable, and not least sensible, addition to Park Place is the Gothic boat-house, at which visitors, who have the privilege of roaming over the grounds, are permitted to land. The really handsome exterior is not belied by the artistic furnishment within, comprising pictures, carvings, and statues. The walk through the grounds, with its surprises of mimic ruins and suggestive emblems, its sylvan glories which owe nothing to the hand of man, and the fairy-like glimpses which owe everything to the bountiful river, is a treat, indeed, of which one never tires, and which every sojourner in these parts should, in duty bound, make his own. From the bosom of the river the white gleams of chalky cliff contrast admirably with the masses of foliage. The residence at Park Place shows well from the second or third meadow below Marsh Lock; but the fields on the Henley side are being converted into brick-yards, and the first view of the town is marred by the coal-sheds, sidings, and ugly little railway station, to which the adjacent block of terrace-buildings cannot be accepted as in any degree a set-off.

The fine old weir which, until recent years, furnished an everlasting object-lesson to young artists at Marsh Lock, has been superseded by a modern arrangement erected near the paper-mill, and worked by a travelling pulley; but on the right bank the brick-mill, house, and exquisitely kept river frontage of its gardens, improve by time, and worthily complete the charms of Park Place; and, zig-zagging across the broad Thames, there remains the wooden bridge by which the barge-horses cross from Oxfordshire to the farther shore and back again without touching land. Underneath the high staging, the river, in alternate pools and shallows, reveals a pebbly bottom more resembling the bed of a mountain-born salmon river than the placid Thames. In the rapid and moderately deep water running from the paper-mill, the patient observer, waiting on a sunny day until the fish have recovered from the alarm communicated by the shadow cast as he took his position, will have favourable opportunity of observing the kind of creatures which inhabit the waters. In the spring months, when the barbel are congregated on domestic cares intent, the almost incredible piscatorial resources of the Thames can be easily understood, and this particular run of water at times appears to be crowded with this sport-giving species.

The district of which Henley is in a sense the riparian metropolis is one of the best along the entire length of the river for the angler, in whose interests we may agree, perhaps, to break off our downward voyage for the moment, in order to complete the information proffered in brief in the first chapter, with respect to the piscatorial capabilities of the river. Although the right of the public to fish in the Thames has been frequently called in question, and threatened with opposition, it remains one of the principal rivers in England free to the general angler. Probably forty or fifty years ago men fished from any section of the tow-path, or with their boats moored in any pool, without let or hindrance. Within the last quarter of a century, however, and especially within the last fifteen years, anglers have increased probaby a thousandfold. A distinct angling literature has been established. The clubs and fishing societies of London alone may be numbered by hundreds, and the increased facilities of locomotion all over the country combine, with other progressive changes, to promote a spirit of sport, and develop the sporting instincts of the people in this innocent direction. One of the results of the multiplication of the angling fraternity, and the consequent hard fishing to which the River Thames has been put, was seen in the evidence given before the Special Committee of the House of Commons during the session of 1883. Prominent amongst the grievances complained of by witnesses who appeared for the general public, was the assertion that waters which had been free to anglers, all and sundry, from time immemorial, were now claimed as private fisheries by riparian owners; and the report of the Committee, as many readers will remember, though it was only an expression of opinion, was rather against than for the anglers. In many of the most important districts of the Thames local Preservation Societies have been established, vested with some sort of control over the fishing, and enforcing, by their bailiffs and keepers, those by-laws of the Thames Conservancy which were framed after consultation with gentlemen representing the different classes of metropolitan anglers. It is only, therefore, in rare instances, that permission to fish is refused to the public, and the system of preservation is acquiesced in by all earnest sportsmen, who do not need to be informed that unless the pastime of angling is con-

BACKWATER AT WARGRAVE: A POOL OF WATER-LILIES.

ducted on strictly fair principles, the Thames, or any other river, would soon be depopulated of its fish.

For angling purposes the River Thames may be roughly divided into three sections. The first comprises the tidal waters, in which the fishing is principally confined to roach, dace, barbel, and an occasional trout in Teddington Weir. Of the coarser fish, incredible quantities have been caught since the regular supervision of the river was undertaken by the local Piscatorial Society of Richmond. The next division is from Teddington Weir to Staines, where the city waters end, and over this the Thames Angling Preservation Society, the most important of its kind in the country, holds sway. The last section comprises all the water between Staines and Oxford, and as I have already intimated, of this Henley is the principal station, or head-quarters.

The trout-fishing of the Thames is probably not what it was in the palmy days when salmon were caught in the river, but it is still surprisingly good, considering the very much-restricted haunts of the fish. It is supposed by many persons who have only a passing acquaintance with Thames trout that it is a distinct species. The fish, it is true, is in external non-essentials different from most of its family, and has, through a long course of residence in the Thames, established certain characteristics of its own. A typical Thames trout, with its deep thick body, shapely head, silvery sides, and fine spots, is an extremely handsome fish, and second to none in its sport-yielding qualities when fighting for its life in a tumbling bay. The difficulty is to catch it. Trout-fishing in the Thames commences on the 1st of April, and terminates in the middle of September; and is chiefly confined to the weirpools. Here, in the foaming and churning water, all the predatory instincts of the species find ample opportunities of practice amongst the delicate bleak and other small fry which love the rapid turbulent streams. Whatever the Thames trout might have been in olden times, it is not to be denied that his representative in these days has no partiality for insect food, of which, however, such a river does not yield an abundance; hence few anglers attempt that most sportsmanlike method of angling for trout—the artificial fly. Failing this, the most fashionable mode is that of spinning with a bleak or small dace, and latterly this has been supplemented by the less commendable practice of live baiting.

In many of the upper waters, as at Henley and Reading, *salmo fario* of the ordinary kind have been artificially hatched and turned into the river. Loch Leven trout have also been introduced, and one of the latest efforts at acclimatisation has been with Great Lake trout and land-locked salmon, sent to this country by the United States Fish Commission, and introduced to the Thames through the National Fish Culture Association and Thames Angling Preservation Society. Whether these interesting experiments in pisciculture will be attended with success time only will prove, but there can be no question that the number of common trout in the Thames have, of late, largely increased, though a greater proportion of small fish have, as might be supposed, been taken.

The principal sport of the Thames, however, must be looked for in what are called the coarse or summer spawning fish, for whose advantage a close time has been instituted between the 15th of March and the 15th of June. The latter date is full early for many of the species. At the same time, the periods at which the fish get into condition after spawning depend so much upon the varying circumstances of the water that the angling public have been, reasonably enough, allowed to enjoy the benefit of any doubt that might have been entertained. The increasing number of steam-launches has in many ways interfered with the pursuit of angling, and the disciples of Izaac Walton entertain anything but a friendly feeling towards the frequenters of the Thames who take their pleasure in other ways than through fishing-rod or punt. The Thames fish have, indeed, many enemies to contend with, and angling in its waters with success becomes a more and more uncertain and difficult art every year. The fish that has deteriorated, most probably, from the introduction of the steam-launch is the pike. The Thames is not, naturally, except in a comparatively few reaches, and at the weirs and mill-pools, a trout stream; but it is precisely the water in which the voracious pike should flourish. The beds of reeds and rushes, the eyots, the deep holes under willow-lined banks, the long straight reaches down which the currents, "strong without rage," maintain their easy progress—these are the natural haunts of *Esox lucius*. But pike-fishing has suffered greatly on account of the pernicious and Cockney system of trailing from the sterns of pleasure-boats and steam-launches. By the murderous flights of hooks, dragged in their wake, without any exercise of skill or attention on the part of the owners of the apparatus, infant fish, too often under the legal minimum of length, are taken. Any pike-fisher who is wise will, therefore, avoid the watery highways which are swept and harried by this legion of pot-hunters.

In the particular district, however, at which we are pausing to indulge in these piscatorial reflections, the troller or live-baiter may find his most liberal opportunities. No steam-launch can push its way up the overshadowed and tranquil backwaters of Hennerton, or round about the islands at Bolney. The skilful pike-fisherman will not only seek such undisturbed retreats as these, but will obtain his best sport by deftly dropping his paternoster fitted with one gimp hook upon a gut trace, and baited with gudgeon or small dace, between banks of weeds, and in those odd and beautiful clearings in the aquatic forests which the practised eye may always find. The Thames, nevertheless, as a pike river, has for some years been a disappointment, and will so continue to be until trailing is prohibited by law.

After the month of October the pike angler has a fairer chance of sport. Simultaneously with the disappearance of the steam-launches and pleasure-boats, from which angling is conducted as a passing amusement, and in utter ignorance of the science, or even rudiments of the art, the decay of the weeds begins. This is the signal for a general exodus from summer quarters by the fish. They sheer off into deep water. The pike, no longer concealed in a thicket of subaqueous vegetation, from which he has, during the summer months, pounced like an insatiable ogre upon the silvery wanderers swimming heedlessly about in search of minute freshwater crustacea and larvæ, takes to the life of a roamer, free from much of the harassing which kept him close, out of the range of roistering Thames excursionists. But it is unfortunate for the pike that the keen sportsman also benefits by this change. The dying down of the weeds leaves him space for the exercise of his skill at the precise time when his game may be taken at disadvantage. Pike-fishing is, therefore, the winter recreation of the angler in the Thames, though, for the reasons indicated, large specimens are rarely killed now.

The perch, most cosmopolitan of fishes in the rural districts of England, the bold biter idolised by schoolboys, whose easy prey under favourable conditions he is certain to be, has almost disappeared from some portions of the Thames. Henley used to be a grand perch preserve, and the late Mr. Greville Fennell, whose angling contributions to literature were chiefly founded upon his observations and experiences in the reaches between Henley and Pangbourne, gave it at one time a first place on the list of good perch waters. But cosmopolitan as the perch may be in its character, habits, and haunts, it is more difficult to rear than many other of the summer spawners, and the peculiar manner in which it hangs its eggs in festoons around the roots and branches beneath water, renders it an easy victim to the rough usages of swiftly-passing traffic. Shiplake hole, and the "tails" (as the fishermen term them) of all the islands mentioned in this chapter, are still favourite places for perch during the winter time, when the steam-launches are in dry dock, though the quality and quantity of the well-beloved zebra of the fresh water have unfortunately declined in the Thames.

The carp family thrive, as ever they did, and in some years are caught in unusually large numbers, rejoicing the hearts of the professional fishermen who have languished for want of customers through a series of depressing fishing seasons. The head of the family is very rarely taken in the Thames proper. Some carp, however, are found in the Cherwell, and by accident, at very rare intervals, solitary specimens are caught in the Thames itself. But these are the accidental wanderers; exceptions proving the rule. Bream are more plentiful, but the most prolific of all are chub, roach, dace, and gudgeon. The popularity to which the Canadian canoe has risen on the Thames is not a little due to the adaptability of the light and elegant boat for chub-fishing. Regulating the drifting of the canoe with one hand, the operator, armed with a suitably short and supple fly-rod, drops down some fifteen yards distant from the overhanging willow-bushes, from under whose branches, close to the loamy or gravelly bank, a lightly-dropped fly of large dimensions will, in the calm of a July or August eventide, seduce the great bronze-coloured "chevin" to its fate, while, in the winter time, artful concoctions of cheese-paste, and other gross baits, directed down stream by a long Nottingham line and the familiar float tackle, will be equally efficacious in the formation of a bag. Roach and dace-fishing, the simplest of angling practices, as conducted from the comfortable floor and chair of a Thames punt, continues to be, as of yore, the most familiar form of the contemplative man's recreation for the average citizen. In the mysteries of fly-fishing, and the ingenious devices invented for betraying the fishes that follow spinning-baits of all descriptions, improvements real and so-called are continually announced, but no change seems to have been suggested for many years in the ancient methods adopted on the Royal River for the capture of barbel by ledgering, and roach and dace by ground-baiting, plumbing, and Thames punt-tackle. Angling in the Thames is a source of untold delight and innocent enjoyment for tens of thousands of persons every year, and long may the day be postponed when the modest privileges of the London anglers, whose opportunities are limited, and whose ambition in the matter of sport is easily satisfied, are reduced or interfered with.

The deeper pool across the river, near the flour-mill at Marsh Lock, used to be a favourite resort of those anglers who pursued their sport from a boat; and the bank from the paper-mill towards Henley witnesses many an exercise of patience from the youthful Waltonian. The utilitarian spirit which has rendered necessary the hideous iron weir above the mill, and which is step by step destroying so many of the gems of Thames scenery, has, however, built a black barricade from the miller's boat-house to the head of the eyot, completely cutting off the communication by water with the further bank. The stream below is narrowed by the two islands in the middle of the channel, and rendered busy by that constant traffic of pleasure-boats which is inevitable in proximity to such towns as Henley and Reading. During the last quarter of a mile the familiar buildings and substantial bridge of Henley have opened to view, and we conclude the voyage to this stage amidst the bustle of boats and boatmen, and a parting glance at the head of Isis as chiselled by the Hon. Mrs. Damer. Water-plants are entwined around the face, which aptly looks in the direction of the river's source.

WILLIAM SENIOR.

HENLEY REGATTA. (*From an Instantaneous Photograph.*)

CHAPTER V.

HENLEY TO MAIDENHEAD.

The Best Bit of the River—Henley—The Church—The "Red Lion"—Shenstone's Lines—Henley Regatta—The First University Boat-race—Fawley Court—Remenham—Hambledon Lock—Medmenham Abbey and the Franciscans—Dissolution of the Order—Hurley—Lady Place and its History—A Strange Presentiment—Bisham Abbey and its Ghost—Bisham Church—Great Marlow—The Church and its Curiosities—"Puppy Pie"—Quarry Woods—The Thames Swans and the Vintners' Company—Cookham and Cliefden—Hedsor—Cliefden Woods—The House—Raymond—The Approach to Maidenhead.

NOTWITHSTANDING the old proverb concerning comparisons, we may venture to assert of this section of the Thames that it is the richest in natural beauties. Though there are spots on the upper part of the river which individually can hold their own with any, there will nowhere be found such a succession of exquisite views of noble reaches of water, of wooded bluffs and slopes, of green meadows and tree-covered islands, of old villages and stately or ancient mansions. There is, of course, nothing between Henley and Maidenhead which can rival the grand grouping of Windsor Castle on its wooded eminence, or the formal magnificence of Hampton Court; neither can the gardens of Kew, or the park on Richmond Hill, be equalled by anything on this part of the Thames; still, it affords us such a series of beautiful views of meadows, woods, and buildings that only between Richmond and Kew can we be induced to hesitate in awarding the palm to the portion of the river which is the subject of this chapter.

At Henley-on-Thames we are on the border of Oxfordshire. From its bridge we obtain not the least striking of the views to which we have alluded. The wider expanse of the upper valley contracts a little as the stream approaches the base of Remenham Hill, whose wooded slopes descend to the neighbourhood of the water. The Thames is deflected slightly towards the left as it commences the curve, in which, a mile or so farther down, it sweeps round the base of the long shelving spur which forms the northern termination of Remenham. On the Oxfordshire side the ground rises more gradually, but perceptibly, from the river bank. Just where the valley is narrowest is the site of Henley. A little farther down the hills recede on this side, and a fertile strath intervenes between their base and the water's edge.

Henley is an old town—indeed, Plot claims for it the distinction of being the oldest town in Oxfordshire—but it makes little figure in history. A conflict between the royal and the parliamentary troops in the "Great Rebellion" is almost the only stirring incident which it has witnessed. Moreover, it has retained fewer relics of ancient days than many places of more modern date. Even its church, which is well situated in the neighbourhood of the river, is not a building of unusual antiquity. The greater part of the fabric is in the Perpendicular style. The tower is even younger, and is said to have been erected by Cardinal Wolsey, so that it belongs to the latest period of Tudor work. Several of the windows have been filled with modern stained-glass, and the interior has been carefully restored, so that the church is not unworthy of its position. Some of the monuments have a certain interest, though no great historical personages have found a grave here. One commemorates Richard Jennings, "Master Builder of St. Paul's Cathedral"; another, Jack Ogle, an almost forgotten humorist of the days of the Restoration; a third, the widow of Sir Godfrey Kneller; and a fourth, General Dumouriez, who ended an eventful life at Turville Park, in this neighbourhood. He was one of those unlucky men who have the misfortune to be too rational for the age in which they were born. A distinguished soldier even in his youth—for by the time he was four-and-twenty he had been wounded almost as many times—he fell under Court displeasure for his liberal opinions. These the Bastille did not eradicate, so that he afterwards became a member of the Jacobin Club. But though he had striven and suffered for freedom, though he had headed the troops of the Directory in a successful campaign in Belgium, he was too moderate in his views to satisfy the fanatics of the Revolution, and, to save his own life, was obliged to put himself into the hands of the Austrians. At last he came to England, where he lived for nearly twenty years the unobtrusive life of a man of letters.

Though Henley has not retained any of the picturesque mansions of olden time, there are several houses, dating from various parts of the last century, which will repay rather more than a passing glance; and the town, as seen from the river bridge, is not without a certain beauty. While these Hanoverian mansions do not afford us the charm of the varied outline and picturesque grouping—the light and shadow—of mediæval buildings, there is a certain stateliness in their strong-built walls and formal rows of windows; and the rich red of their brick façades, especially when relieved by the green tendrils or the bright flowers of climbing plants, is not without its attractions from its warmth of colour. Of these mansions —for they are almost worthy of the name—Henley contains some good examples; and some bow-windowed houses, perhaps of slightly earlier date, are in pleasant contrast with their stiffer outlines, and give variety to the domestic architecture. The Berkshire side of the river also is not without its contingent of attractive residences. On the higher ground are two or three handsome mansions; at the bottom of the slope are many pretty villas—all modern. The bridge itself, a five-arched stone structure, is by no means the least adornment of the town. It, too, is a work of the last century, being built about the year 1787, from the design of Mr. Hayward, a Shropshire architect. He died during the progress of the work, and greatly desired, it is said, to be buried beneath the centre arch of the bridge. This singular place of sepulture—almost rivalling that of Alaric—was out of harmony with the spirit of the age, so, as the next best thing, they buried him in the neighbouring churchyard, and set up a fine monument to his memory.

Close by the bridge is the "Red Lion" Inn, a hostel of note now, as it has been for long years past; for on a pane in one of its windows Shenstone wrote the well-known lines:—

> "Whoe'er has travelled life's dull round,
> Where'er his stages may have been,
> May sigh to think that he has found
> His warmest welcome at an inn."

A sentiment which, though perhaps not very complimentary to English hospitality —or indeed to any hospitality, as the author obviously does not limit himself to our own island—has been endorsed, as Boswell tells us, by Dr. Johnson, who also, in his time, made trial of the "Red Lion." At any rate, Shenstone would have written more guardedly if he had been welcomed by the clerk at the counter of one of the great American hotels. An interview between one of these gentry and Dr. Johnson would make a good subject for an "imaginary conversation," except, perhaps, that it would be too brief.

Henley is generally a quiet enough town, though the increasing fondness for river-side amusements gives to it a certain briskness through all the summer-time; but it has one epoch of thrilling excitement, one brief period of dense crowd and ceaseless bustle, in the early part of July, at the time of its regatta. If the Universities' race between Putney and Mortlake is the aquatic Derby, Henley races are the Goodwood meeting of the Thames. The inns, the lodgings, the private houses, are full of visitors; house-boats are moored on the river, tents pitched in the meadows for those who enjoy the delights of camping out, excursion trains disgorge their thousands, boats of every description bring their contingents from various localities up and down stream. The "Fair mile," the famed approach to the town on the

HENLEY, FROM THE TOWING-PATH.

Oxford road—the special pride of Henley—has no rest from the stream of passing vehicles, and its trees are powdered with their dust; the streets, the meadows, the bridge, every "coign of vantage," are crowded; the usual itinerant accompaniments of an outdoor festivity are there in abundance, and the whole place is noisy with passing vehicles, shouting throngs, vendors of "c'rect cards," and other wares. The course is rather less than a mile and a half in length, from an "ait" below Fawley Court, which bears the name of Regatta Island, to the bridge. So the town itself becomes the theatre in which the interest of the aquatic drama is concentrated. The banks may be said to blossom with artificial colours, for as it is summer-time the "bright day brings forth," not the serpent, but the daughters of Eve in their smartest dresses and their most brilliant of parasols. Beauty and fashion are there, for a day or two at Henley make a pleasant change in the London season, when its gaieties begin to pall a little, and the streets of the metropolis are at a July temperature. Here may be seen subtler harmonies and the delicate blendings of tints that indicate the handiwork of some mistress of the art of dress; there the more glaring colours and gaudier contrasts that mark the efforts of

REGATTA ISLAND.

the shorter purse and inferior taste; but even to these distance lends enchantment, and all unite to form a variegated border to the river and make a flower-bed of the meadows. The men, too, don brighter colours than is their wont, for boating uniforms are in the ascendant. The river is alive with craft of all kinds—skiffs and dingies, tubs and boats of every degree—and the officials find it no easy task to clear the course for each race. The interest is not, as at Putney, concentrated on a single contest; the "events" are many, the chief, perhaps, being the Ladies' Plate, the Grand Challenge Cup, and the Diamond Sculls. These also are not settled in a single race; usually there are two or three heats, in order to reduce the number of competitors, before the final struggle. The interest of the Henley contests also affects a wider circle than the inter-University race. The colleges of Oxford and Cambridge, which have taken the lead on the Isis or the Cam in the annual races, send their representative "eights" or "fours," the best oarsmen of the London clubs put in an appearance, and one or two of our public schools now commonly send a boat, and not seldom carry off a trophy from Henley. Thus these races have a special interest for fathers and mothers, for "sisters, cousins, and aunts;" and a visit to Henley is not without its attraction for those by whom the good things of this life are held in esteem, for luncheons and various comforts for the inner man—and woman—are by no means forgotten.

We may recall to mind in passing that the course at Henley was the scene of the first aquatic contest between the Universities of Oxford and Cambridge. It took place at the beginning of the long vacation, on the 10th of June, 1829, late in the afternoon. Contrary to expectation Oxford was victorious. The race of course was rowed in the old-fashioned heavy boats, without outriggers, which were then termed "very handsome, and wrought in a superior style of workmanship. . . . The Oxford crew appeared in their blue-check dress, the Cambridge in white with pink waistbands. Some members of the crews on both sides afterwards became men of mark; four of them have risen to high positions in the Church. In the Oxford boat rowed W. R. Fremantle, now Dean of Ripon, and Christopher Wordsworth, the venerable Bishop of St. Andrews. In the Cambridge boat rowed Merivale, the historian of Rome, who is now Dean of Ely, and George Augustus Selwyn, the first missionary bishop of New Zealand, who after years of arduous labour in that distant field of Christian enterprise was transferred to the bishopric of Lichfield. There he laboured earnestly at work not less in amount, and more exhausting in nature, than that of the colonial mission-field, till he was called away to his rest." On the Berkshire side of the bridge at Henley a road climbs the steep slope, which may well be followed by any who desire to obtain a wider view of the neighbourhood—stately trees, grassy slopes, now and again a villa with its garden, brighten the nearer distance; below lie the valley and the town. In one place the bank by the road-side is steep and broken, the red soil contrasting pleasantly with the rich green of the foliage. There is a walk also at the base of the hill, which should not be forgotten, where the path leads along a level strip of meadow, dappled in the spring season with innumerable flowers, to the little church of Remenham, with its remnants of Norman work, its exceptionally pretty lych-gate, and its carved porch. Its situation, with the river on one side and the wooded slopes on the other, is not the least picturesque in the Valley of the Thames.

On the meadows below Henley, and on the left bank of the Thames, is Fawley Court, a mansion built by Sir Christopher Wren, but subsequently enlarged. The grounds extend from road to river, and their fine aged trees enhance greatly the beauty of this reach of the Thames. The present house occupies the site of an old manor-house, which was plundered by the Royalist troops at the outbreak of the Civil War. The owner, Bulstrode Whitelock, has left on record a pitiful account of the wanton ravages committed by the troopers. They consumed, or wasted, a great store of corn and hay; they tore up or burnt his books and papers, many of them of great value; they broke up his trunks and chests, stole whatever they could transport of his household goods, and destroyed the rest; they carried off his horses and his hounds, killed or let loose his deer, and broke down his park palings—"in a word, they did all the mischief and spoil that malice and enmity could provoke barbarian mercenaries to commit." We have heard often of the devastation wrought by the Roundheads; it is well to remember that the Cavaliers were by no means guiltless. Remenham village, with its little church, already mentioned, nestles below the slope opposite to Fawley Court, and lower down, on the Buckinghamshire side (for we have now crossed the county boundary), comes Greenland House, opposite to where the Thames makes its sharpest bend. This fared even worse in those unquiet times. About two years later than the incident just

FAWLEY COURT.

related it stood a siege of six months, when it was held by the Royalists against their opponents, and did not capitulate till it was almost knocked to pieces. Some traces of the works raised during the siege still remain, and when the house was enlarged, about a quarter of a century since, quite a crop of cannon-balls was dug up.

Sweeping round the eastern side of the Berkshire slopes the Thames is checked by Hambledon Lock and its islands—well known to fishermen, the reach above being noted for pike—by Aston Ferry, where the river begins to strike out into the more open part of the valley, by Culham Court and the islands below, till it approaches a place well known to the pleasure-seekers of the present day as a sort of half-way house between Henley and Marlow, and as the fittest site for a picnic.

Pleasantly situated on the level meadows in the valley on the Berkshire side, and backed by the wooded uplands which are now some little distance from the river, is

ASTON FERRY.

Medmenham Abbey, a place of more note since its suppression than in earlier times. The convent was founded not long after the Norman Conquest, when the owner of the manor bestowed it on the Abbey of Woburn, in Buckinghamshire, which he had recently founded, for the endowment of a separate but subsidiary house. Medmenham does not appear ever to have become wealthy, and never made any figure in history, except that the abbot was epistolar of the Order of the Garter, a distinction which one would not have anticipated for a place so humble. The report of the Commissioners at the time of the suppression of the monasteries is curiously negative. It had at that time only two monks, "Servants none—Wood none—Debts none—Bells, &c., worth 2l. 1s. 8d. The house wholly in ruins, and the value of the moveable goods only 1l. 3s. 8d." A poor piece of plunder, certainly.

As this statement would lead us to suppose, not much of the original conventual buildings now remain. Even of those parts which bear an ancient aspect, some are only imitations of the last century, when Medmenham enjoyed a certain amount of celebrity. At that time the abbey, which after the suppression of the monasteries had been converted into a dwelling-house, was the property of Francis Dashwood, Lord le Despencer. He determined to found a society, which was called after his first name—the Franciscan Order. It was, however, anything but an Order of Poverty. The number was twelve, in imitation of a band to which these men were the most opposite possible, for the old Latin lines—

"Exue Franciscum tunicâ laceroque cucullo
Qui Franciscus erat, jam tibi Christus erit"

—are the very last one would think of applying to this Order of Debauchery. Great mystery was observed; the workmen who prepared the building were brought down from London, secluded as far as possible from any communication with the people of the neighbourhood, and then conveyed back as mysteriously as they had come. Very few servants were kept in the "abbey," and these were not allowed to wander beyond the monastic precincts, or to hold any intercourse with the neighbouring villagers. Still, though there were no penny papers or "own correspondents" in those days, though "interviewers" and "special commissioners" had not been invented, some rumours got abroad as to the sayings and doings of the new fraternity. It is to be hoped that they were exaggerated, that the author of "Chrysal" has

MEDMENHAM ABBEY.

over-coloured the picture; but that these Franciscans carried out to the full the Rabelaisian motto, "Fay ce que voudras," inscribed over their portal, there can be little doubt. Their rites and ceremonies appear to have been profane parodies of those of their predecessors, their lives in keeping with their religion. Among the band were numbered the Earl of Sandwich, Bubb Dodington, Wilkes, and Churchill. Society seems to have been rather scandalised, but we do not read that the Franciscans suffered any social penalty. Happily, after a time the Order was dissolved, under what circumstances it is not exactly known. One version, perhaps legendary, is that a disappointed member secreted a large monkey in a chest in the hall prior to one of their great festivals. At a particular stage of the ceremonies there was an invocation to the Evil One. At this moment the treacherous monk pulled a string and lifted the lid; Pug sprang upon the table, and then leaped through the open window. The revellers, mistaking their kinsman for their master, thought matters were getting serious, and so held no more merry meetings.

The house is at present a pleasantly situated inn, with farm buildings attached; ivy mantles picturesquely some of the old walls, and the tower, an "antique" of the last century, looks well when not too closely examined. Fine aged trees add greatly to the beauty of the place. The village lies back from the river at the foot of the bluffs, and is reached by a lane, bordered by some of the old-fashioned free-growing hedges which, though not much favoured by modern farmers, are such a delight to the wayfarer. Of the many sequestered spots in the Valley of the Thames, Medmenham village is by no means the least attractive. A wooded slope rises steeply at its back, the little church is half buried among trees, its cottage gardens are bright with flowers, and more than one of the buildings is ancient and picturesque. A farmhouse on the upland above is said to be the successor of one which occupied the site some eight centuries since, and there is an old-world air about the whole place, as though generation after generation of its simple inhabitants had lived and died, apart from the turmoil of the outer world; hearing of stirring events, of battles, of changes of government, even of the dethronement of kings, and of civil strife, as of things which altered but little the even tenor of their lives, and only came home to them when, like bad seasons, they raised prices or

BELOW MEDMENHAM.

lowered wages. In such places generation follows generation with little note of change. The son grows up to manhood, and lives as his father did before him; takes his place on the farm when the old man retires, first to his easy-chair by the fireside in winter, and at the cottage door in summer, and then to his long resting-place in the churchyard; the young man, in his turn, becomes the father of sturdy boys, begins to stoop a little, and to show the signs of advancing years, till at last he too sinks down into the "lean and slippered pantaloon," and then follows his forefathers to the silent land. These quiet days now seem nearly ended for our country—machinery, steam, electricity, have so quickened the pulse in all the great centres of national life that there is a responsive thrilling of the nerves even in the most remote extremities. The old order has changed, yielding place to new. We have gained much, but we have lost something, and can appreciate, from their increasing rarity, the calm of these little nooks and corners of England, where the scream of the steam-whistle, or the bellow of the "siren," does not scarify the ears; where the voice of the costermonger is not heard in the land, and no excursion train disgorges a crowd of noisy revellers; where factory chimneys do not blacken the air, nor heaps of chemical refuse disseminate their fetid odours.

Below Medmenham some more islands vary the course of the Thames, and on the high ground upon the left bank is Danesfield. Woods surround the house and clothe the slope. Here flourish holly, box, and yew—trees, it is believed, of indigenous growth; descendants, very probably, of those which covered all the uplands, when men were few in England, and many a mile of unbroken forest separated the scattered settlements. A curious relic is said to be preserved in the house—a withered human hand, which was discovered among the ruins of Reading Abbey. This is believed to be identical with the supposed hand of St. James the Apostle, presented to that establishment by Henry I.

Hurley comes next, with its islands and locks, interrupting the even tenor of the river, with Harleyford House, backed by sloping woods, on the opposite shore. Hurley is another old-world place, for it too carries back its history to the days of the Conqueror, when a convent was founded here. A former writer on the Thames makes this a text for some sarcastic remarks:—"The fascinating scenery of this neighbourhood has peculiarly attracted the notice of the clergy of former periods, who, in spite of the thorny and crooked ways which they have asserted to be the surest road to heaven, have been careful to select some flowery paths for their own private journeyings thither; among which ranks Hurley, or Lady Place, formerly a monastery." This was founded by Geoffry de Mandeville, a comrade of William the Norman on the field of Hastings, to whom fell a share of the plunder of England. Parts of the church belong to that which he erected, and within its walls Edith, wife of Edward the Confessor, was buried. A group of farm buildings still incorporates portions of the ancient monastery, the chief one being the refectory. But the house called Lady Place, which once occupied another part, has a more important position in history than ever belonged to the

Benedictine convent, which, perhaps, was somewhat thrown into the shade by its annexation to the great Abbey of Westminster. After the Dissolution the site of the monastery of Hurley was purchased from the family to which it had first been granted by Richard Lovelace, who had been a companion of Drake on one of his expeditions. He built a fine house "out of the spoils of Spanish galleons from the Indies," and this, in the year 1688, was the property of his descendant Richard,

BISHAM ABBEY.

Lord Lovelace. "Beneath the stately saloon, adorned by Italian pencils, was a subterranean vault, in which the bones of ancient monks had sometimes been found. In this dark chamber some zealous and daring opponents of the Government held many midnight conferences during that anxious time when England was impatiently expecting the Protestant wind."* In acknowledgment of this the house was afterwards visited by William III. The Lovelace title became extinct in the year 1736, and Lady Place passed into other hands. The purchaser was "Mrs. Williams, sister to Dr. Wilcox, who was Bishop of Rochester about the middle of the last century. This lady was enabled to make the purchase by a very remarkable instance of good fortune. She had bought two tickets in one

* Macaulay, "History of England."

lottery, both of which became prizes, the one of £500, the other of £20,000." The last person to live at Lady Place was a brother of Admiral Kempenfelt. Concerning him a curious story is told in Murray's Handbook. The brothers had each planted a thorn-tree, in which the owner of Lady Place took great pride. "One day, on coming home, he found that the tree planted by the Admiral had withered away, and said, 'I feel sure that this is an omen that my brother is dead.' That evening came the news of the loss of the *Royal George*." The house, which con-

BISHAM CHURCH.

tained a fine inlaid staircase and a grand saloon, its panels "painted with upright landscapes, the leafings of which are executed with a kind of silver lacker," was pulled down and the more valuable part sold in the year 1839; but grass-grown mounds mark the site of the historic vaults; and the old cedars and other fine trees in the enclosed meadows are memorials of its former splendour.

Bisham comes next, a spot of rare attractions. Between the wooded hills and the river there is a broad and fertile strath, the very place on which, in ancient times, monks "most did congregate." Accordingly they soon got hold of a goodly estate at Bisham, and that grey old manor-house standing among groves of stately trees some little distance from the Thames marks the site, first of a house of the Templars, then of an Augustinian Priory. The latter had about two centuries of tranquil existence, for it was founded in the year 1338, by William Montacute, Earl of Salisbury. The last prior submitted to the change, adopted the tenets of the Reformers, and became Bishop of St. Davids. Moreover, he took to himself a wife, who bore him five daughters, each one of whom had a bishop for her husband. His memory is not held in great honour in the annals of St. Davids, for he cared more for money than for the good of the see. After Bisham passed into the hands of secular owners it becomes better known to history. Henry VIII. made a present of it to his discarded spouse, Anne of Cleves, and she, by royal permission, exchanged it with Sir Philip Hoby for a manor in Kent. He was the last Englishman who was legate to the Pope at Rome, and, like many others of his nation, never left the place alive. His brother, Sir Thomas, who succeeded to the estate, was ambassador to France, and he also died abroad. In Queen Mary's time the Princess Elizabeth was committed to his charge, and she spent a considerable time within the walls of Bisham. What she thought of the place and of her keeper may be inferred from a graceful compliment which she paid him on his first appearance at court after her accession. "If I had a prisoner whom I wanted to be most carefully watched, I should intrust him to your charge; if I had a prisoner whom I wished to be most tenderly treated, I should intrust him to your care."

The house, which now belongs to the VanSittart family, is a picturesque old structure of grey stone, with pointed gables, mullioned windows, and a low tower; portions of it, for instance, the tower and the hall—once part of the convent chapel—are remnants of Montacute's abbey; but the larger portion of the building is later than the date of the suppression of the religious orders, most of it being late Tudor work, due to the Hoby family. Bisham is said to have its ghost. Lady Hoby, wife of Sir William, "walks" in one of the bedrooms, appearing as the duplicate, in opposite tints, of a portrait which hangs in the hall, and engaged in "washing her hands with invisible soap in imperceptible water," in a basin which, self-supported, moves on before her. This is the cause which disturbs her rest: "She had a child William, who, being a careless or clumsy urchin, kept always blotting his copy-book; so the mother did not spare the rod, and spoiled the child in a physical sense, for she whipped Master William till he died." The author of "Murray's Guide-Book to Berkshire" states, as a curious coincidence, if not a corroboration of the story, that on altering the shutter of a window "a quantity of children's copy-books of the reign of Elizabeth were discovered pushed into the rubble between the joists of the floor, and that one of these was a copy-book which answered exactly to the story, as if the child could not write a single line without a blot." Bisham has also its secret chamber, an indication that it was built when political struggles had their real perils.

Bisham Priory in former days—perhaps owing to its connection with the Montacutes, Earls of Salisbury—was the burial-place of several distinguished men, whose monuments once adorned the priory chapel, but have disappeared since it became a dwelling-place. The following list of such interments is testimony to the perilous life led by the aristocracy of those days:—" Thomas, Earl of Salisbury, who died at the siege of Orleans in 1428; Richard Neville, Earl of Salisbury and Warwick, beheaded at York in 1460; Richard Neville, 'the king-maker,' killed at the battle of Barnet, 1470; his brother John, Marquis of Montague, killed at the same battle; and Edward Plantagenet, son of the Duke of Clarence, beheaded in 1499, for attempting to escape from confinement." "The paths of glory lead but to the grave" was a true saying in the days of old. Between the foeman's sword on the one hand and the headsman's axe on the other, a goodly proportion of our nobility came to untimely ends.

The Hobys rest in the parish church. This is beautifully situated close to the riverside. Grand old trees cast their shadows on its graveyard and overhang the walk by the brink of the Thames. The grass-grown plot studded with graves is almost merged with the trim garden of the rectory, the flowers in which brighten the view and pleasantly vary the greens of the foliage and of the grassy meadows. These, too, are bright enough in spring-time, when they are dappled with its golden flowers, before the taller herbage of summer has begun to wave in the soft wind.

The tower of Bisham Church—the part most conspicuous from the river—is of very early date—a rude and rather curious piece of Norman work, which may be older than the days of Stephen, when the Templars first came to Bisham. The body of the church is also picturesque, but it has been so greatly restored—even to rebuilding—of late years that it is no easy task to separate old from new. At the present time its most interesting features are some of the monuments of the Hoby family, especially those to the two brothers mentioned above. The widow of the second brother, Thomas, had the bodies of both brought back to England for burial at Bisham; and being a lady learned even for those days, when people did not, as in later times, suppose that a woman made the better wife for being as ignorant as a scullery-maid, she wrote them an epitaph in three languages. The concluding lines on her husband's monument appear to express a willingness, under certain circumstances, to be consoled even for the loss of such a paragon :—

"Give me, O God! a husband like unto Thomas,
Or else restore to me my husband Thomas!"

She seems to have considered that the first part of her prayer was granted, as the second could hardly be expected, for before the year was out she married Sir Thomas Russell. But on the whole the interior of Bisham Church will not detain the visitor for long; he will care rather to linger in the churchyard and its neighbourhood. It is pleasant to pass up and down by the riverside under the shadow of the trees, to gaze upon the noble sweep of the Thames, and over its fertile valley plain, to seek some quiet spot which commands a view of the grey walls of Bisham Priory and the beautiful trees in its park. Even the village is in keeping with the rest of the scene, and is brighter and prettier than is usual, and this is saying much; for though, as a rule, English cottages cannot compare in picturesqueness with many that we see on the other side of the Channel, the frequent poverty and monotony of their design is often atoned for by the creepers which blossom profusely on the walls, and the flowers which make the strip of ground in front one living posy. But in

GREAT MARLOW, FROM QUARRY WOODS.

Bisham not a few of the cottages are picturesque. For some reason or other, partly, perhaps, owing to the absence of mechanical industries, the towns of the south and west of England are commonly, and the villages almost always, more attractive than in the north, and this disparity, as regards the latter, becomes still more marked when we cross the border; for a Scotch village often attains to the extreme limit of dreary ugliness.

Turning aside from the woods of Bisham and sweeping away yet farther into the broad valley plain, the stream of the Thames brings us to the quiet market-town of Great Marlow. A suspension bridge crosses the river, and the gardens and houses on either side afford many scenes of quiet and homelike beauty. Just below it are a weir and a mill, not without a certain picturesqueness; and from a distance the spire of the church, rising from among groups of trees, enhances the attraction of the scene. In this little Buckinghamshire town, before the railway approached its outskirts, life must have passed peacefully, not to say sleepily; and even now, as it lies off the main line, it does not strike us as a place for over-stimulation of the nervous system. The river, during a considerable part of the year, would still be almost as fit a scene for a poet's musing as it was when Shelley resided in the town, and wrote the "Revolt of Islam," spending much of his time dreamily floating in a boat upon the Thames. The most conspicuous feature in Great Marlow, as has been already said, is its church, which stands near to the river and the bridge. Unfortunately, it is one of those where "distance lends enchantment to the view." The present building was erected in the year 1835, on the site of an older one. Whatever this may have been, it could hardly have been so ugly as the present structure. The style may be called Gothic—that is to say, the architect had in his mind some of the English parish churches of the thirteenth or fourteenth century; but it is the Gothic of what we may call the pre-Victorian revival, and about as like what it supposed to imitate as the "English as she is spoke" of the ingenious Portuguese is to our mother-tongue. Efforts have been made, and we believe will continue to be made, to improve it. For instance, the church was constructed for galleries; these have been pulled down—at some inconvenience, we should think, if a fair proportion of the population goes to church; and the interior has been divided by means of the usual arches into a nave and aisle. These, as they come to an end before they reach the roof, have at present a rather forlorn aspect, and, as there is no particular merit in their design, scarcely justify their existence. It is intended, we are informed, to rebuild the whole structure piecemeal; but as the original fabric appears to be in no danger of premature decay, it is a question whether it would not have been better to accept its ugliness, and employ the very large sum which must be expended before the work can be completed for other and more directly useful purposes. The church, however, is not wholly without interest, as it contains one or two "curiosities." Of these, one is a portrait of the "Spotted Boy," the work of Coventry, in 1811. The lad was one of Richardson's "exhibits," and died at Great Marlow. He was a negro, but was mottled with white patches on body and hair—as if he had been imperfectly operated on with soap after the manner of the advertising placards. In fact, he was a parallel example in the human race to Barnum's famous white elephant. The picture might by some be deemed more appropriate to the walls of Madame Tussaud's galleries than to those of a church; nevertheless, so long as it is there, it should be hung where it can be seen. At the present time, the removal of the gallery staircase has resulted in "skying" it most effectually. A good instance of modern mediæval absurdity may be seen in a monumental brass erected to the memory of a lady who died so recently as 1842; for in the inscription the words *charitie* and *mercie* occur as written. More interesting, and in its way quaint, is the monument to a doughty Englishman, Sir Myles Hobart, who once represented Great Marlow in Parliament. He was a steadfast opponent of the Court party in the troublous days before the Great Rebellion, and, on one occasion, with his own hands locked the door of the House during the reading of a protest against certain illegal taxes. For this he was, of course, imprisoned; but it is pleasant to read that the Long Parliament voted a considerable sum to his family as an acknowledgment of his services and a compensation for his sufferings. A bas-relief indicates the manner of his death, which was the result of an accident. His horses ran away down Holborn Hill, upsetting the coach, and fatally injuring their master.

Great Marlow is in truth a town of unusual antiquity, for it is heard of before the Norman Conquest; but an old monastic barn by the bridge, and some fragments of an ancient building in the town, which is called the Deanery, are all that remain from mediæval times. Of the latter, the most conspicuous remnants are two windows, with tracery of a rather Flamboyant character, which are incorporated into an old house, now undergoing "restoration." In short, the lions of the town will not long detain the traveller, although he will be tempted to look rather longingly at some of its substantial houses, with their bright and pleasant flower-gardens.

HENLEY TO MAIDENHEAD.

There is a circumstance connected with Great Marlow, beneath the dignity of history indeed, which, however, as we are writing of the Thames, must not be passed over in silence. In former days—and perhaps still, for we do not wish to make experimental proof—the simple and apparently purposeless question, "Who ate the puppy pie under Marlow Bridge?" sufficed to throw the bargee of the Thames into a state of mind which could only find adequate expression

A PICNIC AT QUARRY WOODS.

in words which more than bordered on profanity. The venom rankling in the taunt is thus explained:—The landlord of the inn at Medmenham had received private information that certain bargemen meditated that night a foray on his larder. He was a humorous man, who had just drowned a litter of young puppies. So he had their corpses baked in a pie, which he placed in the larder, and did not sit up to keep guard. The larder was robbed, the pie was carried off and conveyed to Marlow Bridge, where the plunderers feasted, as they supposed, on young rabbits.

Below the weir, where the Thames is parted by willow-covered islands, are some pleasant nooks for the artist who loves riverside scenery, and quiet spots where he may pursue his work without the presence of a small circle of gaping bystanders. Brothers of the angle also find much employment near Marlow, as the fishing is noted. Taunt, in his useful little "Guide to the Thames," tells us that he saw a trout weighing eight pounds, which had just been caught near Quarry Woods, and that in the hostel called the "Anglers" is one stuffed, which is reputed to be the largest that has been taken in the Thames.

From Great Marlow weir and locks the Thames sweeps back through level meadows to the foot of Quarry Woods. There is now a pleasant diversity in the scenery. On the left the level plain continues, over which we glance backwards to the spire and houses and trees of Great Marlow, and sideways for a longer distance to where the grey, stumpy tower of Little Marlow is almost concealed by foliage. But on the right bank the steep wooded slope of the ancient valley which runs at the back of the groves of Bisham is now approached by the Thames, whose stream for a time hugs the foot of the declivity, and gives us a foretaste of what is to come at Cliefden. At one place the dense woodland is interrupted by a pretty cottage, and an old chalk-pit has been utilised as a part of its garden. A pleasant retreat this would be from the time when the woods begin to brighten with the first buds of spring until they are dappled with the many tints of the dying foliage of autumn. But these nooks and corners by the Thames are no longer, as they would have been a generation since, suited for the abode of an anchorite. All through the summer day there is now little solitude to be found on this part of the river. Boats laden with pleasure-seekers pass and repass—from skiffs and dinghies to steam-launches and house-boats—and there are not seldom obvious signs of Londoners at play. Still, there are quiet, dreamy hours when all the charm of the scene can be enjoyed—most of all in the earlier months, when the flowers are at their brightest, and the verdure is at its freshest; when the dweller in the city is still tied down by duty or by the desire of gain to the crowded streets, and must be content with extensive views of chimney-pots.

After a time the river once more deserts the shadows of the wooded slopes and strikes out again into the open plain. A reach begins when the islands are passed, pleasant when the wind is favourable to those who love sailing. The surface of the water lies open to the breeze, and from bank to bank it is a little wider than usual. In other respects the scene, after the beauty of the last stage, becomes a little monotonous. The chalk downs have receded and their slope has diminished on the right; on the left they are still distant. There are but few trees by the river bank, and the meadows on either side are level and uninteresting; even the embank-

A GROUP OF SWANS.
(From an Instantaneous Photograph.)

ment of the railway, which we are now approaching, is a prominent and not attractive feature in the landscape. The bridge, however, for a railway bridge, is not unpleasing; and when we have left behind both it and some works on the nearer side the scene once more brightens. Just inland is Bourne End Station, where the branch line from Great Marlow joins that which runs from Maidenhead to High Wycombe and Thame.

No one can journey a mile or two along the Thames without noticing the swans, which add so much to the beauty of the royal river, as they "float double, swan and shadow," on some quiet pool, or come ruffling up towards some passing skiff, in defence of their young. A sheet of water is hardly complete as a picture without some swans floating upon it. White specks in the distance, forms of exquisite grace and purity of colour in the foreground, they give a harmony to the composition and add to the scene the interest of life. The moorhen and coot are too inconspicuous as they lurk under the reeds, or swim hurriedly across the open water. The swallow and the kingfisher, brilliant though the plumage of the latter, dart too swiftly by to produce any lasting impression on the mind; but the swan sails slowly along, and lingers here and there, in harmony with the traveller, who is seeking only to drink his fill of nature's charms.

The abundance of swans on the Thames is due to the fact that they are carefully tended. They are not to be reckoned among *feræ naturæ*; indeed, though other species are chance visitors, the "mute swan" is never, strictly speaking, a wild bird in England; but they are private property, the Dyers' and the Vintners' Companies being among the principal owners. Keepers are appointed to look after them, especially in the building season, when they are in some danger from predaceous animals, and more from predaceous persons. Still, they can take pretty good care of themselves, for the cock bird is very fierce in defence of his nest or young, and can deal formidable blows with his pinions, although, if he has succeeded in breaking a limb, as popular report asserts, the sufferer's bones must have been rather weak. The nests are generally built on the "aits," where the osier beds afford a quiet retreat and a good foundation for the capacious structure. This is constructed roughly of twigs and reeds, and raised some little height above the ground. In former days, when *fay ce que voudras* was a motto adopted by City Companies more easily than in the present, they used, as sole conservators of the River Thames, to make an excursion annually in their barges, with all due ceremony and festivity, in order to count and mark their swans. This process was called "swan upping," corrupted generally into "swan hopping." The swans were caught and examined, sometimes not without a good deal of trouble, for a strong old cock bird did not submit himself very willingly to the physical suasion of the "swan crook." Cygnets were marked on the bill with the special symbol of the Company to which the parent birds belonged. The Vintners' Company mark was two nicks, whence, by a slight corruption, came the curious inn-sign of the "swan with two necks." "Swan upping" began on the Monday after St. Peter's Day, just at the time when a water excursion would be most pleasant, in the full warmth of summer, and before all the spring brightness had passed away from the foliage. At an earlier date the birds appear to have been regarded as royal property; and in Hone's "Every-day Book," under the heading of July 12, is the reprint of a curious tract published in 1570, entitled, "The Order for Swannes." It is here enacted that all private owners must compound with the King's Majesty for the right to use a mark; while penalties, commonly fines of thirteen shillings and fourpence, were inflicted for stealing the eggs, for unlawful carrying of swan-hooks, and the like; but the erasure or counterfeiting of marks entailed a heavier fine and a year's imprisonment.

Abney House, below the bridge, is one of those places by the river which must often set the wayfarer coveting, in despite of the decalogue. Climbing plants of many kinds mantle with flowers and with leaves, large and small, the verandas and walls of the house; and their green foliage is in pleasant contrast with its red-coloured bricks. The smooth-cut lawns are green even in the hottest season, and are interspersed with living bouquets of bright-coloured flowers. The shrubberies are adorned and the lawns are shaded with many a rare tree, such as cedars and conifers of diverse kind, which by the side of the English river call up memories of their distant homes in far-off lands.

The woods of Cliefden are now in view in front of us on the right, and henceforth remain a conspicuous feature in the landscape, though it is yet some time before our boat will be gliding along in the silence of their shadows. There is a pleasant reach below the Bourne End bridge, during which the views are more varied and the riverside is less monotonous than on the part which we have left above it. The willows growing by the stream are always pleasant to the eye as they whiten in the summer breeze; there are sure to be tufts of flowers here and there by the waterside, and if these be wanting we need not weary of the woods of Cliefden, on the high chalk escarpments, as they stretch away inland on our left.

The ivy-mantled tower of Cookham begins to show in front, as we approach one of the prettiest spots on the Thames. If it be summer-time, there is evidence that this opinion is held by many. There is no lack of boats on the river; here is a house-boat moored by the shore; in yonder meadow some small white tents proclaim that two or three parties are "camping out"—all being direct indications that the neighbourhood of Cookham has many admirers. Tent life may be all very well in fine weather, but its charms on a rainy day must be more than dubious. Granted the most studious habits, granted that power of immediate concentration upon some absorbing treatise—let us say " the philosophy of the uncreated nothing"—which few possess—at any rate, in holiday time; granted a companion of great but not too provoking amiability, and yet we will undertake to say that a tent will seem, at the end of a day's steady rain, to be rather cramped quarters, and the inmate's thoughts will turn regretfully homewards. Excitement may no doubt be found sometimes when the rain detects shoddy workmanship, and begins to drip upon the floor, or there is a battle with a gale of wind; but an incident of this kind, though a variety, is not always a pleasant one.

HENLEY TO MAIDENHEAD.

Below Cookham Bridge, a light iron structure, the river broadens out before it splits up into channels, in a way that is rather perplexing to new-comers. On the left hand is the original main channel, which takes a great bend outwards towards Hedsor before curving back to pass under the shadow of the Cliefden

COOKHAM.

woods. Then comes "the cut," with its locks—an artificial canal made to avoid the circuit and difficulties of the old channel. Beyond this are the entrances to two smaller channels, one leading to Odney weir, the other entering the Thames some distance below Cliefden House. In the neighbourhood of Cookham it is often hard to say whether the foreground or the distance is more beautiful. Here the ancient fabric of the church, with its ivy-clad tower, rises from its trim churchyard, surrounded with aged trees, some of them little more than huge trunks, which still retain enough vitality to support a short but thick output of branches. Here is an attractive hostel by the waterside. Here are the narrower arms of the river running up invitingly by the side of pleasant gardens and under the shadows of giant trees—places where the idler may linger for a long summer afternoon in some shady nook. Contemplative pursuits appear to be much in favour near Cookham. Fishing for roach out of a punt beguiles the time, and the excitement is of the mildest form, one, probably, from which few persons, however highly strung their nerves, would be debarred. An aroma of botanic origin, but not attributable to any flowers, sometimes steals over the water, to announce that the boat, half hid among the bushes, is not untenanted, and that the occupant is a victim to the herb denounced once by an enthusiastic divine as "the gorging fiend." Here is a student of books, but the volume bears a resemblance to the literature of railway stalls rather than of the academy. Here is a devotee of the brush. He, at least, is at work, but in a leisurely way, as if he entered too fully into the spirit of the picture to spoil it by over-much intensity. In short, Cookham is one of the prettiest, pleasantest, laziest spots that the peripatetic traveller could find within a two hours' journey from Charing Cross.

Cookham Church, which has just been mentioned, is almost hidden by the bridge and by houses from the prettiest part of the river, though well seen higher up the stream. Its low tower is partly covered with ivy; the body of the church is of various dates, the oldest part being Early English. It contains several modern stained-glass windows and old monuments, especially brasses. The cook of Queen Eleanor, wife of Henry III.; the "master clerk of the Spycery, under King Harry the Sixt," have their tombs within the church; a modern monument by Flaxman commemorates the death by drowning of Sir Isaac Pocock, and a bas-relief by Woolner adorns the tomb of Frederick Walker, the well-known artist.

In the distance, to the left of Cliefden, and seemingly forming with it one demesne, lies Hedsor Park, the seat of Lord Boston, with its imitation castle, which would be improved, pictorially speaking, by a judiciously administered dose of dynamite. Hedsor overlooks the old course of the river, but is not approached by the traveller on the Thames, who has to follow the new cut, the only navigable channel. There is nothing attractive in the house, which is in the modern Italian style, and is hardly worthy of the magnificent situation it occupies. The tiny church, which is within the park enclosure, and has been beautifully restored, contains some monuments of the Irby family. Dropmore Park, with its noted pinetum and fine gardens, lies still farther back, another creation of the same reign. The house was erected and the grounds were laid out by Lord Grenville, at the beginning of the present century, about the time that he was Prime Minister.

The chief point of interest about the new cut—which, as might be expected, has rather too much of the Dutch canal about it to attract the traveller just fresh from Cookham—is that, in making it, a number of skeletons, with swords and spears of Roman workmanship, were found entombed together; indicating that these meadows had been the scene of some long-forgotten conflict.

At the lower end of the new cut we pass through a lock into the main channel of the Thames, a short distance below a weir, and at the very foot of the Cliefden woods. It would be difficult to find a fairer scene on any river within the limits of our island, and not easy did we take a wider range over the surface of the earth. On both sides Art has been called in to the aid of Nature; but that aid has been only bestowed where it is a boon. The level island on the right hand has been

A CROWD IN COOKHAM LOCK.

converted into a beautiful garden, where clusters of bright flowers stud the greenest of lawns, and trees from distant lands are mingled with those of native growth. On the opposite shore the hand of the gardener is less conspicuous, and his art, though the more subtle, has been concealed. The chalky upland, which for some miles past has formed a marked feature in the scenery, and has bounded our view in front, now descends to the river brink in steep slopes, sometimes almost in cliffs. Between the foot of these and the water, only here and there does a narrow strip of level land intervene. On two or three of these a picturesque cottage has been built, and the brightest, gayest, trimmest of gardens planted; but the slope itself is one mass of trees and brushwood, through which, though very rarely, gleams forth a little crag of the white chalk rock. All the trees of England seem to have congregated on this bank: there are hazel and maple and thorn; there are ash and oak, and beech and elm; there are chestnut and sycamore, and, especially at this upper end, the brighter tints of the deciduous trees, and of the broad-leaved evergreens, are dappled by the sombre hues of Scotch firs, with their ruddy trunks, and of ancient yews, very possibly lineal descendants of trees among which the ancient Britons hunted, before ever a Roman galley floated on the Thames. For Cliefden Woods, though doubtless they are in part the result of the gardener's art, are very probably a relic of the primæval forests which once covered so large a part of England. As in the Kentish Weald, this rough and broken ground must always have been waste, and there trees would take root, from the time that the slope first was furrowed out by the river, and there would be the "lurking-place of wild beasts," in days when the huntsman wore skins for clothing, and pointed his arrows with chipped flints. Down by the river's brink what a wealth of beauty is often to be found; the waterside plants grow strong and free, pink willow-herb and purple loosestrife, yellow fleabane and St. John's wort, with numbers more which it is needless to mention; while the bank above is green in summer with many a herb, and bright in spring with many a flower. No trim shrubbery this on the Cliefden steeps; nature is left to wanton at will—nay, even to struggle for existence. Ivy and briony and wild bine festoon and sometimes half smother the trees, while the traveller's joy creeps and clings in masses so profuse that from afar it seems to flicker like grey lights among the green shadows.

From this position we cannot see the mansion, but from time to time as we pass down the stream it comes into view, standing above the slope on the edge of the plateau. Its absence is a boon rather than a loss; its clock-tower, indeed, as it rises above the hills, occasionally forms a pleasant addition to the view; but the house is not particularly striking in itself, and the design is wholly unsuitable for its position. That requires a building of irregular outline and broken, but well-conceived sky-line. This magnificent site, above the great river cliff, ought to have been crowned with a group of buildings, whose outline should suggest a cluster of hills. Yet the design of Cliefden House could readily be imitated with three or four packing-cases. It was a great opportunity, such, for instance, as that of the architect of the Parliament Buildings at Ottawa not only had, but also seized; but here, as is the rule, it has been wholly wasted, for to find an architect who has also the feelings of an artist is rare indeed. Since the Middle Ages they have been seldom more than learned master-masons. So we shall look as little as possible at Cliefden House, and as much as possible at its woods, and be thankful even for the tiny mercies of its clock-tower. The present house occupies the site of an earlier mansion, which was destroyed by fire, as was that which it succeeded. The destruction of the first house, in the year 1751, may be used to point a moral against reading in bed—at any rate, by the light of a candle. One of the maid-servants, while indulging in this practice, fell asleep, the candle set the hangings on fire, she woke up in too great a fright to do anything to extinguish the flames, and in a surprisingly short time almost the whole of the mansion was destroyed, but little of the furniture and few of the pictures being saved. This house had been erected by the notorious George Villiers, second Duke of Buckingham, whose duel with the Earl of Shrewsbury is among the memories of another part of the Thames; and when the latter fell wounded, it was to the shelter of this mansion that the guilty pair went off in triumph. Time, however, brought its revenge, when Villiers died "in the worst inn's worst room":

> "How changed from him
> That life of pleasure and that soul of whim!
> Gallant and gay, in Cliefden's proud alcove
> The bower of wanton Shrewsbury and love . . .
> There, victor of his health, of fortune, friends,
> And fame, this lord of useless thousands ends."

To him, as owner of Cliefden, in course of time, succeeded Frederick, Prince of Wales, father of George III. Of him

> "Who was alive and is dead,
> There's no more to be said,"

except that through him the national air "Rule Britannia" is associated with Cliefden. Thomson the poet had been taken into favour by this prince at a time when he was, to some extent, the patron of literature. Thus the masque of Alfred was first performed within the walls of Cliefden, and into this masque "Rule Britannia," composed by Dr. Arne, was introduced, and has alone escaped oblivion.

This house, which appears to have been a stately structure, was destroyed, as has already been explained, and it was rebuilt in the present century by Sir G. Warrender, from whom it was purchased by the Duke of Sutherland. Another great fire occurred in 1849, after which the present house was built. The gardens are very beautiful, but the walks through the groves which mantle the slope—through the dense vegetation and trailing undergrowth—are in their way not less attractive. The cliff runs by the riverside for more than a mile, unbroken except at one spot, rather beyond the house, where a glen, now forming part of the gardens, winds down to the riverside, and affords an easier access to the terraced plateau above.

Though less favourably situated for prospect or for health, there are, as we have said, homes of no little beauty on the opposite side of the river. Of these the most conspicuous bears the name of Formosa, and so far as its gardens are concerned it would be difficult to find one more appropriate. To apply it to the house would be flattery of which few would be capable. White Place, which obtains its name from the colour of the stone of which it is built, lies back from the river. It, too, like Cliefden, is connected with the memory of Villiers; and its avenue of elms is reputed to be haunted by the ghost of a "white lady without a head," who was one of his victims.

It is difficult to describe the beauty of this part of the river, because it does not so much consist in notable features as in a series of exquisite combinations of subtly varied forms, and in delicate harmonies of colour. There is, of course, the one great effect of wooded slope and of flowing stream, which differs but little from place to place; but there is in addition, at every step, some novel harmony of its minor features—fresh drapery of the aged limbs of trees, a new contrast of sombre yew boughs with the bright green of the sprouting beech or the tender tints of

THE LANDING-STAGE, RAY MEAD.

the maple, or of the darkling water beneath the shadow of the wooded bank, with the sparkle of the sun on the ripples of the stream. There float a pair of swans, white as snow; there darts a kingfisher, a flying emerald; there the lilies speckle the stream with gold; there the tall willow-herb forms a pink-tinted fringe to the river, and with its summer splendour alleviates our regret for the many-coloured carpets which in spring-time overspread the meadows.

Beyond Cliefden, where the plateau begins to slope more gently towards the plain, the river is broken up, and its scenery is pleasantly varied by a group of low islands densely clothed with willows. Boulter's lock forms a new feature in the view. Near here is Taplow Court, which has a most attractive garden. So also have smaller houses near the river; even some mills, which would be tolerable but for their chimneys, bedeck their bank-sides with flowers. To praise these villas would be only to repeat a formula; enough to say that if passers by break the tenth commandment in regard to these little arcadias, it is not for want of temptation; indeed, a casuist might argue that the owners were not morally justified in affording such an opportunity for coveting. In extenuation, however, they might plead that they so much increased the beauty of the river, and enhanced the general

TAPLOW WOODS.

gratification, that they might be forgiven for causing lapses in particular cases. The guide-book states that the saloon of Taplow Court was built in imitation of Kirkwall Cathedral. This must be a curiosity; it sounds almost as attractive as a bedroom built in imitation of the catacombs. For this, however, the present owners are not responsible. The house was erected by the Earl of Orkney, one of the Duke of Marlborough's companions in the great European wars.

Now houses begin to thicken on the river bank, and boat-sheds are dotted on the strand. The view of the landing-stage at Ray Mead will give an idea of the appearance of this part of the Thames, when the pleasant summer weather brings good times to the boatmen. Maidenhead on the one hand, Taplow on the other, straggle—vaguely, in the latter case—down to the riverside. The Thames is crossed first by the seven-arched stone bridge that carries the high road; secondly, by the single arch of brick, one of Brunel's bold designs, that supports the Great Western Railway. The former has been for long the site of a bridge across the Thames—at any rate, from a date prior to the reign of Edward III. At this spot there was once some smart fighting, when the Duke of Surrey, brother of Richard II., held the bridge against Bolingbroke's troops all through one winter's night, so as to cover the retreat of his friends, himself at last stealing away without molestation. Except for this, and for being the place where Charles I., when fortune had deserted him, met his children, after a long separation, Maidenhead is nearly in the blessed condition of a place that has no history. It has been asserted to derive its name from the fact that the head of a British maiden, one of the eleven thousand virgins martyred with St. Ursula, at Cologne, was kept here; but the etymology is as legendary as the maiden, the true derivation being Maiden hythe, as there was here a wharf, or "hythe," for timber in olden times. The town is not in any way remarkable. Though its streets are less busy than in the old days of stage-coaches and post-horses, it has a well-to-do look, and there are not a few pleasant residences in its outskirts; but it is very destitute of attractions for the antiquarian. The parish church is modern, having been rebuilt about sixty years since; but that of Boyne Hill will afford satisfaction to those with whom the movement in favour of ritualistic development finds favour, and as a work of the architect is far superior to churches of the earlier part of the present century. The only building in Maidenhead which carries us back to days earlier than the last century is a block of almshouses, which, though plain, has a rather picturesque appearance.

At the end of our journey we look back on a view, more artificial, but hardly less pretty, than most of those which we have seen. Railways are often deservedly execrated, but it may be doubted whether something may not be forgiven to the Great Western for the singularly attractive view which its bridge affords. The riverside between the two bridges is occupied by well-built houses, with lovely gardens and shrubberies. Green lawns, brightened with beds of flowers, groups of shady trees, villa residences of not unpleasing design, and an island on the river, combine to form a view that is not readily surpassed within an equal distance from London.

T. G. BONNEY.

CHAPTER VI.

MAIDENHEAD TO WINDSOR.

Maidenhead—Bray—Jesus Hospital—The Harbour of Refuge—Frederick Walker—A Boat-race—Monkey Island—The River—Surley Hall—Boveney Lock—Eton—Windsor—St. George's Chapel—The Castle—Mr. R. R. Holmes—James I.—Surrey—The Merry Wives of Windsor.

HE next scene in our shifting panorama of the gentle river will be the fair stretch of bank and stream which extends from picturesque Maidenhead to the winding shore from which rise the proud towers of Royal Windsor. From its source, at which a little bright spring bubbles up with a low, softly singing sound, from amid stones and moss and grass, until it becomes merged into the immensity of ship-bearing ocean, our Thames, unhasting but unresting, flows for ever onward between source and sea.

"Thames! the most lov'd of all the Ocean's sons
By his old sire, to his embraces runs,
Hasting to pay his tribute to the sea,
Like mortal life to meet eternity."

And Denham adds a wish—

"O, could I flow like thee! and make thy stream
My great example, as it is my theme;
Though deep, yet clear; though gentle, yet not dull;
Strong without rage; without o'erflowing, full."

One of the charms of Thames is, that the calm river "glideth at its own sweet will." It is no straight-cut, level, mechanical Dutch canal; but it winds and curves, it widens and it narrows, according to its own caprice and delight. And then, through what scenery it wanders! Locks are erected, in order artificially to check its full-flowing stream; but then Thames subdues even locks to himself, and makes them—especially the old wooden ones—singularly picturesque. Man is sometimes too many for him, but Thames, when allowed to have his own way, will tolerate nothing about him that is not lovely as himself. I love to think of the dear old river through all the seasons, and under all aspects. I fancy the sun-bright day, and then "the dark, the silent stream" of evening, when the untrembling shadows are so deep and full; when the belated boat is itself a creeping shade, hardly seen, but regularly audible through the sound of its beating rowlocks, and the splashing fall of its rowing oars. And who can ever forget the cool freshness of the dewy summer morning, before the sun's "burning eye" gleams upon the shining surface of the watery sheen? Then the still, dusky stream becomes "clad in the beauty of a thousand stars;" and then, perhaps, moonlight sleeps silverly upon flood and banks, on trees, and on dreamy habitations of slumbering men. The back-waters are then all mystery, the birds have all gone to roost; sleep and silence rest upon the resting river, the peace of night is over all, and still the gentle river glideth ever to the sea.

The Thames has, however, undergone one disastrous change. It has become crowded, noisy, vulgar. Its beauties remain what they ever were; but its

BRAY CHURCH.

character has deteriorated. Gone are the pure peace, the cool calm, the tranquil seclusion which—say twenty years ago—rendered it the most charming haunt of the lover of Nature, of the poet, too much in populous city pent, who sought an alternative of mental repose in most lovely and most quiet scenery. An aged ghost, restlessly revisiting the *cari luoghi*, must often find a change, great and sad, in the old places in which life was lived in love and joy; and he who knew the tranquil Thames in the old time, long ago, must find, in its brawling loudness of to-day, a change which renders sad the heart. I remember it when there were no steam-launches. Now, Captain Jinks, of the *Selfish*, too often troubles the water, as he, with his friends, enjoy themselves on the pure river which they pollute with their presence, and disturb with their rowdyism. I am credibly informed that one Sunday no less than nine hundred pleasure-craft passed through Boulter's

Lock. To what secret ait can the river nymphs now fly for rest and delicate delight? Yes, the dear old river is sadly changed indeed, and our joy in it is lessened and lowered; but its own inherent loveliness is almost unspoiled; and we have to console ourselves by thinking, with Coleridge, that—

> "We receive but what we give,
> And in our life alone doth Nature live."

Shaking off dull thoughts, let us stand for a moment on Maidenhead Bridge, which was built in 1772 by Sir Robert Taylor. Maidenhead, or Maidenhithe, is now almost the central point of those pleasure-lovers who frequent the Thames. Looking up to Boulter's Lock, which suggests very pleasant memories of Mr. Gregory's charming picture of it, we see on high the white mansion of Cliefden, embowered in a wealth of thickly clustering noble woods, which slope downwards from hill crest to river bank, and present a long sky-line composed of every shade of ever-

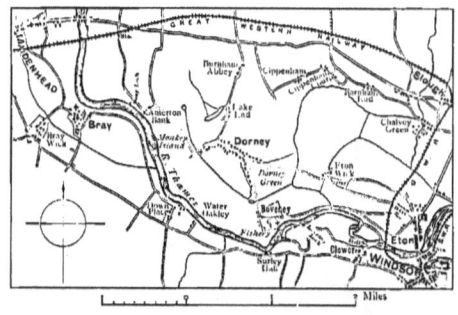

MAIDENHEAD TO WINDSOR.

varying greenness. Nearer to us, on the right, rise from out the thick leafage the turrets and spires of Taplow Court. On the left is the ivied Bridge House; on the right the old hostelry, the well-known "Orkney Arms." In the golden days of Mr. and Mrs. Skindle the house was white, but it is now of brand-new flaring red-brick, thickly pointed. Before us, where the channel slopes round towards the Lock, we see a large ait of alders. Gardens, flowers, trees, adorn the land, while the water is crowded with boats and punts. Looking towards Bray, the view is spoiled by the railway bridge, which is a leading case of engineering *versus* the picturesque. The railway bridge need not, however, mar our pleasure much, for shall we not soon row under it on our way to Windsor?

And now our boat awaits us at Mr. Bond's landing-stage, and we will start on our pleasant little smooth-water voyage. In a few moments we float under that wide-arched railway bridge which had hidden Bray from us; and we see on our right hand an old church tower rising apparently from out a cluster of tall, gaunt, windy poplars. We must land at Bray. The church has been severely restored, but it presents specimens of the historical architectural sequence of Early English, of Decorated, and of Perpendicular. It contains good brasses (particularly one of Sir John Foxle and his two wives), which range in date between 1378 and 1594;

and it is celebrated for its only too well-known vicar, who enjoys all the popularity which attaches to comic baseness. Stone and flint are largely used in this church; but we eagerly pass on from the churchyard to seek the Jesus Hospital, founded in 1627 by William Goddard, as a refuge for forty poor persons. This beneficent refuge is a very picturesque quadrangle of one-storey brick almshouses, and the quadrangle encloses garden plots planted with flowers. It seems to be well maintained and well cared for.

But the Jesus Hospital has an interest which transcends its own picturesqueness and exceeds almost its own value. It is the scene selected by a young painter of genius—Frederick Walker—as a suggestion for his noble picture, the "Harbour of Refuge." Ruskin says, "A painter designs when he chooses some things, refuses others, and arranges all;" and Walker has chosen to do away with the gardens, and to fill up the quadrangle with a lawn, a statue, and a terrace. The old chapel he rightly retains. He has sacrificed fact to the higher truth of ideal art. Walker was emphatically not one of those many painters who have mistaken their vocation. He was a true and an original artist. He saw a poetical suggestion in this retreat for poverty and for age; and in the tender sadness of summer evening after sunset he has placed a mower, whose scythe, like that of Time, is sweeping down things ripe for death. The night, in which no man can work, is about to fall; and a few figures, chiefly of sad, of aged, worn-out men and women, are waiting until the angel of Death shall gather them to deathless peace. The sentiment, the poetry of the picture, are most touching. Scene and hour are felicitously selected, and the humanity which belongs to this pictorial drama is finely conceived. World-wearied, life-worn creatures, old, weak, poor, and sad, with the flicker of faint life just lingering tremblingly, are those on whom the painter has laid stress. They stand upon the low, dark verge of life, the twilight of eternal day; and soon, very soon, shall they relax their weak grasp of life, and go hence, and be no more seen. Such was the idea that dominated Walker, and he has realised his idea. There is infinite pathos in this work of tender melancholy and of exquisite loveliness. I saw the window from which the painter made his study of the place—a study of fact which his genius afterwards so nobly idealised. As we row away from Bray, my mind is full of the picture and of its painter; and a reminiscence of my dead friend, the gentle artist, Walker, rises in my thought: a reminiscence which will, I hope, fitly find a place here.

At the time during which Fred Walker was staying at Cookham I was very frequently rowing on the river. A dear old friend, Mr. E. E. Stahlschmidt, was my constant companion, and we were much in the habit of using a light outrigged pair-oar, which was a fairly fast, though rather a crank boat, and which could carry a sitter who could or would sit still and steer. We were staying at the "Orkney Arms" at Maidenhead, and Fred Walker inhabited a cottage at Cookham. It was arranged one day that we were to row our boat from Maidenhead to Cookham, and were there to take Walker on board, and then to row him to Marlow and back.

The day was one at the end of May in 1866. How well I remember that row! It seems to me that the same sun is shining to-day that shone upon that day, so vividly does the fair by-flown time come back to me. We passed by the Cliefden Woods, by Formosa, and through Cookham Lock, and then rowed down the narrow arm of water which is opposite to Hedsor, and joins the main stream a little below Cookham Bridge. No doubt as to whether Walker were ready. The slight, active figure was dancing about with delight as he hailed us. He was full of joy, was quivering with excitement. The day was warm and fine, but the sky was grand with towering cumulus cloud-masses, which might change to nimbus clouds, and then mean rain. We went ashore, and strolled about the pretty, quiet, old village, looking at, amongst other things, the churchyard in which the great painter who was that day so much alive now rests in death.

At length we were ready for our voyage, and we entered the boat. Walker was steering, I was pulling stroke, and my friend was bow. It happened that while we were stopping at Cookham, a randan boat was also waiting there to start. This boat put off just when we did, and when both boats reached the broad, open water, the randan proposed a race to Marlow.

Both my friend and myself would have treated such a proposal for a scratch race with extreme contempt; but not so our coxswain. His keen nature always craved excitement, and he eagerly accepted the randan's challenge. I told him that it was all nonsense, and not worth doing; but race he would. I then warned him that a race to Marlow was a long one, and that I should pull a slow stroke, so that he must not be surprised if the other boat got a long way ahead. I knew that my bow could pull steadily any stroke that was set him. We were both rowing a good deal at that time, and were in decent training. The preposterous race commenced; a thing that would have been comic, but for the intense eagerness and feverish excitement of our eminent but nervous little coxswain. His eyes grew large, he breathed short, his face was pale. If something of moment had depended upon the race he could not have been more in earnest. One annoying result of his mental condition was, that he kicked and stamped about, and rocked the boat. I cautioned, and entreated him to sit quietly; but I preferred a request with which he could only with difficulty comply.

As I expected, the randan started pulling with all its might; and soon went away from us. Poor Walker was in despair. He saw the other boat apparently gaining fast, and he was seized with twitchings. In a voice weak with anguish, he implored us to "wire up, you fellows; wire up! Oh, you don't know how far they're getting ahead. For Heaven's sake, pull all you know!" He was depressed and dismayed, and was really unhappy. I could not talk much, I could only growl out an occasional adjuration to "be quiet!" an injunction with which he complied the better because he thought that we were losing fast. I continued pulling very steadily a long stroke of about twenty-nine, and was well backed up. The other boat still went ahead, and poor despondent Walker was almost in tears.

A mile or more is traversed, and I begin to fancy that I can hear the rowing of the other boat. Presently Walker's frantic joy tells us that we are gaining upon them, and he urges us to furious exertions. Of his counsels we took no notice, but pulled steadily the old, long stroke. Then we began to draw level with the antagonist, and soon I saw a bit of the boat. We never looked up or altered the stroke; but Walker chortled aloud, and could not restrain some expressions of exulting and emotional *persiflage*. We tried in vain to dissuade him. He was too excitable for such self-repression. At length I found my oar pulling level with the

SURLEY.

scull of the randan, and a little later I was up to the bow oar. The randan put on a spurt, but we drew quietly away, and had passed them when the boats were a couple of miles from home. Walker's triumph was irrepressible; his laughter was long and loud, and I thought that he would have tumbled overboard. He *would* mock at the other boat. When we were well clear of the randan, I saw that they were pumped out, and were splashing wildly. No further danger from them. We increased our distance gradually, passed through the lock, and had finished our luncheon at the "Compleat Anglers" before the defeated randan arrived. During our meal, we were looking at the scene of Walker's delicious picture, the "Marlow Ferry." He was delightedly elated at the result of the little race—kept talking about it with the eagerness of a happy child, and admitted gleefully how efficient the long, quiet stroke was, especially in a hard four-mile pull against stream. He was happy that day.

Afterwards we walked out along the banks which look on Bisham Woods. Then the born painter forgot the race, and became absorbed in his deep, reverent love of nature. Supporting his chin upon his hands, he lay down on the warm, dry turf, and his large eyes dilated as he gazed, with inner rapture, upon that lovely scene, glowing in the light of such a perfect sunny day. He had no gift of expression—except with the brush in his skilful hand. He could never find expression in words. I remember hearing him murmur then, as he gazed long and lovingly upon a calm beauty that he could feel so well, "Opalescent!" That

BOVENEY LOCK.

was all he said; but his spirit had drunk in the joy, the peace, the glory of the scene and time. He was probably seeing a picture; though he did not lean to painting full sunlight. His cheek even then was hollow; his large eyes were dangerously bright; his whole aspect expressed his ambition, his self-consuming art-soul, his terribly if exquisitely-strung nervous system, and fatally excitable temperament. We did not then foresee that the frail frame, so sensitive and delicate, would fail so sadly and so soon. His was, indeed—

> "A fiery soul, which, working out its way,
> Fretted the pigmy body to decay,
> And o'erinformed the tenement of clay."

Now, by the sweet banks of the fair river that he loved so well, the gentle and gifted painter sleeps his last sleep beneath the shadow of the old Cookham church tower.

The Thames is all the dearer to those who have often seen it with Frederick Walker. A sacred memory is blended with the river's loveliness.

We rowed him back that day to Maidenhead, and afterwards to Cookham. For some time after he spoke often, and spoke joyously, of the little scratch race which I have now endeavoured, with a sad, yet soft regret, to recall to my own recollection, and to bring to the sympathetic knowledge of my readers. Those who knew the man will ever love him; those that did not know him personally may well love the ardent, strenuous painter for the sake of his pure and gentle art.

The Thames, following Sheridan's advice, flows ever between its banks—and then what banks they are! As our boat glides along the sliding stream, we pass by many a fair and stately home of ancient peace; we pass many a smooth lawn and garden gay with flowers; we pass by rushes, willows, aits; we pass noble woods, and full meadows in which the rich grass is studded with white and yellow flowers, while sunlight is softly speckled by the calm shadows of lofty, feathery elms. The tall elms have thick clusters of foliage glowing in sunshine, and beneath these bright leaf-clumps sleep deep hollows of soft shade. Yes, our Thames is emphatically a summer stream. We row by reeds, the home of swans, the haunt of moor-hens; by islets which bear alders, osiers, weeds, and rushes. Reflected in the water is the purple colour of the wild foxglove, while the many bank flowers are interspersed with meadow-sweet, with loose-strife, and with broad dock-leaves. On the shining surface of the bright, calm water float lovely lilies, white or yellow, which are connected by long, wavy stems with roots which hold firmly to the ground at the very bottom of the river. We pass the turbulent mill-stream, and the foam-fretted weir; we see picturesque eel-bucks and shady backwaters. We wind and curve with the ever wayward flood, and we find but few stretches which fail of beauty or are wanting in peace. The Thames is the chosen haunt, too, of pleasant painters and of pretty women; and to this choice combination the grateful stream lends a charm as great as that which it receives from such artists and such girls. Truly, our Thames is almost too fair to be looked upon except on holidays.

Following the law of natural or elective affinities, fair women are attracted by the fair river. I think that I never was upon the Thames without seeing some specimen, or specimens, of female loveliness and grace. Pretty girls belong as naturally as the swans do to our Thames. It is a singular fact that natural objects of great charm allure to themselves suitable women. Art does not, as a rule, draw to itself much feminine youth and charm; but the Thames emphatically does so. Look at that boat which we have just passed. What loveliness and love in those two young, graceful girls who are being rowed, while one of them—the one in the boating hat—steers. What eyes those were which they rested for a moment upon our passing craft! Which do you prefer? the one in the blue serge frock, or the taller one in the white robe? I don't know; I could not decide; but I do know that we shall probably meet with more distracting charmers before our little voyage shall cease. Girls often steer very well, and sometimes they row, especially with light sculls, very admirably. I have known pretty young ladies who sculled deliciously, and who lent to the exercise a distinctively feminine skill and grace.

The Thames is essentially a summer river; always with the reserve of the delight of the sad and splendid hues of autumn in the woods. The aspect under which the river shows to least advantage is that of a bleak, grey day, when a coarse, cold, blusterous wind is blowing loudly. Like a pretty woman, the fair Thames should never have its surface serenity disfigured by passionate turbulence or wrinkled by debasing anger. A rough, cruel wind disturbs the characteristics, and distorts the appearance of the pure silver stream. The gentle, peaceful river should ever be smiling and be calm. There is less objection to the sullen grandeur of a heavy storm, dark with thunder, squally with rain, while a gust of fierce wind sweeps beneath the sombre cloud-heaps, and lashes up the troubled water. Yet the Thames should preserve a chaste and delicate quiet. Sunny stillness, the majesty of soft repose, are its true characteristics. In brutal, cheerless weather, it looks like a fair face degraded by ignoble pain. Its sweet essence should not be outraged by vulgar fury.

Now, just as we come to Monkey Island, a hush falls upon the sunshine, a soft shade overspreads the heavens; and a summer shower, dinting the glassy surface of the water with dimpling rain-drops, falls gently, and ceases soon. Shine out, fair sun! And so it does again, till joy returns with sunlight and with warmth. "Man's delight in God's works" can find rapture in nearly every phase of nature. Monkey Island is an inn built upon an islet. It comprises a pavilion erected by the third Duke of Marlborough, in which certain monkeys are cleverly depicted by one Clermont. We need not land there to-day.

Our boat floats ever downward with the stream, and we pass Down Place, Oakley, the Fisheries, until we reach Surley Hall, an inn much frequented by Eton boys, who come here for refreshment at that happy age in which it is possible to lunch off olives and toffee. The river, at this part of it, is not distinctively beautiful. Soon the channel is bifurcated, one arm, over which is written the ominous warning, "*Danger,*" leading to the Weir, while the other arm conducts to Boveney Lock. Once through this lock, and we are in the region of Eton and of Windsor. Eton Chapel is on our left, while before us, growing greater as we near it, Windsor Castle rises upon the sight in ever grander proportions. Eton is the swimming and rowing school, and we are passing the Brocas and those memorable playing-fields which have trained so many boys into men of mark and leading. We will halt for just a glimpse of Eton, and will stroll through the picturesque quadrangles of noble fifteenth century brickwork; will glance at the stone chapel; at the hall, library, and masters' houses; thinking of Henry VI., who, in 1441, founded Eton, after having studied the statutes of William of Wykeham at Winchester. Eton is no longer a school for indigent scholars, as it was meant to be, and was, when William of Waynflete was its first master. But we cannot linger long at Eton, and we shall see again the chapel from the North Terrace of Windsor Castle. Let us re-embark. Our voyage is nearly over. Already the landing-place at Windsor; and lo! the great castle towers just above us.

Windsor Castle is the noblest regal pile, the most splendid palace castle in Europe; but it is seen to the best advantage when regarded from a distance, and contemplated in its totality. Perhaps the grand, irregular castle, which is perched upon a height, looks finest when seen from below; and the towering mass of royal buildings certainly appears at its best when seen from afar. There is no finer view than that from the river. When the castle is seen from within, and when its detail is looked at closely, there is much that is disappointing; and the chief architectural blot is the abominable restoration of Sir Jeffrey Wyattville, who was the architect of George IV. and of William IV. Wyattville is credited with some respect for the interior, but his external architecture is wofully bad. He has made the great quadrangle in the Upper Ward a most dreary thing. His uniform, conventional Gothic is mean and ugly in the extreme; and Wyattville adopted the hateful system of pointing his stones with black mortar. For an illustration of the bad effect of this evil work, it is sufficient to compare any of Wyattville's restored towers with the recently and well-restored Clewer or Curfew tower, in which white mortar is used, so that the stones are not cut off into squares set in black borders. It were devoutly to be wished that a competent Gothic architect should get rid of the traces of Wyattville's fatal work. The cost would be great, but no expense could be too heavy for restoring such an historical building to architectural beauty and value.

On my visit to Windsor—a visit made for the purposes of this article—I had the singular advantage of being conducted over the castle by my friend, Mr. R. R. Holmes, Her Majesty's librarian. This courteous and cultured gentleman has in the royal library curious old plans of the castle in various stages of its creation; and no one can speak with more authority about the great palace which Mr. Holmes loves so well and knows so thoroughly. I cannot too strongly express my gratitude to him for his invaluable assistance, so pleasantly rendered.

It is, of course, impossible in these narrow limits to present any complete picture of Royal Windsor. Such a subject cannot be exhausted in such an article. I can only suggest a few points of interest, and merely endeavour to place those readers who may visit the Thames in my wake in a position to obtain some hint and glimpse of part of the romance, at least, of our most royal castle.

Windsor is associated with the records of all the reigns since the Conquest. Its annals cover the Court life of all the centuries since the Norman came to rule in England. It is linked with all our history, and old as it is, it is ever young with the glow and poetry of romance. The Saxon kings had a palace at Old Windsor, and Edward the Confessor kept his Court there, but the site of that old royal dwelling cannot now with certainty be determined. William the Norman moved to New Windsor, but there is nothing visible in the present castle of an earlier date than

WINDSOR CASTLE FROM ROMNEY LOCK.

Henry II. The ancient entrance, with the sheath of its portcullis plainly visible, shows clear traces of Henry II.'s work.

The great royal builders at Windsor whose works follow them, and are still extant, are Henry II., III., Edward III., IV., Henry VII., VIII., and Queen Elizabeth, who was the foundress of the splendid North Terrace. All Charles II.'s work has been swept away by that Wyattville who did so much disastrous "restoration" for George IV. and William IV. In Windsor was founded by Edward III., 1348, that noble and royal Order of the Garter which sprang from a king's chivalrous homage to a pure and lovely lady. William of Wykeham was, in 1356, surveyor of the king's works, and dwelt in that Winchester tower which has been so sorely spoiled by Wyattville; and Chaucer, also under Edward III., was "clerk of the repairs" at Windsor. One loves to fancy the sweet, cheerful old poet riding, on some fair and fresh May morning, through the royal park, while birds were singing and May blooms blossoming, looking lovingly at nature, his lips wreathed with a serene and sunny smile. Chaucer would, I think, ride gently on an ambling palfrey in preference to backing a mettled steed. One likes dearly to connect images of Chaucer and of Shakspeare with Windsor Park.

The St. George's Chapel is an ideal chapel for such a palace-castle as Windsor. It is at once sumptuous and romantic, picturesque and full of colour. It is a chapel for kings and knights; especially for kings who were also knights and warriors. It is also the royal chapel for the kingly Order of the Garter. The Choir is fitted up with the stalls of the members of the Order, and above each stall hang banners, helmets, crests, and swords. The chapel was built by Edward IV., and was commenced in 1172, the architects being Bishop Beauchamp and Sir Reginald Bray. All such old ecclesiastical edifices are sacred to the dead as well as precious to the living. Beneath the feet of those who worship there to-day rests the dust of kings; of those who, in by-past times, worshipped also here while they were in the land of the living. The tomb of the founder, Edward IV., is here; and the chapel covers the graves of Henry VI., Henry VIII., Jane Seymour, and of Charles I. In 1813 the grave of Charles was opened, and the few who looked upon the remains of the beheaded king could easily recognise the face and form of Charles Stuart. Every visitor to Windsor is, however, sure to see St. George's Chapel; and true it is, of this building, "that things seen are mightier than things heard." We leave our readers to the delight of seeing this poetically regal chapel.

John, King of France, taken at Crécy, David II. of Scotland, captured at Neville's Cross, were prisoners of state at Windsor, and were probably immured in the King John's tower; but the romance of imprisonment in the castle centres round two other figures, to which it seems worth while to devote some little space.

Two romances of chivalry and captivity are intimately connected with Windsor. One is that of a king, the other that of an earl. The king is James I. of Scotland, the earl is Surrey. After the murder, by starvation, of his eldest son, the Duke of Rothsay, Robert III. was minded to save and to educate his second son, James, by sending the youth to France. The Scottish ship was captured by English vessels off Flamborough Head, and the young prince was taken prisoner. This occurred on the 13th of March, 1405. King Henry said, "his father was sending him to learn French. By my troth! he might as well have sent him to me. I am myself an excellent French scholar, and will see to his instruction."

And so began a gentle and generous captivity, which was certainly of advantage

ST. GEORGE'S CHAPEL, WINDSOR.

to the poet king. The prince was provided with masters, and had every luxury and indulgence. James was trained in all arts and arms, became a scholar and a cavalier, and benefited by contact with culture and civilisation.

While at Windsor, love came to James in a shape of singular romance and charm, and he lived actually the adventure which Chaucer had devised for his Palamon and Arcite. The fair Emilie was doing observance to May, and so—

> "She romid up and down, and as she liste
> She gathrith flouris party white and rede,
> To make a sotill garlande for her hede;
> And as an aungel hevynly she song."

The "grete Tour, that was so thik and strong" was, we hear, "evyne joynaunt to the gardyn wall;" and Emilie was walking and singing in the garden, while the imprisoned Palamon gazed from his high dungeon window in the tower,

> "And so befell by aventure or caas,
> That through a window, thik of many a bar
> Of iron grete, and square as any spar,
> He cast his eyin on Emilia;
> And therewithall he bleut, and cryid A!
> As though he stongin were unto the herte."

Then Arcite gazes from the same window—

> "And with that sight her beauty hurt him so,
> That, if that Palamon was wounded sore,
> Arcit was hurt as much as he, or more."

So far a royal poet's fancy. Now take a kingly fact; the fact being also told in song—

> "Now there was maid fast by the Touris wall
> A gardyn faire, and in the corneris set,
> Ane herbere greene, with wandis long and small
> Railit about, and so with treeis set
> Was all the place, and hawthorne hegis knet.
> * * * * *
>
> "And therew' kest I doun myn eye ageyne,
> Quhare as I saw walkyng under the toure,
> Full secretely, new cumyn hir to pleyne,
> The fairest or the freschest zoung floure
> That ever I sawe, methot before that houre,
> For quhick sodayne abate, anon astert,
> The blude of all my bodie to my hert.
>
> "And though I stode abaiset tho a lyte,
> No wonder was; for quhy? my wittes all
> Were so ouercome wt pleasaunce and delyte,
> Only through letting of myn eyen fall,
> That sudaynly my hert became hir thrall
> For ever, of free wyll; for of menace
> There was no tokyn in hir suete face."

Then follows a young poet's lovely and rapturous description of the fair vision, who was, indeed, the Lady Jane, or Johanna Beaufort, daughter of John, Earl of Somerset, and granddaughter of John of Gaunt.

Their loves prospered, as they deserved to do, and James married Lady Jane, at Windsor; the hero of Agincourt being then our king. James was crowned King of Scotland in 1424, and the lovely lady that he loved so well was his Queen.

In 1437, at the Abbey of Black Friars at Perth, James, who strove in vain to rule his turbulent and brutal nobles, was murdered by Sir Robert Graham; what time the heroic Catherine Douglas tried to bar the door against murder, and had her fair arm broken, while the brave and loving Queen was wounded by the assassins.

Surely this royal love romance may give us sweet and tender fancies as we gaze upon the gardens and the towers of Royal Windsor. The story is a true Thames episode.

We may glance for a moment at another noble captive in Windsor—at the Earl of Surrey, likewise a poet; and, indeed, the poet who was the first writer of English blank verse. Impetuous of temper, heady of will, the gallant Surrey developed a lawless ambition which, on the 21st of January, 1547, led him to the Tower block. He was, says Mr. Robert Bell, "formed out of the best elements of the age, and combined more happily, and with a purer lustre than any of his contemporaries, all the attributes of that compound, and, to us, almost fabulous character, in which the noblest qualities of chivalry were blended with the graces of learning and a cultivated taste. It might be said of him that he united in his own person the characteristics of Bayard and of Petrarch;" and yet all these fine qualities led only to the scaffold.

Surrey is connected with Windsor because he was educated and spent his youth there, together with the Duke of Richmond, base son of Henry VIII., who married Surrey's sister; and, further, because, in his day of misfortune, he became a sad prisoner in the castle in which he had spent so many joyous youthful days. Surrey was contracted when he was sixteen, and was only twenty when he became a father. He married, in 1535, the Lady Frances Vere, daughter of John, Earl of Oxford; but romance links his name for ever with that of the fair Geraldine, who was the Lady Elizabeth, daughter of Gerald Fitzgerald, ninth Earl of Kildare, and of Margaret, daughter of Thomas Gray, Marquis of Dorset.

Geraldine, when Surrey died at the age of thirty, was but nineteen years of age. It was the fashion of that day for gallants to wear the sleeve of a mistress of the imagination; and Surrey's passion for the fair young girl was probably fantastic and partly feigned. The fair child was an adopted ideal of a knightly poet's passion. In the only one of Surrey's poems in which he speaks openly of her, he says:—

> "From Tuscane came my lady's worthy race;
> Fair Florence was sometime their ancient seat.
> The western isle whose pleasant shore doth face
> Wild Camber's cliffs, did give her lively heat.
> Fostered she was with milk of Irish breast:
> Her sire an earl; her dame of prince's blood.
> From tender years, in Britain doth she rest,
> With kinges child; when she tasteth costly food.
> Hunsdon did first present her to mine eyen;
> Bright is her hue, and Geraldine she hight.
> Hampton me taught to wish her first for mine;
> Windsor, alas! doth chase me from her sight.
> Her beauty of kind; her virtues from above;
> Happy is he that can obtain her love!"

The Lady Elizabeth was one of the ladies in attendance on Mary Tudor, and Surrey probably went with the Duke of Richmond to Hunsdon, there to visit the

Lady Mary; and on that visit he first saw Geraldine. Later, when a captive, Surrey sings—

> "So cruell prison howe could betide, alas,
> As proude Windsor, where I in lust and joye,
> With a kinges son my chyldysh years did pass,
> In greater feast than Priam's sonnes of Troy."

The contrast was indeed cruel between Windsor as a palace and Windsor as a prison. Scott, like a true poet, lays hold of the romance of Surrey's reputed love for Geraldine, and entrusts to "Fitztraver of the Silver Song" that almost matchless ballad which tells how the wise Cornelius, across the ocean grim, showed to the gallant Surrey the vision of the peerless Geraldine. Who, asks Walter Scott—

> "Who has not heard of Surrey's fame?
> His was the hero's soul of fire,
> And his the bard's immortal name,
> And his was love, exalted high
> By all the glow of chivalry."

Our fancies are stirred, as we gaze on proud Windsor, by the thought of Surrey.

What a view it is from the round and regal tower, over which, when our loved and honoured Queen is in state residence, floats our "glorious SEMPER EADEM, the banner of our pride!" Let us mount with our artist to the summit of the keep. The view is wide and winsome. Surely earth has not many things to show more fair, though the prospect is distinguished rather for soft beauty and placid loveliness than for grandeur or for wildness. I have stood on the summits of Mont Blanc, the Matterhorn, and of many another Alpine peak, and know well that there are outlooks in nature more sublime, more austere, more soul-stirring; but yet this English landscape (which includes twelve fair counties), so peaceful, soft, and smiling, has its own distinctive charm. The bright river, gleaming in sunlight, winds and stretches far through all the calm scene. There stand stately English trees; and the view includes broad green meadows, hedgerows, low, gentle hills in the far purple distance. It is a typical English landscape scene; and then over all there is to-day the splendour of a serene summer sky, and the glory of fantastic, sunlit cloud masses.

I took a great interest in finding out the exact sites of the imprisonment of those two noble and romantic captives who suffered imprisonment here. Kings John and David were, as we have seen, probably immured in the King John's tower, which is a prison. The Round Tower, as Mr. Holmes points out, never was a prison, though a wooden pillar is (if you ask for it) absurdly shown as the one to which Prince James was chained. The prince chained! Why, his was an honourable and a most gentle captivity. He was rather guest and pupil than prisoner; and Mr. Holmes leads, with no uncertain step, to a chamber (now used as a bedroom) in the second floor of Edward III.'s tower, which was the room of the prince, and in which is still the window from which the poet-prince looked into the garden—not now existing in its former state—and saw the Lady June. This point may be considered as set at rest: and it is very pleasant to be able to identify James's chamber, to look out of his window, and to see, with the eye of imagination, the fair sight that he saw in the garden beneath. Of the place of Surrey's captivity no record or tradition remains, but the ambitious earl, who strained after the crown, was surely held in more rigid confinement than was James. Surrey was possibly immured in the King John tower. And now we leave the castle; but before we quit Windsor we will stroll into the park, and try to summon up some fair fancies connected with Shakspeare and with Elizabeth.

It is a charming legend—even if it be only a legend—which tells us that Queen Elizabeth (*El Iza Beata*) when at Windsor, commanded Shakspeare to write a play in which Falstaff would be shown in love—that is, in such "love" as he was capable of—and that the result of the royal order was the *Merry Wives*, surely the most genial, and the fullest of human humours, of all comedies. It seems certain that the first version of this "most pleasaunt and excellent conceited comedie" was written very rapidly, because the second version is so much longer, and differs so widely from the earlier play. In consequence of the prevalence of the Plague in London, the Court, in 1593, lay long at Windsor, and there can be little doubt that the first version of the play was commissioned and acted at Windsor in that year. It is a delightful theme for the imagination to picture a sunny morning on which the royal cavalcade rode through Windsor Park. Essex and Southampton were there; and Shakspeare, no doubt, rode, for a time at least, by the bridle of great Elizabeth. The words then spoken between queen and poet we cannot recall; but as we read the lovable *Merry Wives*, we enjoy some of the results of the conversation. So genial is the comedy that even Sir John's base humours do not excite moral indignation. He fails so hugely. Beaten, baffled, and befooled, the merry but honest wives have the laugh of him, and we feel the spirit of the pleasant jest when Mrs. Page proposes to

> "Laugh this sport o'er by a country fire;
> Sir John and all."

We fill in fancy some street in Windsor with those quaint old gabled houses, set in fair gardens, in one of which Master Page, in another Master Ford, lived; and can reconstruct the "Garter" Inn;[*] we see the fields near Windsor in which the mock duel does not come off, and we can imagine to ourselves the farmhouse near Frogmore at which pretty Mistress Anne Page was feasting. We can follow the basket of washing to the "whitsters in Datchet Mead," and can almost recognise the muddy ditch by the Thames into which the unchaste knight was thrown. As regards the question whether the date of the play be the time of the wild Prince and Poins, or the contemporary day of great Elizabeth, one little passage makes

[*] The old "Garter" Inn stood in the High Street, nearly facing the Castle Hill, and adjoining the site of the present "White Hart" Hotel. "Annals of Windsor" (Tighe & Davis).

the point clear. When Falstaff wants to use the chimney as a hiding-place, Mrs. Page says, that they always use to discharge their birding-pieces up the chimney. Now, in the days of Henry IV., there were no "birding-pieces," while the part country gentleman, part opulent burgher of Elizabeth's time, especially in such a place as Windsor, would possess a fowling-piece. The old characters of Pistol, Bardolph, Nym, are only used because they seem to be the natural hangers-on of Falstaff; and we may safely assume that we are reading, or seeing, a comedy of manners belonging to Shakspeare's own day.

Did the Queen, Shakspeare, and the Court ride by that oak of Herne the hunter, who was
"Sometime a keeper here in Windsor forest"?

Round that oak, in Windsor Park, occurred the last revenge of the merry wives, after which foul old Sir John is bidden to
"Serve God, and leave your desires, and fairies will not pinse you."

We may enjoy a most delightful time of fair fancies in Windsor forest while we think of those
"Spirits, which by mine art
I have from their confines call'd to enact
My present fancies;"

and the charming crowd of characters in the dear *Merry Wives* fade like an insubstantial vision, as we unwillingly leave our ramble in the park, and saunter down, with imaginations sweetly and subtly stirred, to our waiting boat. It is good to leave Windsor with minds filled with creations of our Shakspeare in his sweetest and his gentlest mood.

H. SCHÜTZ WILSON.

PROCESSION OF THE BOATS, ETON.

CHAPTER VII.

WINDSOR TO HAMPTON COURT.

Leaving Windsor—Eton, its History and its Worthies—The College Buildings—Windsor Park—The Long Walk—The Albert Bridge—Datchet and Falstaff—Old Windsor—"Perdita's" Grave—The Tapestry Works—The "Bells of Ouseley"—Riverside Inns—The Loves of Harry and Anne Boleyn—Magna Charter Island—Runnymede—The Poet of Cooper's Hill—Fish at Bell Weir—A Neglected Dainty—Egham and Staines—John Emery—Penton Hook—Laleham—Dr. Arnold—Chertsey—The Lock and Bridge—Albert Smith and his Brother—Chertsey Abbey—Black Cherry Fair—Cowley the Poet—A Scene from "Oliver Twist"—St. Ann's Hill—Weybridge—Oaklands and the Grotto—Shepperton Lock and Ferry—Halliford—Walton—The Scold's Bridle—Sunbury—Hampton—Moulsey Hurst and its Sporting Associations—Hampton Court Bridge.

HE course of the beautiful river as it glides onward from Windsor to Hampton might provoke many a quaint historical conceit, as other rivers have done less aptly. We drift on the tide of Thames, as on the tide of Time, away from the Norman to the Plantagenet, from the Plantagenet to the Tudor; and it is the life of England that we can scan as the waters flow past scenes which, through all the mutations of the ages, through all the seasons' difference, year after year, are ever freshly, strongly, characteristically English. From the Conqueror's steep-throned stronghold; past the ait and meadow associated deathlessly with the solemn declaration that "no freeman shall be seized, or imprisoned, or dispossessed, or outlawed, or in any way destroyed"—"nor condemned, nor committed, to prison excepting by the legal judgment of his peers or by the laws of the land;" past more than one memorial of the mild king whose piety and love of learning founded Eton College; and still onward, far onward, along the fluvial current of history, till we reach the stately and substantial record of another and a later Henry, last of that royal name, and opposite in all respects of character, temperament, and will to the weak and gentle Plantagenet. From Windsor, then, and from Eton's distant spires; from the playing-fields, the bathing-places, the Brocas, and Firework Eyot; from the fisheries, too, which exist on the same spots they occupied eight hundred years ago—for wherever a fishery or a mill is named in Domesday Book, there it will generally be found, as of yore—we turn reluctantly, and not until a lingering look has been cast back on the old familiar scenes. Our way is on the river, or by its side, the towing-path being a track which pedestrians may follow with pleasant ease. But here and there the land may win us astray; and at the very commencement of our jaunt there is more to interest us ashore than afloat. Not that the river hereabout lessens in charm. On the contrary, its winding beauty is almost at its height. But that very beauty half depends on prospects which lead our thoughts inland; and inland we must consent, therefore, to be led.

Eton is a well-worn theme—but can never be outworn! The Royal College of the Blessed Virgin, whose assumption is depicted in the centre of the collegiate seal, was founded in the year 1440, after Henry's visit to Winchester, whence came Eton's first head-master, William of Waynflete. "It was high time," says Fuller, "some school should be founded, considering how low grammar-learning then ran in the land." The original endowments were for "ten sad priests, four lay clerks, six choristers, twenty-five poor grammar-scholars, and twenty-five poor men to pray for the king." There are now on the foundation a provost, vice-provost, six fellows, three conducts, seventy king's scholars, ten clerks, and twelve choristers. Beside these, there are above seven hundred scholars—Oppidans—who are not on the foundation. One of the masters of Eton College is illustrious through all time in which the English language is studied, as the writer of our earliest comedy, or the earliest which has come down intact to modern days. Not later than 1551, as critics and scholars are mostly agreed, did Nicholas Udall write his "Ralph Roister-Doister," which, in plot and dialogue, is immensely superior to John Still's "Gammer Gurton's Needle," supposed to have been written a few years—perhaps as many as fourteen—afterwards. The greatest of literary rarities in the library of Eton College is a copy of Master Nicholas Udall's right merry conceit, which is divided, in what would be considered as modern orthodox form, into five acts; is constructed with comic art of uncommon excellence; contains thirteen characters, some of them powerfully and distinctly marked; and exhibits the manners of the middle order of people at that day, the scene being laid in London. All that we can reasonably guess concerning Nicholas Udall as a teacher of youth is, that he was one to help in setting that early example of severity which was long afterwards followed as a sacred tradition of public-school custom and discipline. Perhaps he had, according to custom of his time, a whipping-boy, or puerile scapegoat, to take on his back the sins of happier pupils. But it is more likely that Master Udall swished without favour all round. Thomas Tusser, who wrote the didactic poem, "Five Hundreth Points of Good Husbandrie," was one of Udall's scholars, and gives hard report of him as a most exacting master. It is somewhat remarkable that the first two writers of comic drama in the language should both have been schoolmasters. John Still, author of the piece of low rustic humour before mentioned, of which the dramatic point is that of the needle itself, found by Gammer Gurton's man, Hodge, in a manner equally startling and climacteric, was master of St. John's and Trinity Colleges, Cambridge, and afterwards Chancellor of the University, and Bishop of Bath and Wells. One of the best things in "Gammer Gurton's Needle" is a song, the well-known—

"I cannot eat but little meat."

Professor Craik was inclined to opinion that "Ralph Roister-Doister" was the later of the two "farcical comedies," as they would now be called.

Men of the world, active in social affairs, as well as clerkly and diligent in the conduct of the school, were some of the earlier provosts and masters. Roger Lupton, whose rebus, the uncouth syllable, LUP, surmounting a tun, is carved over the door of the little chantry which contains his tomb, built the great tower and gateway, when he was provost in Henry VII.'s time and later. Sir Thomas Smith was a Secretary of State and a well-known diplomatist in the reigns of Henry VIII., Edward VI., Mary, and Elizabeth. Sir Henry Wotton was conspicuous both as a writer and a statesman, having been an ambassador of James I.; nor is it necessary to say that as an angler he was the companion of Izaak Walton, by whom he was beloved and praised, notably as an "undervaluer of money." Francis Rouse, Speaker of the Barebones Parliament, saved Eton from confiscation, and founded three scholarships. All these men might have been famed in other paths than those of learning had they never seen the college they influenced and benefited. Other illustrious provosts and head-masters, though not so versatile as to have influenced worldly affairs and the state of the nation in any direct way, or to have written freely and jovially for the "inglorious stage," have left a mark on their time which is more than merely scholastic. Such were Sir Henry Savile, reader to Queen Elizabeth, and one of the greatest scholars of her learned reign; Thomas Murray, tutor and secretary to Prince Charles; Dr. Steward, clerk of the closet to that prince after his accession to the throne; Nicholas Monk, brother of the Duke of Albemarle, and sometime Bishop of Hereford; Richard Allestree, Canon of Christchurch, who built the Upper School; and the late Dr. Hawtrey, famed for his elegant scholarship as well as for his success as head-master. The "ever memorable" John Hales (whose name, brilliant at one time as that of a keen theological controversialist, might in this age be forgotten but for Milton's well-known sonnet), Bishop Pearson, Bishop Fleetwood, Earl Camden, Dean Stanhope, Sir Robert Walpole, Sir William Draper, and Archbishop Longley were all, as boys, on the foundation of Eton College; and other celebrities educated there —some of their names carved on the old wainscoting—were Edmund Waller, Harley, Earl of Oxford, Lord Bolingbroke, the great Earl of Chatham, Lord Lyttelton, Thomas Gray, Horace Walpole, Wyndham, Fox, Canning, Henry Fielding, Admiral Lord Howe, the Marquis Wellesley, the Duke of Wellington, and Henry Hallam.

In the notes to Collier's map of Windsor, published in 1742, an etymology is assigned to Eton which is not clearly demonstrated, if, indeed, it be demonstrable. "Eton," we

ETON, FROM THE PLAYING-FIELDS.

read, "is so called from its low situation among the waters; for Eton is the same as Watertown, but, as they are running waters, and it is a gravelly soil, it is observed that no place is more healthy than this." Few buildings, indeed, are more happily situated than this venerable pile of old red-brown brick and Caen stone, marked by the characteristic architecture of three centuries. The old part of the college, begun in 1441 and finished in 1523, consists of two quadrangles, in one of which are the chapel and school, with the dormitories of the foundation-scholars, while the other contains the library, provost's house, and lodgings of the fellows. It is, of course, the chapel—a fine example of early Perpendicular, resembling in outline King's College Chapel, at Cambridge—that gives dignity and distinction to all pictures of the place, from whichsoever point you take your view. The spires of this beautiful structure are those which "crown the watery glade," and are conspicuous above the quaint turrets of the surrounding buildings seen from afar. Many are the views of Eton which are commended, each as being the best, by different persons. The curving railway-run from Slough gives a continuous succession of changes not to be despised; but undoubtedly the riverside is best. Gray's distant prospect was from the north terrace of Windsor Castle. Mr. David Law chose Romney Island for his standpoint when he made the sketch for one of his finest etchings. But in truth the buildings group well everywhere, as they are seen from a distance; the crowning glory being always the pinnacled chapel. It is scarcely to be doubted that Henry VI., who laid the first stone lovingly, and with meek emulation of the noble foundations of William of Wykeham, at Winchester and Oxford, in his mind intended that the structure, perfect as it now appears, should form only the choir of a magnificent collegiate church. To the beautiful building, as we see it, he would have added a nave and aisles, grandly vaulted, as the strength of the buttresses sufficiently indicates the chapel itself was meant to be. But the troublous years that closed his reign prevented the fulfilment of those designs; and it was left for the present century to bring the interior more worthily into accord with so fair an outside. Bird's bronze statue of the royal founder, erected by Provost Godolphin, in 1719, stands in the centre of one of the quadrangles. There is a look of cheerful gravity in the brick fronts of the college buildings. The elaborate quaintness of the chimneys, the sedate solidity, whether of plainness or of ornament, give a pleasant character to these quadrangles, in the larger of which, containing the bronze statue of Henry VI., is as picturesque a clock-tower as any architect in soul might wish to see. Here, on the opposite side, the hand of Sir Christopher Wren is denoted in the fine arcade supporting the Upper School. The second and smaller quadrangle, called the Green Yard, is surrounded by a cloister, and in it is the entrance to the Hall, a curious apartment, with a daïs, and with three fireplaces, which were long panelled in and lost to memory.

But we must not lose ourselves too long within the College and its precincts, lest the attractions of the library, the provost's lodgings, the election-hall, and the new buildings erected in the Tudor style, make us oblivious to our riverside ramble. It is on the stream itself, and on well-known spots along its banks, that Eton manifests her vigorous training in various traditional ways. The river is constantly covered with boats, and its proximity to the College has given Etonians that proficiency in swimming and rowing of which they are justly proud, and which they maintain by practice and prize-giving. Chief among the bathing-places of note is Athens, with its Acropolis, famed for "headers." On the Fourth of June, now the Speech Day, loyally instituted in celebration of the birthday of King George III., a procession of boats from the Brocas to Surley Hall is the event of the afternoon; and the evening closes with a display of fireworks. There are many old Etonians whose memory goes back to the Montem, abolished forty years ago. It was a triennial celebration, held on Whit Tuesday, and has been the subject of many a picturesque description, not the least vivid and truthful of which was a dramatic sketch by Maria

Edgeworth, intended, like other of her charming essays and tales, such, for example, as "Barring-Out," for the delectation of youth. Eton Montem partook somewhat too strongly of the old saturnalian character for modern tastes; and it was at the instance of a head-master that the old custom was discontinued, not without aid from Government and opposition from young and old Etonians. The last Montem, in 1846, was conducted with such maimed rites as to be a mere shadow of olden heartiness and gaiety; but in its jolliest days, which were in the reign of "Farmer George," there was no doubt something really salient in the mock-ceremony on Salt Hill. It was "for the honour of the college" that the boys, handsomely and expensively dressed in various fancy costumes, levied contributions on all and sundry passengers along the Bath Road, past the mound, believed to be an ancient tumulus, which bore the name already mentioned. The money called Salt was gathered for the captain or senior boy to defray his expenses at Cambridge, whither he was going as king's scholar. When Henry VI. resolved on founding a college at Eton, he incorporated two small colleges or hostels at Cambridge, one of which he had instituted two years before. Thus originated King's College, to which, as Lambarde says, "Eton sendeth forth her ripest fruit." The scholars are required by the statute to be "indigentes," but of course this provision has long been a dead letter.

To take farewell of the College of Eton is usually to take farewell of the town, in which, as guide-books say, there is little interest; though, forsooth, interest may be wanting where we may find pleasure of the less exciting kind, not soon to die away. At least, there is in Eton unaffected substantiality of old-fashioned building, taking the form of its own age, and not stealing outward conceits from any other. Some of the houses of professional persons and private inhabitants, tutors and others, opposite the College, with its chestnut-trees, are staid, and even venerable. As Dickens said of the old unspoilt pier at Broadstairs, they have "no pretensions to architecture," and are "immensely picturesque in consequence." Hereabout is the well-stored shop of the librarian and publisher, never lacking custom in term-time. Eton College has its literary "organ," and lives in hope of a Canning to immortalise pages which meanwhile are not deficient in sense and style, as how, indeed, should they be? In this publication are duly chronicled events which now are scarcely eventful, but which will make history some of these days. The doings in playground and on river, at football and cricket, rowing, swimming, and diving, are here registered, to the satisfaction of "wet bobs" and "dry bobs," as the boys whose varying athletic tastes lead them in different directions are called. There is no house of public entertainment in Eton which is distinguished by the modernised term, hotel; but there are some decent inns, the most comfortable of which is reported to be the "Christopher." It is a direct continuation of Eton into Windsor across the bridge connecting the Berkshire with the Buckinghamshire town, the town being to all intents and purposes one. Windsor Great Park—thus designated in distinction from the Home Park of 500 acres, which adjoins the Castle, and is the enclosure which contained Herne's Oak—should now be seen. It is separated from the castle precincts by the high road, and by part of Windsor. Apportioned here and there to farms, it still comprises a clear area little short of 2,000 acres, forest-like in much of its scenery, and abounding in walks and drives, from which a herd of deer is a frequent addition to the regal beauty of the prospect. When, for purpose of ridicule and burlesque, the title Duke of Shoreditch finds its way into modern literature, it may be called to mind that the first man so dubbed was a Londoner named Barlow, who, at one of the great archery meetings held by Henry VIII. so excelled all the Buckinghamshire yeoman, that his Majesty forthwith gave him, half in pique, half in pleasantry, the mock style and honour. For three miles the Great Park is traversed by the elm-bordered avenue called the Long Walk, at the far end of which, set up high on an eminence known as Snow Hill, stands a colossal equestrian statue in lead, by Westmacott, of George III., in classic costume.

The fine perspective, with its countless noble trees, was planned by Charles II., and finished under William III. Only by accident, fortunate or unfortunate we hesitate to say, was it that the quaintly beautiful gate, built at Whitehall by Holbein, "with bricks of two colours, glazed, and disposed in a tesselated fashion,"* and taken down in 1759, did not take the place now occupied by the leaden statue of George III. The materials of the Tudor gate, carefully preserved, were brought hither by the Duke of Cumberland, with an intention which was frustrated by his mortal sickness. It was as well, after all, that a civic gate was not set up in a sylvan park, however stately. Transplanted monuments seldom, if ever, find congenial surroundings. The Duke of Devonshire, in quite recent years, declined the offer of the fine water-gate of York House, built for Duke Steenie by Inigo Jones; and there it stands to this day, elbowed into obscurity by the Thames Embankment. No Cavendish was ever yet so wanting in a sense of the fitness of things as not to feel that a river edifice, designed as a point of landing and embarking, would be out of place as a portal of a mansion in Piccadilly. Wherever Temple Bar may be erected, it will be an incongruity and an anachronism, serving only to turn men's minds fretfully to the incongruous pile of maimed heraldry, portrait-sculpture, allegorical confusion, and vulgar commonplace, in stone and bronze, built up by Mr. Jones—Horace, not Inigo—in the middle of the road over against the Law Courts. Not as completed according to its original plan does the Long Walk in Windsor Great Park now appear. It was a walk, and nothing more, for Charles II. was a pedestrian. And as a walk it remained till 1710, at which time the carriage-road down the avenue was constructed, a new footpath having meanwhile been made for Queen Anne, which to this day retains its old title, the Queen's Walk. Royal residences and olden sites, and monuments relating to royalty, distinguish Windsor Park and its neighbourhood.

Our way now lies past Frogmore, and over the river again, by the Albert Bridge, to Southley, where we set our faces up-stream, going back a little on our course to

* Pennant's "History of London."

visit Datchet. Another iron bridge, higher up—named the Victoria Bridge—would have taken us thither more directly; though we must then have missed the park and its scenic associations. But if it were only to see the Datchet of Shakespeare, the Datchet lane, and Datchet mead by which Ford's men carried Falstaff to the river-brink in the buck-basket, and there canted him into the water with the foul linen, we could as well have remained on the Berkshire shore. There might, indeed, have been a wooden bridge between Windsor and Datchet in Elizabethan days; but it is

THE ALBERT BRIDGE.

most likely that the name of Datchet, bridge or no bridge, applied to spots on both sides the river, and that Datchet Mead was a piece of low land between Windsor Home Park and the river. As such, it is mentioned by Mrs. S. C. Hall, who agrees with other writers in supposing that Falstaff and the foul linen were tilted, according to Mistress Ford's injunctions, into the muddy ditch among the whitsters, close to Thames side. But Falstaff, both in his soliloquy at the "Garter" Inn, and in the account of the affair to "Master Brook," distinctly says he was thrown into Thames, the shelving shore of which river saved him from drowning. The real Datchet on the Buckinghamshire side could not have been intended by Shakespeare, unless, by a poetical licence, he brought Datchet over to Windsor. The topographers and the Shakespearean annotators alike have been content to slur this point, leaving their readers to reconcile the doubts concerning which all modern authorities, or such as ought to be authorities, are silent. The nearest elucidation is yet afar off. We find it in a note by Malone on Dennis, who had objected to the probability of the circumstance of Falstaff's having been carried to Datchet Mead, "which," says Dennis, "is half a mile from Windsor." This would refer, certainly, to a mead on the Berkshire side, and not in the parish of Datchet, in a county separated by the Thames. Mrs. Hall was doubtless right in placing Datchet Mead between the towing-path and Windsor Little Park; but it is a pity she was not more explicit. The muddy ditch named by Mistress Ford in the play was probably that which, being covered over in Queen Anne's time, was used as a drain, and came to be called Hog Hole. It was destroyed when the embankment was raised to form a

OLD WINDSOR LOCK.

foundation for the Windsor side of Victoria Bridge. Both from this bridge and from its fellow, lower down, good views are obtained of Windsor Castle. At Datchet, no longer so pretty and picturesque as it was half a century ago, is an old church of Early English and Decorated styles, in which Queen Elizabeth's printer, Christopher Barber, is buried; as also Lady Katherine Berkeley, daughter of Lord Mountjoy. Above Datchet, Izaak Walton used to fish, sometimes with his friend, Sir Henry Wotton, the provost of Eton before mentioned.

Albert Bridge, with its long, flattened, Tudor arch, spanning the river at one bound, bears a miniature resemblance to the design of the bridge at Westminster, and is light and elegant, though of a modern taste, which lacks the picturesqueness and simplicity of the old objects on the river. The span, however, adds safety to the navigation, especially in these times of steam-launches, the most unpopular and best-hated craft on the Thames. Like other ills, we have come to tolerate them for a certain one-sided convenience, esteemed by the selfish, the lazy, and the fast.

All pleasant quiet on the river, as, indeed, on the shore, is a thing of long ago. Idlesse, dreamy solitude, *could* only be enjoyed by the few, and *can* never be enjoyed by them. In coupling, or rather in identifying the fast and the lazy, we may, by hasty thinkers, be suspected of a contradiction. There is none in what we have said. The lazy are often restless in their inert desire to be conveyed swiftly from place to place; for they have no energy for idling. To rush, screaming on, with their hands in their pockets, and no motion of their own, is the height of bliss to such people, and this is the enjoyment which a steam-launch affords. Yet the unpopular vessel has a popularity of its own. Riverside folk in the mass, from the Duke of Westminster to the poorest toiler who profits by the Early-closing Movement and the Bank-holidays, all join in decrying the rowdy intruder—the " 'Arry " of river craft. But " 'Arry " is all-pervading, and multiplies himself with astonishing exuberance and rapidity. There is more of him every day; and there are more and more steam-launches, for all the outcry against them.

Old Windsor, whatever may have been the state to which it attained when young or middle-aged, is now only a village, and not, so far as appearances go, a very old one. Like the schoolboy's knife, which had first a new blade, then a new spring, then another new blade, and then a new handle, it has been transformed by successive renewings. It is by a new road from New Windsor that this Old Windsor, which is much the newer of the two, is reached. We pass the Prince Consort's model farm to get at the one bit of antiquity. This is the church. It is very picturesque, and derives a certain venerable suggestiveness from the yews and other old trees surrounding it. But it is not of remarkably early date; and restorations have robbed it of more age than it could well afford to lose. Trees are the best friends that Old Windsor can boast. They keep it warm and green and comely. The by-road leading to the church is not often trodden, except by those who really deserve the name of church-goers. It is not a show-place; that is one thing in its favour. It has a green and tranquil churchyard; that is another. The name best known to people who have read the modern history of Old Windsor Church is a whispered name; the tomb which bears it is a neglected tomb; few go out of their way to pace that little cemetery; fewer still find the grave of Mary Robinson. Hush! it is a name that the kind will be kindest to forget. No loving care kept the tomb from being overgrown with nettles for fifty years; and it is too late now, even were it seemly, to vex the ghost of poor " Perdita." That was the lady's romantic designation. She had played the character in Shakespeare's *Winter's Tale*. There was a Prince Florizel who wore a chestnut wig, a frogged and fur-collared frock-coat, tight breeches, and silk stockings, and cut a very elegant figure on the throne of England—" George the Magnificent," he is called by Mr. Thackeray. He invented a new shoe-buckle. It was five inches broad, covering almost the whole instep, and reaching down to the ground on either side of the foot. "A sweet invention!" exclaims the satirist. And a pretty Prince Florizel, truly, whose head was so full of such matters as these, and whose heart was so choked with egregious vanity that, having "kissed and fondled poor Perdita on Monday, he met her on Tuesday and did not know her." She sleeps, now, peacefully in the tree-shaded churchyard of Old Windsor—she and her daughter, Maria Elizabeth Robinson, both of literary fame, the inscriptions tell. What did they write? Poetry, was it? There is a tombstone also here to the memory of a shepherd, named Thomas Pope, who died half a century ago, aged ninety-six, having been "cheerfully laborious to an advanced age." On the same stone is recorded the death of "Phoebe, wife of the above," aged ninety years. Their fame was not literary, nor their work of the poetical kind. Their bodies, nevertheless, are buried in peace, and their names, merged, it may be, in "the long pedigree of toil," live for evermore.

But though Old Windsor is reduced to an insignificant suburb of New Windsor, it was the royal dwelling-place when all was forest around, and when the solitary chalk hill, standing up from a tree-covered clay tract on the riverside, was uncrowned by feudal masonry. That was before the Norman took hold upon England. At the Conquest, Old Windsor was a manor belonging to the Saxon kings, who are conjectured to have had a palace here from a very early period. We may fancy theirs to have been a less splendid court than that of "George the Magnificent." When Edward the Confessor—who afterwards presented the manor to his newly-founded monastery of Westminster—ruled England from this spot, a few serfs and swineherds dwelt sparsely in huts among the thick woods. The site of the Saxon palace can only be guessed; but antiquaries have surmised that an old farmhouse which stood west of the church, and near the river, surrounded by a moat, probably marked the place. When the Conqueror, having obtained the land from the monastery, by fair bargain, as appears, built a fortress on the eminence which we now call Castle Hill, the palace at Old Windsor remained a palace. It is probable that the first Norman castle on the neighbouring mound was simply a defensive work, with no convenient residence, till Henry I. completed additional buildings. Thenceforth Windsor Castle was Windsor Castle indeed; and little is heard in history of Old Windsor. The manor passed from hand to hand, each tenant for a time holding from the king by service. One man, with lance and dart, for the royal army, was all required. Since the fourteenth century, the holding has been on lease from the Crown. Tapestry works, which of old were maintained at Windsor, and fell into disuse, have in late years been revived. Looms were set up in buildings specially adapted to the industry, which was initiated by foreign workmen, the art of tapestry-weaving in England having quite died out. One of the artists engaged in supplying designs was the late Mr. E. M. Ward, R.A., and the application of the modern tapestries to household decoration was mainly encouraged by Messrs. Gillow and Company, who have employed these hangings with great effect in the royal pavilion, each succeeding year, at the South Kensington exhibitions. Keen interest in the revival of this artistic and dignified class of manufacture was taken by the late Duke of Albany, under whose patronage an exhibition in furtherance of the scheme was held in Windsor Town Hall. An early and munificent encourager of the work was Mr. Christopher Sykes, M.P., whose town mansion was richly adorned with the Windsor tapestries.

No traveller bids farewell to Old Windsor without paying his respects to one of the best of the riverside taverns, the time-honoured "Bells of Ouseley." Perfectly free, at present, from modern revivalism, and from all manner of conscious style, is this genuine old inn, separated by the high road from the river bank. Its quaint bow-windows, one on either side of a porch entered by way of a steep flight of steps—the wholesome dread of unsteady topers—are just of the period and fashion to captivate an artist in search of the picturesque; nor can we look on this unspoilt hostelry without thinking of Mr. Leslie, Mr. Boughton, or Mr. Tissot. In France, a village cabaret or auberge, humbler than this, would yet be far more advanced in the art of public entertainment. "They cook very well at these places," is a remark you frequently hear in Normandy, Picardy, or Champagne, from the lips of culinary judges, versed in all the intricacies of Parisian gastronomy; and if the unpretending inn be near a trout-stream, be sure you may have a dish fit for a prince, and within the means of a woodcutter. Were some enterprising cook to lease a cosy tavern like the "Bells of Ouseley," and introduce a really high-class *cuisine* on a choice but simple scale, the place would be talked about in a month and spoiled in a year, at the end of which time the proprietor might be either a rich man or a bankrupt. Let us take our pretty, rustic riverside resort, for rest and refreshment, as we find it. Fine cookery would drive out honest companions whom old Izaak—who had a face like an elderly pike, but was a right good fellow for all that—would have drawn into profitable talk; for at the "Bells of Ouseley" you meet anglers and bargemen from whom much is to be learnt, if you go the right way to get hold of them. On the left as you enter is the tap, often crowded; on the right a bar-parlour, in which the company is more select. Of old the "Bells" had the reputation of being a house of call for "minions of the moon," as Falstaff called them, or "knights of the road," to choose a later phrase, such as the authors of "Paul Clifford" and "Rookwood" would have applied to the same order of gentry. But the landlord does not, in these days, give stall and fodder to nags of suspicious character, like bonny Black Bess. The old stone stable is oftener occupied by steeds that consume neither oats nor hay; and the highwaymen are not such as wear crape over their faces, or carry pistols like demi-culverins, or dance minuets with ladies they have plundered, but are in fact only members of a bicycle club. Under that old roof with its odd chimneys, standing against a background of greenery, there are jolly ghosts, you may be sure; for the grimmer goblins that have haunted the "Bells" in time when gibbets were plentiful, and when every one of these evil trees bore its rotting fruit, that swung and creaked in the night-wind, were laid long ago in a red sea of steaming punch, by boon companions of those who, as the phrase was, "suffered." The fishing at the "Bells" is good. Capital chub and dace are taken with the fly, and gudgeon are plentiful as blackberries.

On the Bucks shore, above the outfall of the little River Colne, which flows into the broader Thames, below Bell Weir Lock, is Wraysbury, a name which has been conveniently adapted to local phonetics from "Wyrardisbury." Over the Colne there is a suspension-bridge; and the river is crossed by the South-Western Railway, which has a station here. Wraysbury Church is distinguishable under its restorations, which appear to have followed in good faith the original design, as a fair example of Early English architecture. It preserves one of those rarities of monumental design of

THE "BELLS OF OUSELEY."

which so largely outnumbering a proportion of village churches have been robbed. This is the brass which represents, in the habit of an Eton scholar at the beginning of the sixteenth century, John Stonor, who is not the only recorded association of this place with the College of the Blessed Virgin Mary. There is Ankerwyke House, a modern mansion, embowered in trees, by the riverside. It occupies the site of a Benedictine priory, which, in its later days of dissolution, was given as a residence, by Edward VI., to a provost of Eton, Sir Thomas Smith. This priory, for nuns of the above-named order, was founded in the reign of Henry II., by Sir Gilbert Montfichet. Of the old religious buildings hardly a trace remains. Ankerwyke House is associated by tradition with the courtship of Anne Boleyn by Henry VIII., who used, it is said, to meet her under a yew-tree, which has since grown to the goodly girth of twenty-eight feet. Great trees of this kind have an exceedingly venerable look, but as a matter of fact their age seldom comes near comparison with that of the oak; and a yew-tree pretending to the age of three hundred and

MAGNA CHARTA ISLAND.

fifty years and more may be looked upon with doubt, at least as reasonable as the scepticism of a Thom or a Cornewall Lewis, directed to the subject of human longevity. Wraysbury is rather a pretty village than otherwise, and we leave it with a wish that it may be spared any loss of its present prettiness for many years to come. An unspoiled path leads to the ferry, by which Magna Charta Island is reached, the lower of two islets in midstream. Topography is at loggerheads as to whether, the barons holding the island, this was the place of meeting between them and King John, or the field named Runnymede was the spot on which the grant of English freedom was signed. Anglo-Saxon authorities derive the name from Rûn, and say that Runnymede means Council Meadow. So that the island and the field on the Surrey shore—for we stepped across the boundary of Berks when we bade farewell to Old Windsor—hold the great historical honour in dispute. We should certainly incline to a decision in favour of the island. It was on the plain level field, such as we now see it, unbroken by hedge or wall, house or barn, that Edward the Confessor no doubt occasionally held his "witan," during his residence at Old Windsor. The Norman barons would have been likely to choose the island, both on account of its association with those very rights they were met to assert, and because it was at a convenient distance from Windsor, sufficiently near for the king, but far enough to prevent any treacherous surprise by his forces. It is, indeed, asserted by early historians that the island opposite the meadow was chosen by the barons, the king having proposed Windsor as the place of meeting. Local tradition, which may be taken for what it is worth, accords with written history. The Charter bears date June 15, 1215; and in that very year John had taken refuge in Windsor Castle, as a place of security against the growing power of the barons. On Magna Charta Island a Gothic cottage has been built by one of the Harcourts, lords of the manor, as an altar-house for a large rough stone, which bears an inscription setting forth that King John signed Magna Charta on that island. Tradition or fancy goes a step farther, and represents the stone to have been the royal writing-desk.

From Runnymede the slopes of Cooper's Hill rise, on which Sir John Denham conferred celebrity, if, indeed, Cooper's Hill did not the rather confer celebrity on him. It is certain that his poem, which disputes the palm of descriptive verse with Ben Jonson's lines on Penshurst, is far better known than anything else he ever wrote. No one thinks of naming Denham without quoting those four lines which Dryden and Pope have lauded, and which remain to the taste of a changed epoch "the exact standard of good writing." Many critics, so to speak, have taken off their hats to the quotation, and have printed it usually in admiring italics. Addressing Thames, the poet says:—

> "O, could I flow like thee, and make thy stream
> My great example, as it is my theme!
> Though deep, yet clear; though gentle, yet not dull;
> Strong without rage; without o'erflowing, full."

Reflective power, almost equal to Wordsworth's, characterised the poetry of Denham; but he can hardly be compared to the modern poet in the quality of description. If he has drawn a pretty good likeness of the river, which does, however, occasionally overflow, and at other times is by no means full, surely the hill is unrecognisable in such portraiture as this:—

> "But his proud head the airy mountain hides
> Among the clouds; his shoulders and his sides
> A shady mantle clothes; his curled brows
> Frown on the gentle stream, which calmly flows,
> While winds and storms his lofty forehead beat,
> The common fate of all that's high or great.
> Low at his foot a spacious plain is placed,
> Between the mountain and the stream embraced,
> Which shade and shelter from the hill derives,
> While the kind river wealth and beauty gives."

Over Denham's gorgeous clouds of fancy, clothing the sides of this Thamesian mountain, some pretty villas, with lawns and gardens, are dotted among the trees, with which Cooper's Hill is beautifully planted. Its gentle slopes, green and gradual, scarcely attain steepness at any point; and the wild hyperbole which makes the airy mountain hide his proud head in the clouds is absurdly misleading. The view from its summit is wide and fair; and the silver windings of the river towards Windsor Castle, which stands up in its pride of strength, are finely shown on the face of the landscape; and if Denham lost himself in picturing a hill, he is at home again in his representation of a plain. Runnymede, as fair a pasture as it was seven hundred years ago, is still unbroken, still sheltered by the hill and enriched by the river. An army might assemble there, as in the feudal days of old.

Bell Weir, and the Picnic, as the island a little above Runnymede is called, are favourite resorts of holiday-makers, anglers among others; and no more delightful part of the river for rest and recreation, as well as for sport, can be found than in the beautiful reaches which succeed one another as the stream winds, now this way, now that, between banks that never lose their charm of interest and variety. On the Picnic, however, picnics are ceasing or have ceased. Liberty, so cheerfully accorded, has, with too many picnic-parties, been turned to licence, and permission to use the island for such kind of pleasure-making had at last to be stopped. If the bold Briton can be brought to see the gracelessness of accepting a grace, and then abusing it, perhaps there may be a renewal, in time, of the concession. Close to the weir, on the Surrey side of the river, is an excellent inn, aptly, though it

RUNNYMEDE.

may be tritely, called the "Angler's Rest." Barbel, roach, chub, and gudgeon are plentiful round the Bell Weir; and trout are often taken. Thames gudgeon run somewhat small, and anglers who do not combine culinary taste and skill with their proficiency in the craft are apt to regard the little fish as no more than a good sort of live-bait. And this, indeed, he is, especially if you are pike-fishing. Arthur Smith, the amiable brother of Albert, who thought a pike, stuffed and baked, very good eating, knew the partiality of this voracious fish for gudgeon. Indeed, Master Jack, who is both a *gourmand* and a *gourmet*—the characters being oftener united than is commonly supposed—will pass by any other prey to get at the silver morsel, which has been called by some human epicures the freshwater smelt. So, by-the-by, will a perch, the only fish that can live with pike on terms of armed neutrality or amicable defiance. Freshwater smelt, indeed! Why, the despised gudgeon, properly cleansed and treated with salt, is, when freshly caught and delicately fried, the smelt's decided superior; and it is perfectly surprising that the former is not more in request—is never asked for, in fact—at London restaurants. Yet, in Paris, at Bignon's, Champeaux', the Café Anglais, the Café de la Paix, the Maison Doré, or at the Cascade in the Bois de Boulogne, or at the Tête Noir at St. Cloud, no judicious diner misses *goujon*, when that fish is to be had, as it generally is. Why must we wait till we go abroad before we think of asking for gudgeon? Why should we pooh-pooh the dainty little fellow? Is it because it is so easy to catch him that his very name has passed into a proverb? Depend upon it, in spite of the ridicule which follows gudgeon-fishing as the facile entertainment of "a young angler," we make a great mistake, and lose many a dainty dish, in this scornful, or at least jocular, disregard of so sweet and delicate a fish.

Gudgeons swim in shoals, are always greedy biters, and, in fact, hook themselves with so charming and ready a will, that ladies and boys have no greater trouble than pulling them out of the water as soon as the hook, baited with a red worm, is dropped into it. No other labour, and

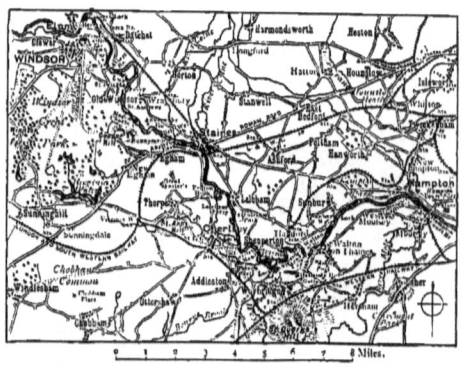

WINDSOR TO HAMPTON COURT.

no skill or activity beyond, is needed. The hook must be small, and the worm must be small also; and the gravelly bottom should be raked, to stir up the aquatic insects and larvæ, and so to summon the confiding fish together. Angling and rowing are not the only pursuits on the river or by its banks. The student of natural history and the landscape-painter, by which, in these days, is mostly meant the same, may botanise to their hearts' content; and, if they care more for popular and poetic than for scientific botany, may be glad to find there are still such beings as country folk, and still such names for flowering plants as codlings-and-cream, which the vulgar call *Epilobium hirsutum*. Call it what you will, this same plant, which is in truth the large-flowered willow-herb, and has a wholesome, but not very distinct or pungent odour, supposed to resemble the scent of apples with cream, as above named, is liked by cattle, and was at one time recommended for cultivation as fodder in wet places where other useful plants will not grow. The true forget-me-not—the *Vergiss-mein-nicht* of the German tale —grows in extraordinary luxuriance and beauty in these fresh grassy places. An amphibious little weed, with red-shaded green leaves floating on the water, and with pink spike-blossoms, called the persicaria, is beautiful and harmless when dancing on the rippled recesses of the river, but a bane to farming when it takes to a life on shore. We are close to Egham now, and may either put up at the "Angler's Rest" or enter the town and seek bed and board at the "Catherine Wheel."

Egham is a small town surrounded by some very pretty country, which, having been, bit by bit, blemished by taste, has received the final blow from the conspicuous benevolence of a millionaire. This gentleman has built a gorgeous palace, nay, two gorgeous palaces, for imbeciles of the superior class. The buildings, taken together, may be about as big as Windsor Castle, and they are as visibly prominent in the landscape, though not precisely with the same effect. A white granite bridge, designed by Rennie, and opened in 1832, by William IV. and Queen Adelaide, connects Egham with Staines, and in these iron times is a positive relief to the eye afflicted by such viaducts, railway and other, as are rapidly spoiling the Thames. Altogether there is a comfortable modern look about both places, their comeliness, such as it is, being entirely due to natural surroundings. Egham is mainly one long street. The church is a plain structure of the negative style of 1820, the tower being a landmark seen from far. There is likewise a chapel of ease; and there are places of worship for different denominations. What more of Egham? Strode's almshouses are in its High Street, and a cottage hospital is healthily placed on Egham Hill. The Elizabethan house of Great Fosters is in the neighbourhood. Egham has an annual race-meeting, the course being no other than Runnymede. Staines, new and manufactural as it now appears for the most part—the "linoleum" works having largely contributed to its industrial aspect—is as old as any place in true English history. Ancient records give the name as Stanes. Modernised though it be, Staines is by no means bereft of all antiquity. The church is venerable; and near it is Winicroft House, a Tudor building which some of the good folk, innocent of architecture and chronology, soberly assign to the reign of King John, who sure enough had a palace at Staines, somewhere or other, and not impossibly on this particular site. One of the earliest bridges in England, preceding the Roman, as may be inferred from the Itinerary of Antoninus, crossed the river here. When the Roman road to the west was made, and a military station formed at Staines, it is probable that a stronger bridge was built; and, as most of the Roman bridges in England seem to have been of wood, supported by stone piers, the guess that Staines, or more properly Stanes, took its name from those relics of Roman occupancy, is perhaps pardonable. Just above the town, at the mouth of one of the entrances of the Colne into the Thames, where an ait is formed, stands a monument worth careful attention. It is a square stone shaft on a pedestal, which again is raised on a base formed of three gradations. This is the ancient London stone, or boundary stone, as it is alternatively called, for it has served during many ages to mark the beginning of Middlesex out of Buckinghamshire, and the termination of the city's jurisdiction over the waters of Thames.

The Conservancy of the river, by long prescription, confirmed by various Charters and Acts of Parliament, was vested in the Lord Mayor and Corporation. Apart from the courts, which were held by the Lord Mayor in person, and with much state, most of the administrative duties have long been performed by a committee of the Corporation, aided by four harbour-masters, an engineer, water-bailiff, and subordinate officers. Till recent times, the Navigation and Port of London Committee, as it was called, held jurisdiction from Staines in Middlesex to Gantlet in Kent, and exerted a strong hand in preventing encroachments on "the bed and soil of the river," or any injury to its banks. The duties also extended to the regulation of the moorings of vessels in the port, the deepening of channels, the erection and maintenance of public stairs, the repair of locks, weirs, and towing-paths, the control of fisheries, and the seizure of unlawful nets. Tolls and tonnage-dues contributed to the revenue on which the Corporation depended for means of executing all these obligations. They had, as one of their public advocates tersely put it, "a surplus below bridge which they were unable to appropriate, and a deficiency above bridge which they had no power of making good." Still, no hesitation or serious shortcoming appeared in their fulfilment of duties. But some years ago, a claim to "the bed and soil of the river" was set up by the Crown. Thirteen years' litigation ended in a compromise. The City consented to acknowledge the title of the Crown, and the Crown consented to grant a title to the Corporation, stipulating, however, that a Government scheme should be embodied in an Act of Parliament. Hence, the Thames Conservancy Act of 1857, which vested the rights and duties in a Board composed of the Lord Mayor, two Aldermen, four Common Councilmen, the Deputy-Master of the Trinity House, two persons chosen by the Admiralty, another person chosen by the Board of Trade, and another by the Trinity House, making twelve in all. By a later enactment these rights and duties were abrogated, and now the jurisdiction of the City over the Upper Thames has altogether ceased.

LONDON STONE.

This is certainly not the place for any argument for or against the deprivation of almost regal authority which the City of London has long swayed. But up-river men, especially anglers, have cause to be grateful for the protection afforded them in the past by the conservators. Staines Deep is a good instance. All the "deeps" on the river are formed for the especial behoof of the angler, who is indebted to their peculiar construction for the abundance of fish that reward his patience, trouble, and skill. A deep is so staked or otherwise protected that no net or coarse process of any description can remove the fish that collect there. Old boats are not unfrequently sunk to prevent the use of nets. All the deeps between Staines and Richmond have been formed on this or some such system at the expense of the London Corporation; and at Staines the never-failing abundance of large roach is due, no doubt, to the careful plan on which the deep is formed. The accessibility of Staines from London makes it exceedingly popular, as is evidenced by the number of boat-houses, and constantly increasing trade of boat-building. The hotels and inns are not spoilt by custom. The little "Swan," one of the prettiest of old-fashioned houses on the river, is just below the bridge on the Surrey side, and really in Egham parish, though boating men generally speak of it as the "Swan" at Staines. Then there is the "Packhorse," on the Middlesex shore, with a good landing-stage. The "Angel and Crown," which is traditionally associated with the Emery family, having been kept by one of that name in the days when John Emery was the recognised and unapproached stage Yorkshireman, is in the High Street. He played Tyke as probably no other man ever played that character; nor was he less effective in the monstrosities of the stage, Caliban being one of his pet parts, and Pan another. He had a fair range of Shakespearean repertory, being a terribly truthful Barnardine, in *Measure for Measure*, and a capital Sir Toby Belch. In some panegyric memorial verses which appeared soon after his death was the line—

"And Farmer Ashfield with John Emery died."

This praise was exaggerated and indiscriminate. The present writer was sitting

STAINES BRIDGE.

many years ago at the "Angel and Crown," in a mixed company of oarsmen, anglers, and residents, when he heard the performance of John Emery as Farmer Ashfield called in question. Somebody had extolled it for its rich Yorkshire dialect. Thereupon a grey-headed old man broke in with a quotation from "Speed the Plough." In the scene supposed to follow a ploughing-match, when Sir Abel Handy's patent invention has been kicked to pieces, and carried off at the heels of the frightened horses, Bob Handy answers his father's question, "Where's my plough?" by turning to the farmer and inquiring the name of the next county. "We ca's it Wilzhire, sur," is the reply. The scene, in fact, is laid in Hants. The grey-haired man was an old actor, and he finished his pertinent reference to Morton's play with the quiet remark that he too remembered Emery, and admired him in Yorkshire parts, but that Farmer Ashfield was *not* a Yorkshire part. With a London audience in Emery's reign all countrymen were Yorkshiremen, just as all foreigners were Frenchmen. We must not leave Staines, where barge-life and riverside character generally may be studied better than at any other spot by the Thames, and the boundary stone without mentioning that this ancient monument bears the traces of its original inscription, dated A.D. 1280, "God preserve the City of London."

Penton Hook, on the Middlesex side, is a horseshoe-shaped piece of water, where the river shoals out a great deal, so that boats going down the rapid shallow run of half a mile will do well to keep in mid stream, so as to avoid grounding on that shore. If you pronounce Penton Hook as you see it written, you may chance to miss being understood. Penty Hook is the common pronunciation, and if without the aspirate, so much the nearer local correctness. Penty Hook Lock has an average fall of two feet and a half. There is a ferry here, by which you may avoid the Hook and its long pull. The bend at Penty Hook is the natural course of the river, and its horseshoe form, enclosing a large meadow, has the lock for a base. For general fishing, Penty Hook has been famed time out of mind; and, though disappointed men are sometimes heard to lament the growing signs that this fishery has begun to be worked out, every season yields many a well-filled creel. The lock is a good thing for those who voyage for pleasure; not that they go through it, but that it leaves them the undisturbed solitude of the ancient passage, by drawing away the hurriers, who think of nothing but of "getting there," wherever "there" may be. This retired and tranquil bend is the haunt of water-fowl, and is a very wilderness of butterflies. One of the countless tributary streams that feed the Thames, the Abbey River, babbles of days when the monks of Chertsey kept their preserves well filled from these productive fisheries. Fine trout are taken here still; the strong barbel gives excellent play; a number-twelve hook, mounted on a single hair, and baited with a gentle, will take roach and dace in any quantity, though a heavy float is necessitated by the force and depth of the stream; he who seeks the big chub should cast his line under any of the overhanging willows; and he who scrapes for gudgeon may choose from twenty pitches, any one of which will give him a day's quick work. Down stream now to Laleham is a short row, or walk, along the Middlesex shore, on which side is the towing-path from Staines as far as Shepperton.

Such charm as may be found in a flat landscape—and it is not small when there are trees and water, red roofs and quaint chimneys, sheep, cows, and an old church—we find at Laleham Ferry, one of the quietest and prettiest spots on the Thames. The nearest railway stations, neither of which is too near, are Staines and Shepperton. This little village was for nine years of work, study, and wedded happiness, the residence of Dr. Arnold, the mild but firm Erastian in most of his

LALEHAM FERRY.

ecclesiastical views, the parental educator, the Liberal in politics but not in party, the Church reformer who clung to the Church not as a priesthood but as a body of believers, the man of thought and man of action. To Laleham, Thomas Arnold went at twenty-four years of age; took pupils, as, since he was twenty, and elected a Fellow of Oriel College, he had done at Oxford; married, and worked on, with the grand idea before him of bringing new life and spirit into our public-school system. It was at the end of his nine years' sojourn at Laleham that he took priest's orders, and turned a corner in his life whence his useful fame began. He was appointed to the headmastership of Rugby School. In those nine years at Laleham, peaceful and happy as they were, sorrows were not "too strictly kept" from Arnold's home. Four of his family are buried here; his infant child, his mother, his aunt, and his sister. It is no matter of mere opinion or dispute that Dr. Arnold and Rugby are associated as no person and place, no school and master, ever were before or since. Illustrious men have indeed raised the high standard of tuition higher than they found it, at other public schools. Their names add lustre to a shining roll. But Dr. Arnold, of Rugby, whose constant longing from his youth had been to "try whether our public school system has not in it some noble elements which may produce fruit even to life eternal," justified his belief and his mission so well, that he not only raised Rugby School to its highest fame, but introduced a great change and improvement into all school-life in England. He trusted much to his Sixth form, his elder boys, whom he inspired with love, veneration, and confidence, so that their recognised authority over the junior pupils was exercised as with a reflected light. He would have no "unpromising subjects," no pupils likely to taint others. "It is not necessary," he said, "that this should be a school of three hundred, or one hundred, or of fifty boys; but it is necessary that it should be a school of Christian gentlemen." All good hearts in time were bound to the firm, manly, sympathetic master, whose devotion to duty was contagious, and whose unceasing interest in his scholars was repaid by their reverence for him. Dr. Arnold's writings are earnest, clear, and independent. The six volumes of sermons, chiefly delivered to the Rugby boys, should be read by all boys, all parents, and all masters. Dean Stanley, his pupil and biographer, collected and republished his tracts on social and political subjects; and in the striking picturesqueness of his "Roman History," in which he adopts the "ballad-theory" of the Prussian historian, Niebuhr, he forestalled a mode of animated illustration, and contrast of ancient and modern events, which is so popular in the hands of Macaulay and Grote.

Nine years of such a life as Arnold's would be enough to confer perpetual dignity on a more important place than Laleham, which contains a population of not more than six or seven hundred souls, and is not honoured and spoilt by a "surrounding neighbourhood" of new wealth, refinement, and education. Not that the village is more rustic—in the depreciative sense—than a village inhabited by people who "have known some nurture" ought to be. There are a few good old-fashioned houses about it, Arnold's being one of them; a solid red-brick house with a large garden. The occupant came to regard the country as "very beautiful." He had always a resource at hand, he tells us, in the bank of the river up to Staines; "which, though it be perfectly flat, has yet a great charm from its entire loneliness, there being not a house anywhere near it; and the river here has none of that stir of boats and barges upon it, which makes it in many places as public as the high road." Laleham House, the seat of the Earl of Lucan, is a plain square modern mansion with a Tuscan portico. The grounds of forty acres are noted for the noble elms, shrubs, lawns, and flower-gardens.

"One spot for flowers, the rest all turf and trees,"

as Leigh Hunt sang, though not of so large and fine a domain.

Some years ago a galvanised monkish movement, led by an Anglican clergyman, who went about town with sandalled feet, a girdle of knotted cord, and a cowl over his tonsured head, made a descent on Laleham, where the poor enthusiast tried to found a monastery, with what temporary noise of local wonderment is now a subject of much forgetfulness. The church at Laleham is small, old, and patched with modern brickwork; and the church across the river, at Chertsey, a mile lower down on the Surrey shore, is square-towered, part ancient and part new. Nothing very old, or noticeable as old, will be seen if we go inside; but we may do this reverence to modern art if not to antique religion, for there is a memorial bas-relief of simple beauty, carved in a Christian spirit by the Greek-souled sculptor,

Flaxman, the subject being the raising of Jairus' daughter. Cattle are feeding on the grass of Chertsey Mead, or cooling themselves in the shallow stream. How different are they from the droves of builders and architects who try to improve the banks of the river! The cattle positively decline all effort at picturesqueness; but they are picturesque, which the new houses or villas, and stuck-up towers and turrets, with all their ornamental pretence, decidedly are not. A hundred years ago was built, by James Payne, the bridge of stone, with seven arches, the high middle arch

LALEHAM CHURCH.

being beneath a pointed summit of the parapet. This bridge, though steep, seems right under the lock, which is built of wood, and has a fall that averages three feet. Ancient and modern both are the intimate associations of Chertsey. Among the modern are reminiscences of Albert Smith, whom even James Hannay, a contemner of comic authorship, allowed to be a writer who was easy to read. He rattled on, with too little thought, it may be, but with a shrewd common-sense and an almost feminine justness of view, that won him friends among his enemies, even if a careless witticism now and then made an enemy of a friend. This was never for long; while it is certain that Albert Smith lived down a great deal of hard and even scornful criticism. He brought round all his old *Punch* companions from whom he had cut adrift; and even the high-toned *Examiner*, seldom merciful to " light writers," pronounced one of his books of travel to be, "frank, genial, and manly." He practised in his early career as a dentist, but soon drifted into magazine work, and amused the laughter-loving public with his "Adventures of Mr. Ledbury, and his friend, Jack Johnson." The man-about-town style of writing was more amusingly and inoffensively exemplified in Albert Smith than in any of his rivals; for with him it was spontaneous—a hearty emanation from personal habit, which had grown into nature. Student-life—medical-student-life, that is to say—in Paris and London gave both incident and tone to his tales and sketches—the

CHERTSEY BRIDGE.

incident being of a practically jocular kind; and the tone, that of rollicking levity. He went a little out of his way to take up historical romancing in his novel on the subject of the venomous Marguerite d'Aubray, Marchioness of Brinvilliers; and Douglas Jerrold went a good deal out of *his* way to assail his "former crony" Albert's dabbling in "arsenicated literature." More congenial, certainly, to his powers of lively common-place were the stories of "Christopher Tadpole" and "The Scattergood Family." He had some dramatic faculty, which took now and then the proper dramatic form of theatric art; and, beside the stage-burlesques collaborated by him with Shirley Brooks, Charles Kenny, Stoqueler, and others, he wrote a few pieces, whereof one was suggested by the famous Chertsey bell, and a romantic legend in connection with that relic of Saxon days. Albert Smith's brother Arthur, a man of singular gentleness, was devoted to him, and spared no pains to please and serve him in a multitude of ways, little and great. The affection which existed between the two was never shaded by difference of any kind. Here is a little story which now sees, for the first time, the light of print:—When Albert Smith was giving his long-lived entertainment of "Mont Blanc," Arthur, his right-hand man in the business of management, took a holiday, and, visiting some glass-works in the north, was so struck with the resemblance of certain waste products to icicles, that he brought a number of specimens away with him, had them mounted like pendants, and, on his return to Piccadilly, hung them in triumph round the eaves of the little *chalet* which formed a prominent part of the set scene. Albert, who would not have damped his brother's delight for all the world, was "charmed" with the effect, and thanked the good Arthur again and again. "I *can't* tell him," said he, secretly, to the present writer, "that the flowering plants, the Alpine heaths, and all, are in full summer-bloom. It would break his heart to be reminded of the little contradiction in the seasons."

As the first religious house founded in Surrey, the abbey, of which there are now few vestiges, gave Chertsey a name of imperishable renown in English annals. We are carried back by the sound to Saxon days, to King Egbert and the sainted Erkenwald, who founded the great monastery at Barking as well as that of Chertsey. Abbot Erkenwald received his first Charter from Frithwald, subregulus of Surrey, nine years after the foundation of this abbey of Cerotæsni, Cerotesege, or Certesyg, as the name last given appears in "Domesday Book." The etymology, then, of our familiar Chertsey is "Cerota's ey," or island. Erkenwald's monastery and church were erected on a grassy ait, formed by the Thames and the little stream now called the Abbey River, or Bourne. When was there ever monastery or abbey built in England, France, or other part of Christendom, but it was near a river, teeming with fish? In nine out of ten cases, the ground has been an island, whatever it may be now. Take Westminster, for instance. It is not, you will say, insulated; but it was, and its name was the Isle of Thorns; and the very first angelical promise in relation to the Saxon abbey was made to the fisherman, Edric, who was told by a supernatural visitant, sent by St. Peter, that a plentiful supply of fish would never fail him so long as he duly carried his tithe to the monks. From that time, quite early in the seventh century, till near the end of the fourteenth, the Thames fishermen religiously paid their tithe of salmon to the abbey; and it is a singular fact that the first violation of the custom was by a priest, the vicar of Rotherhithe, who denied his tithe until the monks of Westminster enforced it by law, protesting that the right had been granted to them by St. Peter, when their abbey was founded. As an instance of the primitive state of society, in the England of the Middle Ages, every bearer of fish to the Abbey of Westminster sat by prescriptive right at the prior's table that day, and could demand ale and bread at the buttery-hatch to be brought him by the cellarman.

Fish, not on fast days alone, but as a constant staple of diet, was one of the needs of monastic life. Nor did the monks and their lay brothers generally wait for tithings from secular piscatory sources, as the fraternity at Westminster seems to have done. Mr. Dendy Sadler has no doubt hit with main truth of history, if with some exuberance of playful humour, the monkish habit of angling. At Chertsey the Benedictine friars of the tenth century left such evidence of perfection in fish-culture as is pleasingly apparent to every Thames angler of the present day; and the salmon-trout nurseries of Mr. Forbes, on the Surrey shore, revive a goodly tradition of the olden time. Pike, perch, chub, bream, and barbel abound near Chertsey and Shepperton, as of yore; but the good monks, let us remember, had the lordly salmon always at hand, as well as the trout, which was too plentiful to suggest any thought of artificial hatching. The once stately abbey, of which all the remains now are the fragments of an arched gateway, part of a wall, and a bit of encaustic tile pavement, occupied an area of four acres, and looked like a town. The Danes pillaged and burnt the place two hundred years after its foundation, murdering the abbot, Beocca, with all his monks, to the number of ninety. It is scarcely possible, even now, to dig deep on the ground without unearthing bones and fragments of masonry, relics either of the ancient Saxon foundation, or of the second and still Saxon convent re-established by King Edgar, in the tenth century.

During successive periods many great men were interred here; but the abbey is chiefly remarkable, as a place of sepulture, for having been the brief resting-place of Henry VI., whose remains were brought thither from Blackfriars by water. It was on her way "toward Chertsey with her holy load" that the Lady Anne encountered crook-backed Richard of Gloucester, as the scene in Shakespeare's play of *Richard III.* vividly represents. Having been interred there with much solemnity, the corse of the murdered king was only suffered to remain undisturbed till the second year of his successor's reign, when Richard caused the coffin to be removed to Windsor. The weak but well-meaning king, whose piety and love of learning may be said to have been too strong for his mental sinew, held Chertsey in high regard and favour, following, indeed, other sovereigns by whom in long succession from Saxon times the abbey was often strengthened and endowed. To benefit a religious institution and the town pertaining thereto was formerly one and the same act, a state of things now hardly comprehended in its full significance. It was to the abbot, in kingly piety, that Henry VI. granted the right of holding a fair on St. Ann's Hill, on St. Ann's Day. The "Black Cherry Fair," as it is called, is now held in the town, and the date is changed from the 26th July to the 6th August. Another great fair, for cattle, horses, and poultry, is also held there on the 26th September, in view of Michaelmas Day, which ancient feast is generally honoured with the goose as a standing dish; for by that time of the year this bird, hatched in the spring, has attained a goodly form and condition, while preserving some of its tender youth. So notably do these considerations affect the fair in September, at Chertsey, that it is popularly designated the Goose and Onion Fair, the sale of geese and onions eclipsing all other traffic, not only as regards poultry, but horses and cattle to boot. As before observed, mills and fisheries survive the changes of epochs with extraordinary vitality. We have seen that Chertsey is still a head-quarters of angling, as it was a thousand years ago; and the abbey mills flourish in modern fashion to this day. More remarkable far is the survival of the curfew bell; for there are fisheries and mills of ancient origin all over England, but the curfew is heard at few other places than Chertsey. Here exists a curious old custom of tolling the day of the month, after a brief pause, at the close of the "knell of parting day." In the tower of the rebuilt parish church, with a peal of six bells, is one that is believed to have belonged to the ancient abbey. There is warrant for the tradition which assigns so venerable an age to this bell; for the Latin inscription

ORA : MENTE : PIA : PRO : NOBIS : VIRGO : MARIA

is in Anglo-Saxon characters.

For a little more than two years, Abraham Cowley, the poet, intending to husband his small fortune, lived at Chertsey, or, rather, continued to exist for a short time. His desire for solitude provoked from Johnson, the lover of city life, a biographical sneer. It is true that the first night Cowley came to his half-timber house at Chertsey—he had desiderated a brick house, by-the-by—he caught a severe cold, and kept his room for ten days; but it is also true that he was an invalid when he came from Barn Elms, whence he was driven by illness. A series of mishaps befell him, which he recounted in a half ludicrous light, in a letter to his friend Sprat; and this letter it is that Johnson recommends "to the consideration of all that may hereafter pant for solitude." Cowley's house, which he only left in funereal pomp and state to be conveyed by water to Westminster Abbey, and there buried, is still sometimes called by its old name, Porch House, from a porch that once projected into the highway, but was pulled down a hundred years ago. In the garden is a fine group of trees, one of which, a horse-chestnut of great size and beauty, sheltered the poet in the short term of his life at Chertsey.

A memorable episode of Dickens's early work of fiction, "Oliver Twist," is graphically connected with this agricultural town, the most commercial establishment in which is a brewery. There were no railways to speak of when Fagin, the Jew that Dickens drew, was redrawn by Cruikshank, and when Bill Sikes, and Nancy, and Toby Crackit, and the Artful Dodger, and Charley Bates, and the bad people generally, seemed as real as, on the other and supernaturally amiable side, Rose Maylie and the rest were creatures of angelic imagination. There is nothing more real in this story, nor in all the stories that Dickens ever wrote, than the expedition by Sikes and Oliver, from Bethnal Green, through Finsbury and Barbican, to the West-end—past Hyde Park Corner, Kensington, Hammersmith, Chiswick, Kew, and Brentford—past Hampton and Halliford, Shepperton and Sunbury—till Chertsey was reached; and, joined now by Toby Crackit, they made their way through the silent town to the scene of the projected burglary. Boating-men know well the landing-stage at the "Bridge House," one of those inns where the comfort is not diminished either by negligence or false pretence. This is the recognised "hotel" of Chertsey; but the "Cricketers," in the Bridge Road, half a mile from the town, is the favourite resort of anglers. Many pleasant walks are still to be found near Chertsey, St. Ann's Hill being within a mile. As the residence of Charles James Fox, the house, with its gardens, lawns, shady walks, and quaint summer-houses, should be seen by all who have the opportunity of visiting it. The old gate of wrought iron, though not by any means extraordinary, nor indeed nearly so elaborate as some examples of smiths' work still to be seen about old parts of Chelsea, Chiswick, and Roehampton, is characteristic and significant of its period. From this gate to the summit of the hill is a short walk which affords a delightful view on a fine day, extending to Windsor on the one side, and to London, a distance of twenty miles, on the other. St. Ann's Well, on the descent, is a sylvan spot, which might better have been left alone than "improved," as it has been. It once looked like what it probably is, a veritable relic of the chapel which has been swept away like its contemporary foundation, the abbey on the marshy island below. St. Ann's Hill is a favourite place for picnics, as also for volunteer exercises and reviews. Returning towards Chertsey Bridge, on our downward Thames journey, we see the wood-crowned heights of Woburn, and presently make or renew acquaintance with the Wey, another tributary of the metropolitan river. The Wey rises in Hampshire, near Alton, where good ale used to be brewed, and indeed continues to be brewed still, in spite of the fact that this national beverage, the wine of the country, is getting more and more into the hands of a few noted brewers, and consequently is more and more "all alike," which is a sad sameness to think of! Time was when small breweries were oftener attached to inns of good repute, and when to taste the ale was a complimentary obligation. It is no question of curious tasting in these times; for you know pretty well what you are going to get when you ask for "a glass of bitter," which is generally good, but somewhat monotonous. What has become of all the country home-brewed, of the ales of different colleges, for example?"

> "I have a friend who loveth me,
> And sendeth me ale of Trinity,"

sang Barry Cornwall. Where is now the good ale, and where are the good fellows who sent it?

> "The wine of life is drawn, and the mere lees
> Is left this vault to brag of."

Through Farnham, Godalming, Guildford, Woking, Byfleet, and Weybridge, through all that country of heath and health, of pine-trees and rabbit-warrens, of scenery that you feel and breathe as well as see, the Hampshire stream flows and grows till it mingles with the waters of Thames below Chertsey, at a mill in the bend of the stream. It is said that the best hay in England comes from Chertsey mead, which also, during a large part of the year, affords right of commonage to neighbouring farmers for their cows; and the milk testifies to the richness of the pasture. It is at Weybridge that the Wey is joined by the Bourne, as also by the Basingstoke canal, and the meeting of the three streams is in a pleasant spot. Weybridge and Oatlands Park are places that hold renown in common. Round about the neighbourhood are country seats, beautifully situate, and two miles south of the town or village is Crockham Hill, from which a transcendent prospect of the whole weald of Kent, Surrey, and Sussex is gained, never to be forgotten. By Chart, or by Westerham Common, the way to Crockham Gap is the loveliest of Surrey walks, and indeed the beauty of the district cannot be overpraised.

At the Domesday survey, "Webrige" was a manor held by the Abbot of Chertsey, and its value was twenty shillings. With other lands, the property pertaining to the ancient abbey, it was annexed by Henry VIII. to the honour of Hampton Court. The estate of Oatlands was acquired by the king in manner following. He was negotiating its purchase when the owner, one William Rede, died, leaving his son John, a child, the heir. A short way was taken by the king to remove all difficulty. He constituted Sir Thomas Cromwell guardian of the boy, and the rest was plain. Very speedily the erection of a palace for Anne of Cleves, the king's intended wife, was commenced, the materials being found in the dismantled monasteries. Stone was brought from Chertsey and Bisham; marble for pavements from Abingdon; while the good red bricks which composed the walls were made at Woking, which name was spelt by the accountants "Okyng," much as it is pronounced at the present day by rustic natives. For his orchards, the king took apple, pear, and cherry trees from the orchards and gardens of Chertsey Abbey. The interior walls of Oatlands Palace were hung with the costliest tapestries of France and Flanders, the floors being covered with "carpets of Turque." But before the casket was ready the jewel had been discarded. Anne of Cleves, on whom Henry bestowed an uncomplimentary epithet, had come, and had proved unacceptable. The bride was divorced, and a new bride was taken in her stead. With the new bride Henry required a new palace. Oatlands was consigned to the keeping of Sir Anthony Brown; and, save that it was made the occasional residence of the Princess Mary, we hear very little more of Oatlands in King Henry's reign. We may fancy it a many-gabled, many-towered, Tudor edifice of red brick, with stone quoins and dressings, ornamental chimney-shafts, and handsome bays, like Hampton Court, in fact, with the same kind of turreted central gate-house in the principal front. There are drawings of it in the Bodleian at Oxford. The foundations are said to have been traced over fourteen acres. Terraces, flower-gardens, orchards, fountains, fishponds, and detached summer-houses adorned the pleasance round the magnificent edifice; and beyond, fenced about by a quickset hawthorn hedge, was the deer park.

An example of wasted labour and misapplied ingenuity, the grotto constructed in the eighteenth century by an Italian and his two sons for Henry Clinton, Duke of Newcastle, may be cited as one of the questionable glories of modern Oatlands. The artificers were twenty years at their work, which cost the Duke, it is said, £40,000; the sum stated in the early accounts being between £12,000 and £13,000. Outside, this egregious sham is built of tufa, which is a volcanic substance, or the calcareous deposit of certain springs. Within, the building has three or four chambers connected by low dark passages, on the ground floor, and one large chamber over all, with an elaborate cupola of satin-spar stalactites. All the inner

SHEPPERTON LOCK.

walls are a mosaic of minerals, shells, and spars of various kinds, blending in many devices, and inlaid with endless patience and skill. Among the many fine specimens of minerals still left, are quartz, crystals, and ammonites of rare perfection. Horace Walpole delivered himself of this criticism on the Oatlands grotto: " Oatlands, that my memory had taken it into its head was the centre of Paradise, is not half so Elysian as I used to think. The grotto, a magnificent structure of shell-work, is a square, regular, and, which never happened to grotto before, lives up one pair of stairs, and yet only looks on a basin of dirty water."

It is evident that Horace Walpole spoke of the upper chamber as the grotto itself; and so it was mainly. This *bizarre* kind of architecture was quite in the taste of George IV., and accordingly, when he was Prince of Wales, and just after Waterloo, he entertained at a supper in this wonderful room the Emperor of Russia, the King of Prussia, and the princes and generals in their train. As being tastefully in accord with a stalactite cavern, lit up by cut-glass chandeliers, the gilt chairs and sofas had satin cushions embroidered by the Duchess of York. Oatlands underwent many transformations; was destroyed by fire in different ages; and has sprung up again and again, like an exceedingly protean phœnix. The only vestiges of its ancient grandeur are the massive gateway and some magnificent cedars by the river. It is curious to think of its many transformations, during the dwindling and declining periods of its history. Once it was a rambling, mock-battlemented structure, in the taste of Strawberry Hill Gothic. A quasi-Italian style has been its later phase, and this remains, in the aspect now presented by the house, which has been converted to the purpose of a residential hotel. Oatlands has a longer story than can be told, even in outline, here. After Henry VIII. abandoned his intention to keep up its royal splendour, it became the temporary abode, at different periods, of Edward VI., Elizabeth, James I., Charles I., and Henrietta Maria—their youngest son, Henry, being born here— Queen Henrietta's second husband, the Earl of St. Albans, and then a succession of nobles, under various terms of tenure, till the Duke of York purchased the property in 1790, when the rococo edifice on the Strawberry Hill pattern of modern antiquity made its appearance, and became the bone of contention between two architects, the inglorious though not mute Pugin and Barry of the time, as we may call them—each claiming the honour of its invention. Greville's Memoirs give us as much as we want of the private life of His Royal Highness in his queer castle, or, if further information be desired, some interesting additions may be found in the "Life of George Brummell, Esq., commonly called Beau Brummell," who passed much of his time there, and whose most constant benefactor, after he had been cut by the Prince Regent, and other summer-friends, was *sa toute affectionnée amie et servante*, the kind Duchess. In justice to one or other of the rival claimants to the glory of architectural design, it may be said that the outside folly of Oatlands, as conceived either by Holland, the architect of Drury Lane Theatre— his best work—or John Carter, more favourably known by his etchings from the Gothic, was redeemed by its interior fitness and stately proportion. An example of the effect produced by transplanted architecture is conspicuous on Weybridge Green. Here, a Cockney wanderer out of his element of Babel life, stands the column celebrated by Gay in his "Trivia."

> "Where famed St. Giles's ancient limits spread,
> An imrailed column rears its lofty head;
> Here to seven streets seven dials count the day,
> And from each other catch the circling ray."

A rover, indeed, was this monument from Seven Dials. First it was taken to Sayes Court, but never erected. Wanting a memorial to the Duchess of York, the villagers of Weybridge picked up the neglected masonry, altered it to suit their purpose, by discarding the dial-faced top and substituting a clumsy crown, and stuck it where it now stands. In the crypt of a small Roman Catholic chapel, facing a fine group of fir-trees on Weybridge Common, the body of Louis Philippe was laid, till the royal remains were taken to France and re-interred in the Orleans mausoleum at Dreux.

Largely frequented by anglers, Weybridge must take care if it desires to retain the favour of boating-men. While the towing-path crosses the boat-yard, and

SHEPPERTON.

dredging is neglected by those, whoever they may be, on whom the duty rests, it is very difficult to avoid grounding; so that many owners have been taking their boats away, as the constant dragging not only scratches but strains them. Shepperton, on the Middlesex shore, is a pretty village, small and quiet, with its chief places of residence hidden away behind trees, or peeping out upon the river. It has a railway terminus, on the South Western system, and is about an hour, that is, nineteen miles, from Waterloo. The deeps afford tolerable fly-fishing in the trout season, and are more frequently fished for jack, perch, roach, and barbel. There are several good swims in pretty equal favour with anglers, to wit the upper deep, the lower deep, and the old deep, east of the creek rails. Besides these, the creek itself is often resorted to. The anglers' inns at Shepperton are the "Anchor" and the "Crown." It is an unspoilt Thames-side village, this Shepperton, in spite of its many pleasure-seeking visitors; a class, to say the sad

truth, apt to disclose a selfish indifference to the pleasure of others. If the
holiday-maker is to be traced by scientific investigation, the marks to be looked
for will be broken bottles, greasy sandwich-papers, and lobster-shell, just as flint
tools and weapons denote other and earlier savages who have lived on the earth,
and have made it as disagreeable as possible for their fellow-brutes. Shepperton
Lock and Ferry are both picturesque in themselves, as well as being foregrounds
of scenery that is charming to the eye nurtured by art. Truly, landscape-painting
has done noble service in fostering the love of nature. Though real beauty must
be above the skill of man to imitate, it is a curious truth that no age in which
that skill has not been exercised has ever left any written records of a feeling
for the grandeur of mountains, or the simple loveliness of woods, fields, and brooks.
Chaucer, you say? Why, Chaucer pictured everything because he had seen it in
pictures; had the very soul of a limner; lived in the sincerest age of art; saw
Flanders and Italy; was familiar with all that was exquisite in the refinement of
courts; and, unless his appointment as clerk of the works at St. George's Chapel,
Windsor, was a gross sinecure, knew how a daisy should look in stone as well as
in nature's finer fashioning. He who imagines Chaucer to have displayed natural
observation without cognisance of art has totally misread Chaucer's time, rich in
actual colour, as in the very dress that distinguished "gentle and simple;" for, as
Mr. Ruskin has said, speaking of "the lovely and fantastic dressing of the thirteenth to the sixteenth centuries" (in the very heart and flower of which period
Chaucer lived), "no good historical painting ever yet existed, or ever can exist,
where the dresses of the people of the time are not beautiful."

Shepperton, hale, green, and old, in its plentiful trees, mostly elms and horse-
chestnuts, has likewise an age in history. It is a noted spot for Roman remains;
it has a church that was venerable and still retains claims to veneration; and it
has a rectory-house older, for the most part, than the parish church to which it
belongs. The dwelling in question is of fifteenth century erection, and is principally of timber, the soundest, the strongest, the most enduring—English oak.
Builders will come back in time to the wisdom of such building, as they are even
now aware of the folly which assumes iron to be fireproof. Halliford is our next
halt, a mile down from Shepperton Railway Station, the nearest to the place. It is
quite accessible enough for anglers, whose interest, if not whose taste, leads them
to a preference of seclusion to racket and noise. Proverbially "jolly," the angler
understands jollity in the Waltonian sense, as, indeed, the most sensible of us all,
anglers or not, understand it. The vulgar adverb, "awfully," does, indeed, too
literally qualify at times the modern adjective. Halliford Bridge was washed
away some years ago by the floods; and now the Surrey and Middlesex shores are
connected by a brick and iron structure which is named Walton Bridge, and
which, having been the occasion of war between Bumbledom on both sides the
river, was painted of two colours, a chromatic difference that greatly increased
the normal ugliness of the design. The most plentiful fish at Halliford are roach

and bream, but there is an abundant variety of others. To distinguish this little place from another Halliford, which is a hamlet of Sunbury, it is sometimes called Lower Halliford. The views along and across the river, every way, are charming; and as we look over to Oatlands, the Surrey hills form a beautiful background; while on one side we have Walton and Ashley Park, and on the other Weybridge. The "Red Lion" is a favourite haunt of anglers, and all who visit the spot by road or river; and other houses of entertainment are the "Crown,"

HALLIFORD.

the "Ship," and Mrs. Searle's. The narrow creek adjacent to the "Red Lion" is in frequent request as a harbour for punts and small craft in general. A little further and we come to Walton, having crossed the river once again into Surrey.

Walton-on-Thames was, in old Saxon days, as its name plainly indicates, a walled town. Etymology apart, the traces of its having been fortified speak for themselves in the neighbouring remains of important earthworks. It is now a village; and, as a village, large; but it is not quite large enough to be considered a town; and of its walls there are no traces above ground. Walton Bridge crosses the river just where there was once a ford that, as relics show, was strongly defended. A little above Walton is the spot at which Cæsar crossed, in the time of his second invasion. It is called Cowey Stakes, and has afforded matter for many an antiquarian discussion. The "Stakes" were driven in front of the bank to repel attempts at landing. Some accounts of them state that they were placed upright in two rows, across the shallow bend of the river, so as to support a bridge. Walton has an interesting church, very old in some parts, but modern in others, with Norman piers, on one of which may be read, deeply cut, the verse ascribed to Queen Elizabeth, when princess, and when it was sought by Mary to entrap her in a heresy regarding the doctrine of Transubstantiation.

> "Christ was the Worde and spake it;
> He took the bread and brake it;
> And what the Worde doth make it,
> That I believe, and take it."

Among the monuments are works of note by Roubiliac and Chantrey, the older sculptor excelling in the bold inventiveness and forcible execution of his work, a superb monument to Lord Shannon. On the left side of the communion table is buried William Lilly, the astrologer, that "cunning man, light Sidrophel," as he figures, or is supposed to figure, in "Hudibras." Another tomb, but it is in the churchyard, not the church, holds the remains of "bright, broken Maginn," who sleeps without a memorial. President Bradshaw's house, at the back of Church Street, is divided into a group of wretched tenements, all in a squalid condition; yet in a room on the ground-floor of one of them, covered with dirt and whitewash, is a carved oak chimney-piece, with coupled columns and a cornice, the room itself being panelled, and an elaborately carved beam crossing the ceiling. There is a tradition that Charles I.'s death-warrant was signed in this room. One of the curiosities of Walton-on-Thames, shown at a house next that of Rosewell, the boat-builder, is a scold's gag, or bridle, few examples of which instrument yet remain in England. This particular specimen bears an inscription which, though now illegible, has no doubt been accurately quoted as follows:—

> "Chester presents Walton with a Bridle,
> To curb Women's tongues when they bee idle."

Chester, according to tradition, was a person who lost an estate through the evil speaking of a loose-tongued gossip, and took this mode of revenging himself on the sex. The bridle is a combination of head-piece and collar, a flat iron projection inside the latter being forced over the tongue, while a slit in the former, which passes over the face and skull, allows the nose to protrude. Not far from the church, on the road to the railway station, is Ashley Park, with its late Tudor or early Stuart mansion of red brick, containing a great hall, which takes the whole height of the house, and a gallery extending throughout its entire length of a hundred feet. The park, a richly wooded demesne, adjoins Oatlands. From St. George's Hill, in the vicinity, the magnificent prospect includes seven counties. The stream at Walton Bridge runs over many shallows, fast on the Surrey shore, and it is not easy to sail round the bend.

WINDSOR TO HAMPTON COURT.

Along a pleasant reach of the river, on the Middlesex bank, lies the village of Sunbury, with three or four boating and angling inns, which are much frequented by pleasure parties, though it is always a marvel to foreigners that the accommodation at these and other Thames-side inns should not be of a higher order. At Sunbury are the rearing-ponds of the Thames Angling Preservation Society; and the fishing of all kinds is excellent, no part of the river affording better sport with the fly than Sunbury Weir, which abounds with trout. The "Flower Pot," the

SUNBURY WEIR.

"Magpie," and the "Castle" are in Thames Street; and the "Weir Hotel" is on the Surrey side. The stone lock and lock-house are prettily placed amid pretty scenery, and there is a good camping-ground. As it is not often that a church is altered to its improvement, justice demands a recognition of the fact that the church of the Virgin Mary, by the river-side at Sunbury, from having been as ugly a brick building as was ever consecrated to public worship, has been rendered sightly by the insertion of new windows and the introduction of a semicircular chancel, and an elaborate Byzantine porch with stone arcades on either side. Till changed to the form we now see, the church appeared as it had been rebuilt in the eighteenth century. The ancient church was of Saxon foundation, dating from the time of Edward the Confessor. When the Orleans family made their retreat in the neighbourhood of the Thames, the Duke of Nemours assisted at the consecration, by Dr. (now Cardinal Archbishop) Manning, of a small Roman

Catholic church a short distance off, prettily constructed of stone, from the drawing of Mr. Charles Buckler. The sailing clubs have made Sunbury a rendezvous, and boat-building is a prosperous occupation.

As we near Hampton, the historical and the "Happy," Garrick's villa comes into view. The watery way, down from Sunbury, is between banks which are flat and uninteresting, osiers hiding the land from those who voyage in boats. Robert Adam, who, in conjunction with his brother James, improved the street architecture

SUNBURY CHURCH.

of London—their fraternal labours being commemorated in the name of that since-spoilt river-terrace, the "Adelphi"—built the Corinthian front of Hampton House, as it was called when Garrick' bought it, though the mansion has since been renamed after the great actor himself. Adam's portico, the salient feature of the house, reaches, with its pediment, above the attic storey. Much is said about the building, its contents and its visitors, by Horace Walpole; for Garrick's dinners, his illuminated grounds, and his night-fêtes, attracted company of the first order. On the lawn, near the water's edge, was and is a miniature Grecian temple, of octagonal shape, with an Ionic porch, the structure being designed for a summer-house, in which for a time was placed Roubiliac's statue of Shakespeare, to be removed, after Garrick's death, to the British Museum. Garrick planted his domain very tastefully, and the trees that have grown to goodly height and umbrageousness since his day, now invest the spot with a dignified grace. For twenty-five years Garrick enjoyed his liberal case and the pleasures of well-chosen society in this home of comfort and elegance; and his wife, who survived him by forty-three years, living to a great age, continued to dwell here, maintaining everything in the same place as when he was her companion. The forget-me-not, commonest of wild flowers in this neighbourhood, finds surely a congenial soil where David Garrick's memory was cherished so fondly and so long. Islands in thick succession dot the stream, and when fishing-rods are patiently extended here and there, the picture is at once socially and tranquilly suggestive. Opposite the town and church of Hampton lies Moulsey Hurst, between the villages of East and West Moulsey. This wide and beautiful meadow, "hard and smooth as velvet," as one of Archibald Constable's literary correspondents describes it, has been degraded in all ways attributable to civilisation. As a race-course, it is probably the vulgarest in the world; and its history is bound up with the annals of duelling and prize-fighting. A letter, very charac-

BETWEEN HAMPTON AND SUNBURY.

teristic of the time, contains the following candid record:—"Breakfasted at Mr. Maule's very early, and went along with him and the Bailie to see the great fight between Belcher and Cribb, at Moseley Hurst, near Hampton. The day was very fine, and we had a charming drive out in our coach-and-four, and beat all the coaches and chaises by the way. We had three hard runs with one post-chaise and four very fine horses, before we could pass it, and drove buggies, horsemen, and all off the road into lanes and doors of houses." Among the gentlemen present, as the same frank-spoken witness testifies, were "the Duke of Kent, Mr. Wyndham, Lord Archibald Hamilton (a famous hand, I am told), Lord Kinnaird, Mr. T. Sheridan, &c. &c., and all the fighting-men in town, of course." These last, we read further, were "the Game Chicken, Woods, Tring, Pitloon, &c. Captain Barclay of Urie received us, and put us across the river in a boat, and he followed with Cribb, whom he backed at all hands. The Hon. Barclay (Berkeley) Craven was the judge." This charming chronicler proceeds to tell us that the odds were on Belcher, but that the hero in question, after a long fight, "was at length obliged to give in." Poor fellow! Modern adherents to the theory that fisticuffs had any early origin in Great Britain may be consoled for the decadence of the "good old national art of self-defence" by the assurance that boxing was a practice which endured little more than a century and a half, if so long, and was learnt from North American savages. Its real antiquity is Greek, the grounds for believing that the Anglo-Saxons, and, after them, the Plantagenets, favoured this form of pugilism being extremely slender. The English prize-fighters of the eighteenth century encountered one another with broadswords. There are other "arts of self-defence" far better entitled to rank as English than boxing. The quarter-staff is one.

On the road to Moulsey from Walton-on-Thames stands Apps Court, or the modern representative of that capital mansion, once inhabited by Mrs. or Miss Catherine Barton, who might have been called a "professional beauty" had the phrase, together with photography, been invented in her day. The manor was

GARRICK'S VILLA, HAMPTON.

bequeathed to her for life by Charles Montague, Earl of Halifax. She was a reigning "toast," and her name frequently occurs in Swift's journal to Stella. Catherine Barton, who was a sort of niece to Sir Isaac Newton, being, in fact, the daughter of his half-sister, has been spoken of as the mistress of Lord Halifax; though it is now pretty clearly established that she was privately married to him, before his elevation to the peerage. She afterwards married a master of the Mint, who succeeded in that office her illustrious uncle. Many other persons of note are historically associated with "delightful Ab's Court," so designated by Pope, in his Horatian epistle to one of its proprietors, Colonel Cotterell. The grounds, like most of the Thames pleasances, contain some grand timber; oaks and elms being conspicuous objects.

A little past Moulsey Lock is Hampton Court Bridge, a five-arched iron structure, by which we take our way to the palace and its famous grounds.

GODFREY TURNER.

THE APPROACH TO HAMPTON COURT.

CHAPTER VIII.

HAMPTON COURT TO RICHMOND.

Hampton Court—Thames Ditton: The "Swan"—The Church—Surbiton—Kingston: The Coronation Stone—Teddington—Twickenham—Eel Pie Island—Petersham—Richmond Park—Approach to Richmond.

HAMPTON COURT is not the stateliest pile upon the banks of Thames. It is less splendid than Windsor, less historic than the Tower; yet it possesses a meed of human interest unique in English palaces. Windsor has its memories of the births and deaths of kings; of proud embassies from Popes and Kaisers while yet the censer was swinging through all England; of sweet brides wedded to the misery which is always lurking behind a throne. The Tower is the most historic building which the world still looks upon—the very kernel of England's history, even as the Chapter House of Westminster is the birthplace of her liberties; in the darkness and silence of its dungeons was matured that intolerance of despotism, that resolve for freedom which began early to mould the modern England; it is a fortress of unending tragedies. Yet Hampton Court, which is newer than either and less historic than either, enjoys a popularity and exercises a charm far beyond that of the two feudal fortresses. The explanation of this which fashions itself when one is in romantic mood, is that the popular imagination is touched by the sidereal rise, the brief glory, and the sudden fall of Cardinal Wolsey—a fall which even the gift of the stately palace itself could not avert. But the footprints of Wolsey at Hampton Court are hard to trace; and it is probable that to the Bank Holiday masses, and to the crowds which stream through its galleries every fine Saturday and Sunday in summer, the most abounding charm of the place is that which Nature, with some assistance from Art, has provided. The terraces, the gardens, the maze, the trim vistas cut through long lines of trees, the Dutch primness and precision of the grounds, and above all the thousand acres of Bushey Park, with its renowned avenue bursting with the tinted blossom which in summer perfumes the air like "an odour sweet of cedar log and sandal wood," are the true delights of Hampton Court.

The old Tudor palace is a significant landmark of the river-side, for it indicates the spot where suburban London may be said to begin. London has a long arm, and the voyager on the silent highway from its source on towards the sea finds, as he nears Hampton Court, unmistakable signs that he is reaching the fringe of a giant population. There is a greater frequency of white villas, glistening in the sunlight, shaded and cooled by the ample foliage which is rarely so green and prolific as on the banks of the southern Thames, the water gently lapping the edges of the shaven lawns. The river is dotted with boats, where before the dinghy and the outrigger had been but occasional; the towing-path is more populated; and—it is a melancholy story to tell—the water begins perceptibly to lose its limpidity. The pollution of the great rivers of the world seems to be one of the ultimate aims of civilisation. Is the Scheldt pure—the weird mysterious Scheldt of the "Flying Dutchman"—the storied Rhine, the classic Tiber, the "blue" Danube? Its immense navigation and the multitudes on its banks put the Thames into worse case than them all; but we trust the time is coming when we shall be more mindful of Nature's lovely heritage, and that if we may never again see salmon taken at London Bridge, neither shall we see banks of festering mud on the very limits of the tide.

Hampton Court has been frequently described as the English Versailles, and there is much reason in the comparison. Alike in history and in human interest, however, Hampton Court is far more attractive than the splendidly frigid palace of Louis Quatorze. It is true that it has few pretensions to magnificence; but it is a compound of history, and the history of people rather than of events. The shades of Wolsey and Charles I. eternally haunt the portals through which so many historic figures have passed. But the ghost of the magnificent cardinal finds everything unfamiliar. Even the Great Hall, so often ascribed to him, was not built until after his death; Sir Christopher Wren's new west front is all strange to him; only in a little wing here and there can he recognise the handicraft of his own architect. Maybe the capacious cellar, with its wine-casks stuffed with broad pieces of gold, which, if we are to believe tradition the cardinal used for a treasury, is still untouched; but where are the five "fair courts" round which the palace was grouped by its builder? Had Hampton Court remained to our day precisely as it left the hands of the Tudor artificers, it would have been a priceless relic of the architecture and the methods of life affected by an English prince of the Church in the early years of the sixteenth century. But Wren has done his incongruous and Nash his clumsy work upon what Henry allowed to remain of the cardinal's design, and Hampton Court, as we know it, has, architecturally speaking, a blind side and a smiling side.

There is no doubt a certain stateliness about the East Front and the Fountain Court; but it is a heavy and monumental stateliness which ill accords with the really picturesque portions of the old palace. Classical symmetry and Palladian regularity are sadly misplaced when joined to Tudor red brick. The style which Wren chose for his additions requires greater space for its effective display than he had at disposal; consequently, the buildings round the Fountain Court suffer from the contracted area of the quadrangle. English brickwork was never better than in the early part of the sixteenth century, and in the buildings erected by Wolsey and Henry VIII. at Hampton Court we have this Tudor brickwork at its best. The cardinal's buildings have upon the outer walls the geometrical patterns which were not uncommon at the time, formed by the insertion of those stout blue bricks which are so potent to resist damp. Of the strictly modern additions and re-buildings, the work of the last hundred and fifty years, it were better to say nothing more than that they, lamentably, still exist.

ENTRANCE PORCH.

It is difficult to obtain a comprehensive view of the palace save from the river. Thence, however, the glimpses of the pile are very varied, the view of the western front being especially charming. The multitude of towers and mullions diversifies the *façade*, very greatly to the disadvantage of Wren's monotonous eastern front; while the interlaced and arabesqued chimneys, the graceful clock-tower, and the high-pitched roof of the Great Hall break the sky-line with a cunning which, although apparently undesigned, is as effective as it well could be. There is something peculiarly appropriate in approaching Hampton Court from the Thames, for in the day of its pride, when the cardinal and his thousand retainers abode there, when Henry retired to it with one or other of his wives, or when his dour daughter Mary passed there her honeymoon with the darkling Philip, snatching a brief leisure from his "Acts of Faith" in Spain and the Netherlands, the river was the silent highway upon which all the world travelled, whether from the Tower or from Westminster. But the approach from Bushey Park, although its historical savour is small, is more attractive almost than that from the water. I never traverse the Chestnut Avenue without regretting that the venerable towers and turrets of the palace do not close in the vista. Such an avenue ought, for the sake of picturesque completeness, to have for objective an ancient country house, gabled, ruddy, and peopled with historic shades. The Diana water is very pretty in its way; but it is not the most effective climax. There are some beautiful avenues in the little park of Hampton Court itself; but there is a Dutch flavour about them which causes them to look less natural and spontaneous than the Chestnut Avenue, which is really only the central of a series of nine, four on each side. Bushey Park, like all other parks, is pretty, but flat; it happily still contains a good head of fallow deer. For nearly fifty years—since the palace was thrown open to the public—Bushey has been a spot of inexhaustible delight to myriads of Londoners, the great majority of whom choose the route through the park to Hampton Court. The novice in woodcraft might imagine that many of the trees had attained a good old age; but so far as those in the nine avenues are concerned, all were planted by William III., the tutelary genius of latter-day Hampton Court, and outside the avenues the timber is neither luxuriant nor remarkable.

By far the most interesting portion of either the new or the old palace is the Great Hall, which, save that it has a new floor and that the painted glass in the windows is modern, is little altered since Henry VIII. built it. This magnificent apartment ranks with Westminster Hall and the Hall of Christ Church, Oxford, as one of the finest open-timbered interiors in Europe. What relation it bears to Wolsey's Hall, the site of which it is believed to occupy, we cannot tell, since no picture of the cardinal's banqueting-chamber has been preserved. But "The Lord Thomas Wulsey, Cardinal, Legat de Latere, Archbishop of Yorke, and Chancellor of Englande," had a nice taste in architecture; and it is tolerably safe to suppose that, beautiful though Henry's hall be, the cardinal's was better. Why the king saw fit to destroy what Wolsey had built there is no evidence to show; probably he was desirous that his own name rather than that of his upstart chancellor should be permanently associated with the place, and the lavishness with which his cypher, together with the rose and portcullis and other heraldic devices of the Tudors, are scattered about the palace, favours the idea. It is a little remarkable that the hall, with the adjoining withdrawing-room, should be disconnected from the other buildings, and that it should not be possible to reach it without passing into the open air. The best view of the hall is obtained, not from the sombre entrance beneath the Minstrel's Gallery, but from the daïs at the upper end, where the high table for noble and princely guests was wont to stand. Its proportions are majestic—106 feet by 40, with a height of 60 feet. The open-timbered roof is elaborately carved and arcaded, and springs, as though naturally, from massive corbels between the windows. At the extremities, where the corbels join the roof timberings, are the graceful pendants characteristic of the Tudor time. The windows blaze with painted glass, all a mass of heraldry and kingly pedigree, while beneath the eye finds rest in the more subdued tones of King Hal's

THE FIRST QUADRANGLE.

tapestries of incidents in the life of Abraham. Who designed this arras and where it was woven are questions which have never been settled ; but there is abundant internal evidence that it is either early Flemish or German work. In the gloomy vestibule beneath the gallery is a series of allegorical tapestries, the most curious of them representing the seven deadly sins in such guise as would suggest that the artist took the idea from the procession to the "Sinful House of Pride" in the "Faërie Queen." Spenser makes

FOUNTAIN COURT.

Gluttony ride upon "a filthy swine"; in this tapestry it bestrides a goat. All this arras is in beautiful preservation, particularly that which deals with the life of Abraham, in which the high lights are worked in gold.

The painted windows of the Great Hall deserve something more than a passing mention. Six alternate windows are filled with the arms and descents of the wives of Henry VIII. ; and it is worth noting, as some indication of the commonness of a Plantagenet ancestry, that each of these ladies was descended from Edward I. The probabilities are, indeed, that even now a large proportion of the English people, above the lower middle class, have in their veins the blood of Longshanks. The seven intermediate windows are emblazoned with the badges of Henry VIII.—the lion, the portcullis, the fleur-de-lis, the rose, the red dragon of York, and the white greyhound of Lancaster. Upon each are his cyphers and the mottoes " Dieu et mon Droit," and " Dne. Salvum Fac Reg." The great eastern and western windows are likewise full of badges, quarterings, and impalements. At the upper end of the south side of the hall is yet another window more beautiful from its pendant fan-tracery than any of the others, and emblazoned with the arms and cyphers of Henry and Jane Seymour, and the arms and cardinal's hat of Wolsey. The daïs characteristic of old time, when distinctions of rank were very palpable, still remains; but the beautiful old flooring of these painted tiles so much used by Tudor builders has gone, although there is reason to suppose that it still existed eighty years ago. A finer apartment for a regal banquet or a stately pageant could hardly be conceived. One would like to believe the legend of Shakespeare representing, in this very hall, before Elizabeth and her somewhat flighty court, the story of the fall of Wolsey; but there is not a tittle of real evidence in its favour. The Withdrawing Room, or Presence Chamber, as it is sometimes called, entered from the Great Hall, is a large, oblong apartment which has apparently been little touched since Tudor times. It has a fine painted oriel, a moulded plaster ceiling, and an ancient oak chimney-piece, into which is let a portrait of the unlucky presiding genius of the place. This chimney-piece is a modern importation from an old house at Hampton Wick. The roughly-plastered walls are covered with tapestries of a wildly allegorical character, considerably older than those in the Great Hall, and less carefully preserved. Above them are Carlo Cignani's cartoons for the frescoes in the Ducal Palace at Parma.

Beyond the Great Hall, the apartments which are shown to the public have little architectural or personal interest. The rooms in which Henry, Elizabeth, and Charles I. lived are all in the Tudor portion of the palace; the series which has been converted into one great picture gallery is in Wren's building, and runs round the Fountain Court and along the eastern front. What this front lacks architecturally it gains to some extent scenically, since it overlooks the geometrical flower-beds, the straight avenues, the long and narrow Dutch canals, beloved of the Stadtholder, the like of which one may still see in the gardens of old world *casteelen* in Holland. Some of the avenues, seen from these upper windows, are very charming and effective, notably that which is closed in by the red mass of Kingston Church. This is not the place to discuss the pictures with which the palace abounds. I shall not perhaps be committing treason if I suggest that they are remarkable more for their quantity than for their quality. There is a sprinkling of pictures of which any gallery might be proud, and some of the portraits, of no artistic importance, are valuable by reason of their personality. Kneller's "Hampton Court Beauties" have acquired a factitious fame, for whatever may have been the charms of the originals, they are assuredly not very obvious here. Poor, indiscreet Queen Mary got herself well hated by the other ladies of the Court whom she considered insufficiently attractive to be numbered among the elect. Perhaps the most famous of all the pictures at Hampton Court is Vandyke's equestrian portrait of Charles I., of which there is a replica at Windsor. Many of the paintings are true memorials of the old palace, and formerly hung in the ancient state apartments. They have the savour of old associations which the rooms in which they are hung lack—memories of times when life was more fitful, more spectacular, and, as it seems to us in this distant age, more romantic than it had become when Dutch William sowed *Je maintiendrai* about the old place, as Henry had scattered his roses and greyhounds and fleurs-de-lis, and all the other heraldic bravery of a century when heraldry was a fine art. Hampton Court is rich in personal history, and many a romantic shade must haunt its Great Hall, there to recall the vanished banquet, when the wine-cup gleamed so red, and bright eyes danced more intoxicatingly than any vintage of Spain or France. Many, too, there be that must still weep out their historic sorrows, and the visionary axe must flash before many a ghostly eye. Here lived Anne Boleyn and Catherine Howard, as well as Bluebeard's more fortunate wives. Edward VI. was born, and Jane Seymour died here. Elizabeth kept her Christmases merrily indeed at Hampton Court; tradition says that she was here dining off a goose when the news came of the defeat of the Armada. It was a favourite residence of Charles I., who passed here some of his happiest and most miserable moments, and hence he fled to Carisbrooke. Both Cromwell and Charles II., who once played a renowned game of hide-and-seek, were fond of the water-side palace; William III. had a passion for it, and in its park met with his fatal accident. The first two Georges stayed here occasionally, but since 1760 it has not been a royal residence. William IV. and his unimportant Queen liked the neighbourhood, and spent much time in the heavy but doubtless comfortable red-brick house at the Teddington entrance to Bushey Park. So long as it endures, Hampton Court will be one of the most interesting of English houses. Attractive in every aspect, in some it is unique, and if it had no other claim to distinction it would always be remarkable as perhaps the very first country house built in England without a moat and drawbridge.

The park of Hampton Court is small compared with the vast chase, covering thirteen parishes, of which Henry made it the nucleus. It is somewhat flat, but is well timbered, and beautifies the towing-path all the way to Kingston. Of the palace from the river I have already spoken. It is in view for a considerable distance towards Thames Ditton; but the glimpse is not so striking as that obtained by the oarsman who shoots suddenly beneath Hampton Bridge and sees the grand old pile full in front of him. Between Hampton Court and Kingston the river is at its most charming hereabouts. Flowing between deep banks, over which the rushes and osiers bend, in summer it is studded to just beyond Thames Ditton with the cool Bohemian house-boat, a veritable desired haven to the heated oarsman. The coquettish window-curtains, the mass of flowers on the flat roof, the whisk of dainty muslin, all go to form one of the prettiest of Thames pictures. The Middlesex shore is fringed with luxuriant hedgerows, quick with life and bursting with blossom, so wide and tunnelled by the boughs of trees that one of Mr. Stevenson's nursery heroes might lose himself amid the interlacements, while imagining that he was stalking the red man in his native forests. On the Surrey side the meadows come down to the water's edge, fringed with rushes and alders. Soon above the trees peeps out the quaint wooden spire of Thames Ditton Church, topping a squat tower of the type beloved of the olden church builders of the Thames Valley. At the river's brink, and under the shadow of the church, is the famous "Swan," dear to the museful angler who delights not in crowds, and loves to make for a charming and unobtrusive stretch of river. With a kindly care for the welfare of the traveller, and not unmindful of other

IN THE REACH BELOW HAMPTON COURT.

considerations, some olden landlord of the "Swan" procured the establishment of the ferry at his very doors. The "Swan" was an important hostelry in the days when Thames Ditton was more in fashion than it happily is now; and it still divides the honours of the spot with Boyle Farm on the opposite side of the road. Dark brown of hue, and not unpicturesque of contour, Boyle Farm stands amid effective masses of foliage, its sloping lawn dipping down to the channel formed by a miniature eyot which screens it somewhat from view. The ample cedars on the lawn contrast well with the older portions of the house which face the water. This pretty spot obtained its name from that Miss Boyle who became in her own right Baroness de Ros, and is mentioned in one of Horace Walpole's letters as having "carved three tablets in marble with boys, designed by herself . . . for a chimney-piece, and she is painting panels in grotesque for the library." By her marriage with Lord Henry Fitzgerald, Lady De Ros became sister-in-law to that ill-starred pair,

Lord Edward Fitzgerald and "Pamela." Time was when Boyle Farm rivalled Strawberry Hill as a centre of gaiety, and its famous "Dandies' Fête," given in 1827 by five young sprigs of nobility at a cost of £2,500, was long a dazzling wonderment to those who are tickled by such things. This was one of the hereditaments which fell into dispute upon the death, without a will, of the first Lord St. Leonards. To the angler it may be that the comfortable old inn is more interesting than Boyle Farm with its Walpolean memories. Many is the wit and the man of letters who, after a day of more or less make-believe angling, has refreshed himself

THE "SWAN," THAMES DITTON.

at the "Swan." Theodore Hook delighted in Thames Ditton, and wrote some stanzas in its praise in a punt one day in 1834; it was natural that with so keen a lover of good living the "Swan" should come in for eulogy.

In the churchyard of Thames Ditton rests "Pamela" beneath a stone which records her original interment in the cemetery of Montmartre, and her re-burial here. Into the stone is let a portion of the marble slab, shattered by a German shell during the siege, which marked her resting-place in Paris. Close by is the grave of the first Lord St. Leonards. The tiny church possesses little architectural interest; but it contains a number of small but curious brasses, which have been removed from their original positions in the floor, and fixed upon the walls, a proceeding which, although it has divorced the memorials from the dust they commemorate, has no doubt tended to the preservation of interesting inscriptions, such as have, in too many cases, been destroyed. A brass which possesses a curious interest is

27

that of Erasmus Ford, "sone and heyre of Walter fforde, some tyme tresorer to Kynge Edward IV., and Julyan, the wife." This worthy couple had a full quiver in all conscience, for the brass bears representations of six sons and twelve daughters. Erasmus died in 1533, and his wife six years later. An even more portentous family was given to William Notte, who died in 1576, and Elizabeth his wife—nineteen, all told. It is hard to imagine such a posterity dying out; yet Notte is assuredly an uncommon name. Few facts in human history are more astonishing

THAMES DITTON CHURCH.

than the rapidity with which names become extinct. Century after century the same names occur upon tombstones and in parish registers, and then there comes a blank which time, instead of filling up as before, only accentuates.

Coming from London, Thames Ditton is the first point at which, in summer, the house-boat, elsewhere ubiquitous, is met with. The charm of the lagoon-like life which the house-boat affords has not lacked eulogists; but who shall justly describe the calm delights of dusk upon the river? It is as undefinable as happiness. The red gleam of sunset is splendidly spectacular; the gloom of dusk upon the water is weird, and a world of mystery seems to reside beyond. The plash of oars continues until the last speck of light has been folded into night; the boats shoot out from the encompassing darkness, ripple past, and enter the farther shadows. Strange fancies enter the imaginative mind, and these gliding boats seem like phantom craft shooting from shadow-land to shadow-land. Sometimes there comes a hissing launch, its lights flashing meteorically across the stream, and throwing their beams in among the rushes and osiers like sudden electric jets, or the fitful gleam of a will-o'-the-wisp. The awakening on the river has something of the idyllic, especially on a Sunday morning, and if the moorings be cast within earshot of church bells. Ditton is a prime point for the disportment of small craft from Kingston and Surbiton, and on fine Saturdays and Sundays the river hereabouts is crowded. All this movement is of course unfavourable for the punt-angler, unless he be astir early or on a day when the water is more or less deserted. In winter, however, when boating possesses charms only for the hardiest of enthusiasts, a good creel can be made within a stone's throw of the "Swan."

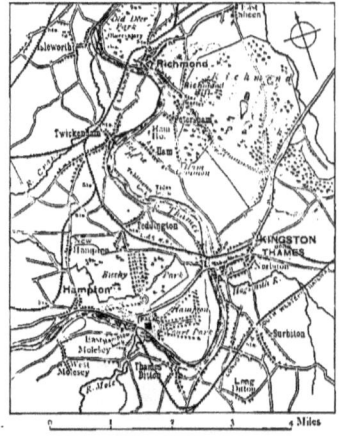

HAMPTON COURT TO RICHMOND.

Between Thames Ditton and Surbiton the river banks possess nothing of especial interest. The broad reach is, however, exceedingly pretty. On the Middlesex shore is a more than usually picturesque towing-path, broad and grassy, backed by the full hedgerow which bounds the park of Hampton Court. On the Surrey side the reeds and alders are profuse, and edge the water almost without break. The river front of Surbiton wears a decidedly foreign air, with its tall white houses, and winding walks and shrubberies along the bank. This esplanade, starting from the water-works, extends for some distance towards Kingston, and is an excellent hint to the local authorities of other water-side suburbs. Surbiton is an interesting spot to rowing men, for it is the head-quarters both of the well-known Kingston Rowing Club and of the Thames Sailing Club. Other interest it does not possess, and everything in and about it is painfully modern. But it is a pretty spot, and being within easy access of London is full of attraction to those who toil and spin daily within sound of the boom of Great Paul. Anything that Surbiton lacks in antiquity its ancient and dignified parent, Kingston, can supply. Kingston Bridge lies pretty well a mile farther down stream, almost at the opposite extremity of the town. The view from the facing bank still has something of the foreign air of Surbiton; but the aspect is Netherlandish rather than French, which the other is. The square red tower of the church, the congeries of tiled roofs, and the quaint little summer arbours in the sloping garden of the river-side hotel, contribute greatly to this effect. The not unhandsome stone bridge, the twin-brother of that at Richmond, which connects

KINGSTON, FROM THE RIVER.

Hampton Wick with Kingston, is a modern successor of a long ancestry of bridges, the earliest of which dated from Saxon times. Civil war, rather than time, seems to have made an end of all the previous bridges save that which immediately gave place to the present. For centuries London Bridge was the only other permanent means of crossing the Thames; consequently, when there occurred one of the frequent commotions in which our ancestors delighted, there was a good deal of competition between the two sides to get Kingston Bridge destroyed first, and so prevent communication between Middlesex and Surrey. In the strifes of the Roses it

THE MARKET-PLACE, KINGSTON.

fared ill, and during Sir Thomas Wyatt's rebellion it was broken down to prevent the passage of the insurgents. Since then, nearly three centuries and a half ago, the bridge has been more tenderly treated.

Kingston is a very interesting old town, and was an important place, and the scene of the coronation of Saxon kings a thousand years ago. It is remarkable as being the last municipal borough on the river, with the exception of the City of London itself. All the other places have to put up with local boards or vestries, or other undignified mushroomy governing bodies. Kingston possesses the real antique thing—mayor, aldermen, town councillors, mace, and all the other symbols of municipal importance, and is duly and rightly proud thereof. Few English towns can boast of such antiquity, and of fewer still can it be said that they have been boroughs since the days of John Lackland. It seems always to have been a loyal town—the result, perhaps, of its ancient regal associations—and much money appears to have been spent by the olden burghers for bell-ringing and other diversions when confusion had overtaken the king's enemies. When the Earl of Northumberland was taken, for instance, the Kingston ringers benefited to the extent of twenty pence—a clear exemplification of the saying that one half of the world lives upon the misfortunes of the other half. When Prince Charles returned from his Spanish expedition in 1624, the joy of the townsmen was so demonstrative that they must needs spend three and fourpence upon the clangour of joy-bells. Doubtless

THE CORONATION STONE AT KINGSTON.

the young prince, who was much at Hampton Court, was well known in the town, and when, after his accession, his troubles pressed thick upon him, the townsmen were loyal to the core. The actual hostilities of the great rebellion began and ended at Kingston, singularly enough. There the first armed force assembled; there, near Surbiton Common, Buckingham and Holland made the last stand for the crown, in which fight Lord Francis Villiers, who is buried in Henry VII.'s Chapel at Westminster, was slain.

Charles I. and Cromwell, however, are mere personages of yesterday in the history of Kingston. Ten hundred and fifty years ago it was the seat of Egbert's brilliant Witenagemot, a couple of generations before ever a king was crowned here. These coronation memories, however, are Kingston's great pride, and almost the only passages in her history of which any material memorial still remains. This memorial is, of course, the famous coronation stone, an irregular mass worn smooth and shiny by a thousand years of rain and friction. It stands finally now in the market-place, railed off in the reverent fashion common to chairs of state by a massive *grille* which tends greatly, no doubt, to its conservation. How many of our kings before the Conquest were crowned at Kingston, and that their consecration really took place upon this particular stone, tradition affords the only evidence. The genuineness of the stone is well enough authenticated for the ordinary believer who does not care to make himself miserable by a course of universal scepticism; but I believe there have been antiquaries (of course they were not born at Kingston) who have ventured to suggest that the evidence is insufficient. Tradition says that seven Saxon kings were certainly crowned here, and that probably others were. Here are the names of the seven, with the dates of their coronation, copied from the pedestal of the stone, with faithful adherence to the spelling affected by Mr. Freeman:—Eadweard, 901;

Adelstan, 924; Eadmund, 943; Eadred, 946; Eadwig, 955; Eadward, 975; Ædelred, 978. Kingston is a bright, cheerful little town, and the inhabitants seem to bear up well beneath the infliction of their terrible Town Hall, of which the sole tolerable points are some very good oaken carvings and some quaint armorial glass, all removed from the old Town Hall when it was demolished. Before the iconoclasts of Cromwellian days wreaked their evil will upon it, the parish church of Kingston must have been internally very interesting. There is reason to suppose that it was rich in brasses; but all that now remains of them are the blanks in the floor left by their removal. There are a few fine monuments, and one ancient brass of considerable interest is to be seen still. It commemorates Joan, the wife of Robert Skern, and her husband. The lady was a daughter of Edward III. and the frail Alice Perrers. After the coronation stone and the church, the only other "sight" of Kingston is the Norman chapel of St. Helen, for many years only a ruin, which is believed to be the successor of a still older building in which "Saint" Dunstan is reported to have crowned King Ethelred. The crooked streets of this old town,

THE ROYAL BARGE.

which disputes with Winchester the glory of having been the ancient capital of England, are made picturesque by many fine old red-brick houses of Jacobean and Georgian date. A generation ago there were standing a number of even older houses irregularly gabled, half-timbered, and barge-boarded; but they have either been demolished or re-fronted. Some of the shops, with painfully modern fronts, have low panelled interiors and carved staircases.

At Kingston the towing-path changes to the Surrey shore, and the river takes a bold sweep towards Teddington. Between Kingston Bridge and Teddington Lock the path is by no means picturesque; but the wooded beauty of the opposite bank diverts attention from the more homely shore. Right away to Teddington, and indeed beyond, is an almost uninterrupted succession of lawns and shrubberies and cool-timbered pleasure-grounds, surrounding pretty riparian villas. Life in summer in these cool veranda-girt pleasure-houses is idyllic. There you may enjoy tennis and boating, fishing and sailing, and drink your fill of admiration of the gaily-cushioned craft as they skim past with their lightsome burden of coquettish

"THE ANGLERS," TEDDINGTON.

muslins and gossamers. Nor is this river-strand to be despised when the winds of "chill October" have stripped the trees and left but bare branches, which look mournful and desolate to those who know only the Thames of sunshine and boating flannels. The stream runs brown, brimming, and turbid, as it swirls along laden with a burden of russet leaves. The angler, happily, can follow his museful sport in all seasons, and to him, as to other contemplative men, the river has attractions in autumn not smaller than those of summer, though different. The best English weather is a fine autumn, good alike for work and play, and the second

STRAWBERRY HILL.

half of autumn is a by no means inauspicious time for the Thames angler, for the river has ceased to be crowded with small craft, and the lumpy water suits better the fish then in season than if it were clear and limpid. A day's angling upon the bosom, or a long stroll by the banks of Thames, has many charms upon a sober October day; and a late autumnal sunset, with its glow fading from across the water and deepening into grey behind the bare poles of the trees, is a thing all of loveliness. Such a sunset, with the soft mist which clouds the banks directly afterwards, is well seen along this reach between Kingston and Teddington, where the thickly-wooded shores shut in the mist, and where the night seems to issue from the weird recesses of the woodland.

Teddington is but a couple of miles, as the river flows, from Kingston, and for the last half mile of the distance the murmur, one might almost say sometimes the roar, of the weir is audible. This same weir is the prime delight of the angler upon the more Cockneyfied portion of the river, and many is the patient piscator who perches himself thereon betimes, and sits at the receipt of finny custom until the gathering dusk renders the enterprise no longer profitable. The old-fashioned carp, that mysterious, long-lived fish which was once, like the peacock, an old English delicacy with which monastic fishponds swarmed, runs to a great size about Teddington Weir, while dace are almost as plentiful as minnows in a brook. Adjoining the weir is the lock, the first in the ascent and the last in the descent of the river. The lock and weir mark, to all intents and purposes, the spot, between sixty or seventy miles from the sea, at which the Thames ceases to be tidal. Henceforth the pilgrim, following the river on its way to the ocean, will see at low water, particularly between here and Kew Bridge, more mud-banks than he cares to count. At such times, too, the sense of smell will, at all events in hot weather, be found to have taken so keen a development that even chloride of lime would be accounted an odour sweeter than that given forth by the nude expanse of festering mud. At Teddington as yet there is happily little annoyance of this kind. To see the little of interest the village affords it is necessary to land at the ferry opposite the "Anglers," an old-fashioned inn which has long been popular with fishermen. At Teddington, be it remembered, is kept jealously locked up, in the custody of Mr. J. A. Messenger, the Royal Bargemaster, the State barge which has descended to her Majesty from early Jacobean times. In form it is graceful and elegant, and in the centre is a covered pavilion for shelter from the sun and rain. It is profusely gilded, and lavishly carved with mermaids and dolphins, while near the figure-head are emblazoned the coronet and plumes of the Princes of Wales, and the badge of the Garter. When he was at Hampton Court Charles I. delighted to spend an hour or two on summer evenings in this barge feeding the swans upon the river. It has not been used since 1849, when her Majesty rowed in state to open the Coal Exchange; but the public had an opportunity of seeing it in 1883, when it was shown at the Fisheries Exhibition. The village of Teddington lies away from the river, and stretches on westward to the gates of Bushey Park. At the head of the main street stands the parish church, a not unpicturesque amalgam of the new and the old. Its architectural interest is small, and the interior is whitewashed, but it contains the tombs of two or three notable people. Of these, "Peg" Woffington, the actress, is perhaps the best remembered. There is a marble monument to her memory which records that, "Near this monument lies the body of Margaret Woffington, spinster, born October 18th, 1720, who departed this life March 28th, 1760, aged thirty-nine years." She was buried in the grave of her infant nephew, Master Horace Cholmondeley, who had died seven years previously. At the end of her wayward career poor Peg could not have found a more peaceful resting-place. The oldest monument in the church is to Sir Orlando Bridgeman, who died in 1674. This descendant and ancestor of a long line of Orlandos was lord of the manor and a legal luminary. He was Charles I.'s Commissioner for the treaty of Uxbridge; and under Charles II. was Chief Justice of the Common Pleas and Lord Keeper of the Great Seal. When the church was being overhauled in 1833 the Bridgeman vault was opened, and Sir Orlando's body was found lying in a lidless coffin. So skilfully had the embalmer done his work that the remains were perfect, even the pointed Jacobean beard being untouched. An express was sent off to fetch the then Earl of Bradford, Sir Orlando's descendant, who thus had the strange privilege of looking upon the features of a progenitor who had been dead 159 years. There are two old and uninteresting brasses, and a tablet to the memory of John Walter, the founder of the *Times*, who died at Teddington. The churchyard is beautifully kept, full of trees and shrubs and climbing plants, which latter have grown luxuriantly over some of the older tombs. Here lie buried Paul Whitehead, the poet, minus his heart, deposited in the Despencer mausoleum at High Wycombe, whence it was most reprehensibly stolen; Richard Bentley, who shares with Walpole the guilt of designing Strawberry Hill; and "Plain Parson Hale," the friend of Pope, who was for more than fifty years the incumbent of the parish.

From Teddington Lock until close to Richmond the stream is undeniably less picturesque than in the reaches described earlier in this chapter. The river is less full of water, and when the tide is out the unsightly and unsavoury mud-banks are always in view. The towing-path becomes stony and arid; the hedgerows filled with poppies cease, and a very matter-of-fact embankment on the Surrey shore has to be reckoned with. Yet the reach between the lock and Eel Pie Island has always been popular, and often in the summer one may see here all sorts and conditions of notabilities disporting themselves at a little water-party. The spot is comparatively near to London, and your amateur boatman, with true wisdom, prefers not to get between two locks. We are coming now to classic ground, where wit and letters, fashion and frivolity, long have reigned. There is not another village in England with literary associations so numerous and august as Twickenham. Pope and Walpole are the presiding genii—neither of them, perhaps, the most genial of genii; but the fairy-like element is supplied by the hosts of feminine friends with whom the two bachelors were wont to philander. Whether as a letter-writer or as an architect, Walpole was vastly diverting; and it is a pity that so little of his brown stucco abode can be seen from the river. Strawberry Hill is the kind of place a mad architect might build in a delirium. We have a side not unlike the west front of Westminster Hall might have been had it been built of lath and plaster; then comes the keep of a Norman castle, flanked by a Renais-

POPE'S VILLA, TWICKENHAM.

sance *tourelle* from Chambord; the whole crowned with crow-stepped Flemish gables from Antwerp, and the twisted and fluted chimneys of a Tudor farmhouse. Then there are wings which aim at imitating these imitations; these, it is fair to say, are due to Walpole's successors. But howsoever astounding the exterior of Strawberry Hill, the interior is far more remarkable. Within, as without, the place bears every mark of having been built by a man who learned his architecture as he proceeded. Walpole leaped gaily over an anachronism, and saw nothing unorthodox in copying a mediæval tomb and fashioning it into a chimney-piece, nor in taking the choir stalls of Old St. Paul's as the model for the bookcases in his library. The internal arrangements of Strawberry Hill are as wonderful as the events recounted in that very Gothic story the "Castle of Otranto." It is a mighty maze without a plan. A long, narrow corridor, leading apparently to nothing, debouches upon a door which, when opened, discloses a large and splendid apartment. It is, indeed, a house of after-thoughts; but, whatever be its crudeness, it is not devoid of value as an early forerunner of the real Gothic revival. Pseudo-Gothic of this pattern was almost as popular towards the end of the eighteenth century as houses built in the guise of Greek temples. Happily, most of the examples have fallen down, but a few, such as that terrible "restoration" of Windsor Castle, still

TWICKENHAM FERRY.

remain. In literary and personal memories Strawberry Hill is far richer than many houses of greater antiquity and of real historical interest. Horace Walpole gathered all "the town" around him in these "enamelled meadows with filigree hedges;" and few are the great names of the last century and a quarter which have not some connection with "the castle," as its builder loved to call it. All the world's familiarity with this *chic* abode of Walpoles, Damers, and Waldegraves excuses me from dwelling lengthily upon its peculiar but undoubted charm. Frances, Countess Waldegrave, made it almost as fashionable as it had been in Walpolean times; and although the bulk of the contents of the house were sold after her death, it is pleasant to know that Baron de Stern, who became the proprietor in 1883, purchased much of the furniture, and that, to some extent at least, the historic continuity of olden associations has not been broken.

A little nearer to Richmond, and so happily placed that it commands the river from below Richmond Hill to Teddington Lock, is the modern and very *bizarre* successor of Pope's Villa. Only a specialist in architectural mania, or a member of the Société des Incohérents, could attempt a description of this astonishing building. It is said to have been erected by a tea merchant, and it certainly looks very much like a cross between a Chinese pagoda and a house of cards. Its lawns and shrubberies are very pretty, and after all there is something to be said for having all the river-side monstrosities gathered into one parish. The house does not occupy the exact site of the original Pope's Villa which, thanks to a too common lack of sentiment, was demolished long ago. The famous Grotto, one of the works of embellishment of the "little crooked thing that asks questions," still remains, but in a damp and mouldy condition, and despoiled of all that rendered it interesting. Pope had no great love for gimcrackeries, and we can in some measure imagine the tenor of the lines in which he would have immortalised the tea merchant could he have foreseen the change a century would bring about. The associations of Pope's Villa and gardens are primarily literary, even as those of Strawberry Hill, at a later day, were fashionable, frivolous, and dilettante. In Pope's time Twickenham was the centre of literary interest in England, and if the Jove who dwelt in this Olympus was querulous and stinging, his genius went a long way towards making lustrous an age in which taste and manners slept. Taste, at least, was still slumbering when Lady Howe considered herself justified in demolishing one of the most famous abodes that have ever been connected with our literary history. It is in the neighbourhood of Pope's Villa that the injury which has been done to the Thames by the mass of sewage sludge that has been so recklessly poured into it of late years first becomes noticeable. Although the effects of the tide are not much felt above Richmond Bridge, the condition of the river hereabouts at low water is lamentable. A broad edging of slimy ooze stretches for some distance from the bank on either side, and when the weather is really hot, and there is a drought of any considerable duration, as happened in the summer of 1884, the odour is hardly that of frankincense. The Thames Conservancy embankment between Twickenham and Richmond will no doubt improve matters somewhat; but it is, to say the least, melancholy that it should have become necessary to so disfigure the Surrey shore. Nor does the presence of unwieldy dredges in these reaches enhance the picturesqueness of the stream, while the new towing-path made with dried mud from the river-bed is an agency of martyrdom.

Behind Eel Pie Island—famous in the annals of angling and sweet in the memory of generations of picnickers—is seen the red tower of Twickenham parish church, architecturally much more interesting than the majority of Thames Valley churches. The ancient building fell down in 1713, and the fact that Sir Godfrey Kneller, who was at that time one of the churchwardens, had a hand in its rebuilding, albeit he was not the actual architect, may account for the excellence of the workmanship. The brickwork is almost as good as some of the best Tudor achievements in that line. Some famous and many interesting people lie buried here and in the churchyard. Pope's own tomb is hidden beneath the seats; but the marble monument which he erected to the memory of his father and mother, and in anticipatory commemoration of himself, is still to be seen on the east wall. In that part of the inscription which refers to himself Pope left blanks for his age and the date of his death; but such is the carelessness which prevails in such matters that these *lacunæ* have never been filled up. Kneller, the courtly painter of so many beauteous coquettes, is also buried in the church. Here, too, sleeps Admiral Byron, the author of the once popular "Narrative of the Loss of the *Wager*," irreverently described by his grandson as " My Granddad's Narrative." Kitty Clive, the charming actress who lived at Little Strawberry Hill, and for whom Walpole had a platonic attachment, is buried in the chancel. Naturally, in a classic village where many tremendous personages have dwelt, Twickenham is full of fine and interesting old houses, mainly of that square red-brick order of architecture which, if not precisely picturesque, is suggestive of comfort and homeliness. The old houses at Twickenham are of the sort in which Thackeray's people lived—still redolent of the charming but indescribable odour of the days of good, harmless Queen Anne. Perhaps the most interesting of them all is York House, immediately opposite Eel Pie Island, in which Anne herself and her sister Mary were born. Lord Chancellor Clarendon, Anne's somewhat plebeian grandsire, lived there, and it is one of the five or six houses in which he is said to have written that monstrous dull book, the "History of the Rebellion." For several years after their clandestine marriage the Duke of York—he who when king made so pitiful an ending of it— lived with Anne Hyde in this house, although it was undoubtedly called York long before then. In the second half of last century, Prince Stahremberg, Viennese Envoy Extraordinary, lived here, and achieved such fame as can therefrom result by a long succession of private theatricals in which a bevy of lovely and high-born dames took part. Orleans House likewise has royal associations, but of a somewhat melancholy kind, as memories of exile usually must be.

Twickenham might pleasantly detain us for a whole chapter; but the wooded slopes of Richmond rise beyond, and tempt us on to "Ham's umbrageous Walks," and the green meadows of Petersham. The river between Eel Pie Island and Richmond Bridge has a charm all its own, which owes much to the associations of the shores between which it flows. The meadows on the Surrey shore are sweet to look upon from the water; but until Ham is approached there is greater interest and variety upon the Middlesex bank. Ham, with its famous "Walks," lies low, and little of it can be seen from a boat. From the towing-path, however, there is to be had a very pleasant glimpse of Ham House, shaded, and, indeed, almost hidden, by splendid elms, some of which, against the pale which divides the grounds from the public path, throw in summer a cool and welcome shadow across the glaring footway. There is something solemnly picturesque about Ham House, as, indeed, there nearly always is about an old red-brick house closely surrounded by graceful, darkling elms. Ham is, in fact, so hemmed in by foliage that

RICHMOND: THE MEADOWS AND THE PARK.

it but narrowly escapes being gloomy. Horace Walpole, who was nothing if not cheerful, declaimed terribly against its dreariness; and to Queen Charlotte it appeared "truly melancholy." It is shut in and almost surrounded by high walls; but a good view of the front may be obtained from the towing-path through the handsome iron gates in the centre of a dwarf wall. These gates are said not to

RICHMOND: THE TERRACE FROM THE RIVER.

have been opened for many years, and the house itself has the appearance of being rarely lived in. Ham is a good example of very early Jacobean architecture. It has a longish front with a slightly projecting wing at each extremity, and is approached by avenues on almost every side. Few country houses in the neighbourhood of London are more interesting either historically or architecturally. Neither within nor without has any restoration been attempted, and there has been only such renovation as was imperatively necessary to prevent decay. It was built, it is said, for Henry Prince of Wales, elder brother of Charles I., who died a mere youth. The actual builder of the house was Sir Thomas Vavasour, King James's Marshal of the Household, and the belief that it was extended for the Prince of Wales is strengthened by the *Vivat Rex*, which, together with the date, 1610, is carved over the principal entrance. Since 1651 it has belonged to the Earls of Dysart. The first Earl of Dysart was, in Hibernian phrase, a Countess, Elizabeth, daughter of William Murray, one of the owners of Ham, and the wife of Sir Lionel Tollemache. After Sir Lionel's death, the Countess married John, Earl of Lauderdale, who, within three years of his marriage, was created Baron Petersham, Earl of Guildford, Marquis of March, and Duke of Lauderdale. John Maitland, the "L" of the Cabal Ministry, and his Duchess, were two of the most infamous creatures of the Restoration. The Duchess was bad because it was her nature to be so; her second husband, whose relations with her before marriage and in the lifetime of their respective first partners had been at the least compromising, was too weak and too easily swayed to withstand his wife's imperious ways. She openly sold the places in the Duke's gift, and it was the opinion of Burnet that she "would have stuck at nothing by which she might compass her ends." The "Cabal" constantly held their councils at Ham House, the Duchess of Lauderdale often being present to sharpen their flagging wits, exhausted by the concoction of shameful schemes for replenishing their own and their master's exchequers. The house was magnificently decorated and furnished out of the spoils of politics, and internally it remains very much as the Duchess left it—full of pictures, portraits, tapestries, rich cabinets of ivory and cedar. In one of these cabinets is preserved a crystal locket containing a lock of hair from the ill-fated head of Elizabeth's Earl of Essex. It had been proposed to assign Ham House as the residence of James II. after his enforced abdication; but that courageous person found it expedient to take himself off, and to weep on a kindlier shore.

It is not alone in summer, when the trees are thick with foliage, and the sun shines cheerily in at the windows, chasing away the memory of Lauderdale, Arlington, Ashley, and their fellows, that Ham House presents a striking appearance. It is the very type of house to make a winter picture, and there is nothing more characteristic of an English winter scene than this historic pile, looked upon from the river-side. The gaunt, bare elms, black against the dull sky, save where the snow has left some traces on less sheltered boughs, the frosted turf, the great iron gates and their tall piers, topped with an edging of snow which has gathered in the corners of the ironwork, a white drift banked up against the rusty hinges and the rarely-drawn bolt, glistening fleecy masses lodged above the door and on every projecting frame and cornice, a long roof hidden in snow, particles of which adhere even to the chimney-stacks—all this makes a picture which should be painted. The village of Ham, with its classic "Walks," is haunted by the towering shades of Pope and Swift, and the gentle ghosts of Thomson and Gay, an appropriate connecting-link between Twickenham and Richmond.

Of Petersham, which adjoins Ham Common, little or nothing can be seen from the river. It has a church which, although architecturally uninteresting, is crowded with old monuments, and with famous and notorious dust. The Duchess of Lauderdale herself was both married and buried here; but she has no monument. George Vancouver, the circumnavigator and the godfather of Vancouver Island, lies here; so do the Misses Agnes and Mary Berry—Walpole's "favourite Berrys." Not the least distinguished of men who have been buried at Petersham within recent years is Mortimer Collins, who is still missed from among the ranks of lighter English writers.

Past Ham the gleaming river winds through the Petersham meadows, with their wooded background of Richmond Park, and the broken, furzy ground near the "Star and Garter." Here the Thames becomes as lovely as it is between Hampton Court and Kingston. The banks are profusely timbered, and a bushy little eyot in the centre of the stream adds to the charm of the view. Where the towing-path for a time ends, near Kew Foot Lane, there begins on each shore an irregular line of water-side villas buried in lilac and laburnum, surrounded by smooth lawns with edgings of geometrical flower-beds, those on the Middlesex side still in Twickenham, which extends quite up to Richmond Bridge. On the Richmond, or rather the Petersham shore, tower high up on the verdant bluff the towers and pinnacles of the "Star and Garter," looking in the distance not unlike a French villa on the heights of St. Cloud. Beyond is Richmond Hill, with its leafy Terrace, adding much to the foreign impress of the scene. The boldness with which "thy hill, delightful Sheen," rises up almost sheer from the water, recalls some more glorified Namur. There is a brightness and a vivacity about this little suburb of St. James's which are rare to find in this stolid island. It is a hard climb up the lovely lane from the river-side to the portals of the "Star and Garter," and the gates of Richmond Park. It is a sweet and toothsome spot this site of the "Star and Garter," which recalls cycles of flirtations and memories of iced champagne. "A little dinner at Richmond" is a heading very familiar to the persevering novel reader, and the scene of these pleasant symposia is of course always this aristocratic hotel at the top of the hill. The delights of Richmond Park, on the opposite side of the road, are of another order. Here we have a vast pleasure-ground, the nearest of its kind to London save Epping Forest. If anything, it is lovelier than Epping, since it has been better cared for, and there has been none of that reckless destruction which has so much marred the forest glades of Essex. Close by the Richmond gate is one of the sweetest bits in this thickly-wooded domain—the old Deer Park. A steep slope, green and timbered, divides this from the higher and more public portions of the park. In this undulating preserve, dotted with stately oaks, is kept a large herd of fallow-deer, tame almost to temerity. The old Deer Park has some retired nooks and lonely glades in which one may surprise the dreamy deer sheltering themselves upon a glaring day beneath the wide branches of the ancient oaks, up to the barrel in fern and bracken and bramble. Scattered here and there are plantations new and old, full of larch and fir as of more stately forest trees, in which abound game of all sorts, but more especially the hopping rabbit and the skimming hare.

The gates of the park open upon the very extremity of the Hill, close to Mansfield House, now an hotel, but formerly a residence of the Lord Chancellor of that name, who once had a redoubtable encounter with the mob, and to Wick House, where sometime lived Sir Joshua Reynolds. Immediately beyond is the Terrace, an umbrageous promenade, dear in morning hours to nurses and their lively charges, and later in the day somewhat of a Rotten Row or a ladies' furlong. From his seat beneath the trees the gazer looks down away to the west upon one of the most lovely sights that earth affords. Between edgings of quivering green, of lighter and of deeper hues, winds a glistening silver ribbon, in and out among villas and townlets, always narrowing, and when the limit of vision is reached, it seems to the straining eye as though meadow and woodland met and stayed all further passage. From this eminence the country lies mapped out as though seen from a balloon; and far away beyond all trace of the river there closes in a dark and swaying mass of foliage, like to one dense land of trees. A thin line of brilliant blossom marks in early summer the Chestnut Avenue in Bushey Park; and were it not that here and there the sun catches a high church tower, a gilded vane, or mayhap a turret of feudal Windsor, it would seem from here above that this wide and lovely stretch of country was still a vast untrodden forest. So great is the height that even immediately below the long, comfortable pleasure-boats loom tiny as toys, while the steamers, happily rarer in these waters than they are lower down, become almost picturesque in the distance. All this gleaming valley, stretching across to the west, is the way we have come; these are the woodlands past which we have rowed, and beneath whose shadow we have rested; those the bends and reaches where we have done some honest, straightforward pulling; those the shaded lawns where we have longed for tennis, the rather for the sakes of the muslin goddesses we furtively watched than for pure love of that Olympian game. That which remains of our course shoots us past the skirts of Richmond town, the river bank coloured and diversified with houses old and brown, and houses new and white, with here and there yet other lawns. One more embowered eyot comes round which the waters, parting, gently swirl. Here lie anchored a lazy barge or two, and as we plash past them into mid-stream again we are full in face of Richmond Bridge, grey, many-arched, and slightly bowed. Beyond the bridge there rises a chain of swelling uplands, all massy with foliage, and dotted with the red and white of lotus-eating villas.

J. PENDEREL-BRODHURST.

RICHMOND BRIDGE.

CHAPTER IX.

RICHMOND TO BATTERSEA.

The River at Richmond—A Spot for a Holiday—The Old Palace of Shoen—The Trumpeters' House—Old Sad Memories—Richmond Green—The Church—Kean's Grave—Water Supply—The Bridge—The Nunnery of Sion and Convent of Sheen—Sir William Temple—Kew Observatory, Isleworth—Sion House and its History—Kew Palace and the Georges—Kew Gardens—Kew Green—Brentford—Mortlake—Barnes—Chiswick—The Boat-race—Hammersmith—Putney—Barn Elms—Putney and Fulham—The Bishops of London—Hurlingham—The Approach to a Great City.

IT would be easy to find spots on the Thames where the natural features—the wood-clad slope, the grassy sward, and the gliding stream—were associated in equal beauty, but it would be difficult to meet with any more picturesque combination of these with the dwellings of men than may be seen on the river reach just below Richmond Bridge. The light-grey arches, through which the Thames flows ripplingly, are backed by the groves of Richmond Hill. On the one hand are the shady gardens and villas that now thickly stud the meadow plain of Twickenham; on the other, the houses of the town, after thronging down to the waterside at the entrance of the bridge, give place to statelier mansions with ample pleasure-grounds. It was doubtless a more imposing sight when the façade of the old palace of Sheen, which these mansions have replaced, overlooked the margin of the Thames; but it can have hardly been more picturesque than now. The low iron railing allows the eye to wander from the path by the riverside to the green pastures and lawns overshadowed by fine trees, to the old-fashioned façade of the "Trumpeters' House," while the more ambitious semi-classical design of Asgill House on the one hand, and of Queensberry House on the other, in closer proximity to the river, give an irregularity to the grouping, and perhaps accord better with the neighbouring town than the unbroken front of the Tudor palace is likely to have done. In its days, also, there were no bridges, and though the railway viaduct below might well be spared, the stone bridge must have improved a view of this kind. At any rate, this reach of the Thames is classic in art and literature; it has engaged the pencil of Turner, and is full of memories of Pope and Gay and Thomson.

The space between the two bridges seems to invite the traveller to linger. Fresh, perchance, from the streets of London, the odours of the underground railway still in his nostrils, the vapour of its smoky streets still lingering in his lungs, a little heated, it may be, and still mindful of towns in his walk from the railway station through Richmond streets to the bridge, he walks rapidly for a brief space along the towing-path, and then perforce halts in another and a new land. If, fortunately, he has chosen a day still early in the summer, before the average Londoner has quite realised that it is time to begin to take "an outing;" if he has arrived on the spot at an hour when the rowdy element, still, unhappily, too prominent among the dwellers in our metropolis, has not yet broken the peace of the Thames by those simian howlings or that loud-voiced blasphemy, which is deemed expressive of pleasure, then he will find it hard to detach himself from this reach of the river, and will imitate the elders of Richmond, two or three of whom he will generally find sunning or shading themselves on the benches near the waterside. Is it not enough to watch the trees almost dipping their branches in the stream, to find excitement in the hovering of a fly over its surface, in the splash of a fish, or even in the tiny swirls of the stream itself? In this dreamy calm which steals so quickly over us we watch the hovering butterfly or the flitting dragon-fly, the little dramas of animal life, their comedy or their tragedy, with an interest that causes the graver issues which we left behind barely an hour ago to fade from the mind. What more do we need after the noise of the streets than this perfect calm of the air, undisturbed save by the faint rustle of the breeze among the leaves, or the twittering note of a bird? What more, after the dusty pavements, than these glimpses of green lawn, of summer flowers in garden-beds or pendant from wall and trellis; of shadowy walks under green trees of Britain or darkling cedars from Lebanon? As we dream, memories rise of a dead-and-gone past—of many an episode of English history which is connected with this little reach of our river, perhaps the most classic ground on the Thames outside the precincts of the metropolis. Kings and queens, many a lord and lady of high degree, many a man on whom genius has conferred a place in history which birth alone cannot give, have loved to linger along this bit of Thames-side, or to float idly on its stream.

> "Their mirth is sped; their gravest theme
> Sleeps with the things that cease to be;
> Their longest life, a morning gleam;
> A bubble bursting on the stream,
> Then swept to Time's unfathomed sea."—KENYON.

But let us awaken from our reveries to dwell more particularly upon the memories called up by the Thames below Richmond Bridge.

Once on a time the pride of Richmond and the glory of this part of the river was its royal palace—a favourite residence of several of the kings of England. There is some uncertainty as to when the manor of Sheen—for that was its earlier name—came into the hands of the Crown; but the first royal owner of the entire estate appears to have been Edward III. He also is said to have been the builder of the palace, although there must have been a residence on its site in the days of both his father and grandfather. Here, in fact, his long reign came to its melancholy end. Within the walls of Sheen he lay, robbed and deserted by courtiers and favourites, tended only by a "poor priest in the house," who found the dying monarch absolutely alone, and spoke words of exhortation and hope to soothe the parting struggle. His body was conveyed from Sheen by his four sons and other lords, and solemnly interred within Westminster Abbey. His grandson and successor lived here for a time, but on the death of his queen, Anne of Bohemia, within its walls, took such a hatred to the palace "that he, besides cursing the place where she died, did also for anger throwe down the buildings, unto where former kings being wearie of the citie, were wont for pleasure to resort." Henry V. rebuilt the palace, erecting a "delightful mansion of curious and costly workmanship." It was a favourite residence of Edward IV., and was held in equal regard by Henry VII. In his days, however, it suffered twice from fires, the first one destroying a considerable portion of the older buildings. Henry rebuilt the injured part, and altered the name from Sheen to Richmond. It is also interesting to learn that architects made mistakes or builders scamped their work even in the courts of Tudor kings, and at peril, as one would have thought, of their ears, if not of their necks; for shortly after the second fire a new gallery, on which the king and his son Prince Arthur had been walking a short time previously, fell down, fortunately without injuring any one. Richmond Palace was the scene of many of the principal festivities in this king's reign; much also of his accumulated treasure was hoarded within its walls. His successor, the much-married monarch, came frequently here in the earlier days of his reign, but the palace fell out of favour in the later, and was the country residence of his divorced wife, Anne of Cleves. Elizabeth, however, greatly liked it, and her last days were spent under its roof. She came from Chelsea to Richmond, in the month of January, sickening of the disease that caused her end, and overcome with melancholy for the death of Essex. She refused to take food or medicine or rest; she would not go to bed, but sat on cushions piled on the floor; a melancholy picture of distress, but with the old spirit left, as when she flashed out upon Cecil for having inadvertently used the words " she *must* go to bed."

The palace was an occasional abode of James, her successor, and his queen, and the residence of their eldest son, the accomplished Prince Henry, " England's darling." Here he died, amid universal lamentation, and his brother succeeded to the expectation of a crown, and the ultimate doom of the headsman's axe. Prince Henry would hardly have pulled down " Bishops and Bells," but his brother secured their downfall and his own by trying unduly to exalt them. Prince Charles, after an interval of some three years, took up his abode at the palace, and Richmond once more awoke, for the new Prince of Wales scattered his money—or rather the nation's money—royally while he played the fool with " Steenie," Duke of Buckingham. After he assumed the crown his visits here became less frequent, and after his execution it was ordered by Act of Parliament that the valuable contents of the palace should be sold. Though inhabited from time to time after the Restoration, it never returned to its former greatness. Much of the building was destroyed before the end of the seventeenth century, and now only a few fragments remain. Old pictures and documents enable us to form a good idea of its ancient splendour, and a right noble and picturesque structure it must have been. On the north side it looked on to Richmond Green, where now may be seen the remnant of its ancient gatehouse; on the south-west it came down to the margin of the Thames. A narrow lane, leading from the Green to the riverside, and emerging opposite to the noble old elm which forms so marked a feature in the view from the river, passes across the site of the court of the ancient palace, and, doubtless, over the foundations of its principal buildings. Roughly speaking, the site of the river façade is now occupied by the three mansions already mentioned, which themselves, as we shall show, are not without a history. The principal of these lies far back from the river; a lovely garden, shaded with trees, intervenes. Judging from the drawings, the buildings of the palace approached near to the waterside, and the space between had rather a barren and desolate look, as though it were left in the rough as a mere foreshore. Now the gardens make the passer-by long to trespass. The owners, however, beneficently (or is it to secure a good view of the river?) keep their boundary fences low. Building on the site of the old palace seems to have begun quite early in the last century. The heavy, but stately red-brick house, with a stone portico, which we have already mentioned, was erected by a Mr. Richard Hill, brother of Mrs. Masham, the well-known favourite of Queen Anne. It bears the name of the Trumpeters' House, from two statues of figures blowing trumpets, which once adorned the façade. The more modern mansion, nearer to both the bridge and the river, stands on the site of the villa occupied by a noted character in the last century, the Duke of Queensbury—commonly known as old Q.—one of the least virtuous and respectable members of the aristocracy in a not too virtuous age. There is a characteristic story quoted by an historian of Richmond which carries its own moral. Wilberforce, when a young man, was invited to dine at the Richmond mansion. "The dinner was sumptuous, the views from the villa quite enchanting, and the Thames in all its glory; but the Duke looked on with indifference. 'What is there,' he said, 'to make so much of in the Thames? I am quite weary of it: there it goes, flow, flow, flow, always the same.'" In his old age he deserted Richmond, having taken offence, it is said, at the inhabitants. There is an open space between the towing-path and the railings of the ducal villa that the Duke

BETWEEN RICHMOND AND KEW.

enclosed and converted to his own use, trusting that fear of his rank, desire to retain his custom, and gratitude for his benefactions—for he was no niggard—would combine to secure the acquiescence of the inhabitants. But these motives proved insufficient; the town commenced an action, which was, of course, successful, and the Duke, deeming its inhabitants ingrates, withdrew to London. There he found occupation in such pleasures as money and rank could purchase—and these could bring more a century since than now. When he became too infirm to move about, he sat on his balcony under a parasol, to ogle the pretty women as they passed by, and died with his bed-quilt strewn with *billets-doux*, which his enfeebled hands could not open—*vanitas vanitatum*. No spot on the Thames, save Hampton Court, is so rich in historic memories as this delightful bit of the river below Richmond Bridge. In the still evening air, as the glow is fading from the west, as the riverside becomes deserted, and the toilers and pleasurers have alike gone home, the ghosts of old times come back, and the actualities of the nineteenth century fade away into the shadows of the past. Tender and pleasant memories would be most in harmony with the scene, and these are by no means absent; the sounds of music and dance are not wanting from the stately walls which our fancy conjures up, nor from the gilded boats which seem to float along the stream. Yet still the more prominent are sad, lurid evenings, presaging coming storm—Edward dying in solitude and dishonour, to leave his kingdom to a feeble fool; Elizabeth, in her overshadowed youth, quitting the palace *tanquam ovis*—to quote her own words; and again, when the brightness of life had passed, dying slowly there, her last hours darkened by many sorrows, not the least being the thought of the unworthy pedant who would take up her sceptre; the parting agony of his eldest son, making way for one whose very virtues were his bane, and whose memory was only redeemed by the mistaken necessity of his execution. These are thoughts tragic enough to darken the recollections of the palace of Richmond, and cast some shadow on a scene which is one of the fairest in the neighbourhood of the metropolis.

We must not, however, pass on from this spot without turning aside from the river to give one glance at Richmond Green, on which the northern front of the palace formerly looked. It is a noble expanse of grass, surrounded by trees, some, probably, survivors of the old elms which bordered it in the days when the Parliamentary commissioners made their visitation; others of more recent planting. Here, where once jousts and tournaments were held—sometimes with fatal result—the lads play at cricket, and the old folk saunter under the shadows of the branches. The enclosing road is bordered with houses of every date, from Queen Anne to Queen Victoria, and among some of the former stands the chief remnant of the palace of the Tudors. It is the gateway of Henry VII.'s structure, a plain four-centred, depressed arch, over which is a mouldering stone still bearing the royal arms. The adjoining house, though modernised, is a part of the ancient façade, and contains a fine old staircase; and the buildings running backward on one side of the courtyard still retain in their walls pieces of the ancient brickwork; and some of their rooms are of interest. One, indeed, is commonly pointed out as that in which Queen Elizabeth died; but the tradition is unworthy of credence. The modern "Queen Annist" will take much pleasure in the contemplation of Maid of Honour Row, a line of houses erected early in the last century. On the stage of the theatre which stands by the Green the best actors of London often appeared, and it is noted as the place where Edmund Kean, stricken by fatal illness while playing the part of Othello, sank into the arms of his son Charles, who was acting Iago. He died shortly afterwards in a small room in an adjoining house, and is buried in the churchyard.

Richmond Church is not without interest, though it is without beauty. There is a much-battered low stone tower, and a body, which dates chiefly from the last century, built of brick, in what may be called the Hanoverian style. It is, however, in good repair, and in excellent order within, and is, at any rate, of more interest than many feeble modern imitations of mediæval work. Several men of note have been buried within its walls, or in its churchyard. Among them is the noted Gilbert Wakefield, sometime vicar of Richmond, one of the victims of the reactionary terror inspired by the French Revolution. James Thomson, the poet, also lies within the walls. Besides these are members of the Fitzwilliam family, who had their residence near the Green, among them being the Earl who enriched the University of Cambridge with a fine collection of paintings and drawings. Lady Di Beauclerk, the friend of Dr. Johnson, with Dr. Moore, the author, father of the hero of Corunna; Mrs. Barbara Holland, also among the well-nigh forgotten names of literature; and many actors besides Edmund Kean have been laid to rest in the precincts of Richmond old church. The increase of the town has caused the building of two other churches, and the institution of a cemetery.

Richmond, though so near abundance of water, has sometimes been in danger, like the ancient mariner, of being without a drop to drink. To use the Thames is, of course, impossible, the present age objecting to dilute sewage; and the supply from other sources has not always been sufficient. A few years ago an attempt was made to obtain a supply from the porous beds which, in most parts of England, succeed to the stiff blue clay underlying the chalk. The result was more interesting to geologists than satisfactory to the ratepayers. As is the case beneath London, this porous stratum was found to be wanting, an upland mass of more ancient rock having evidently interrupted the sea beneath the whole area now occupied by the London district, and the boring tools pierced for nearly seventy yards through more ancient beds, till at last the unprofitable task had to be abandoned.

The stone bridge, which we have already mentioned, is a comparatively modern institution. The Act for its erection was obtained in the year 1773, and prior to that the Thames had to be crossed in a boat. Local chronicles tell us that there was much disputation and some heart-burnings before the site was determined. The design is good, and the light grey of the stone contrasts well with the verdure of the trees and the darkling water of the Thames. The railway bridge—an iron structure—is a doubtful addition to the scenery of the river; like many another institution of modern times, a railway is of unquestionable utility, but the less we see of it the better. However, we may honestly say of this that it might easily be a greater eyesore. Beyond it houses of a substantial character, and their pleasant gardens, continue to border the left bank of the river, but on the right bank the scene quickly changes, and we could fancy ourselves dozens of miles away from the metropolis. The slope of the elevated plateau, which forms so marked a feature from Richmond Bridge, and is climbed in part by the town, has now trended inland. The Thames has struck out for the middle of the shallow valley, along which its present course meanders, and is bordered now on either side by an alluvial plain. This, on the right bank—the inner side of the curve formed by the stream—is occupied by the extensive property belonging to the Crown of England, of which the

SION HOUSE.

more northern portion—that known as Kew Gardens—is the more familiar to the London public.

We come first to the more secluded part—the old deer park, a great expanse of meadow-land, dotted, often thickly, by groups of fine trees. This was an appendage of the ancient palace of Richmond, or Sheen, of which mention has already been made. It is separated from the towing-path and causeway, which runs along the riverside, by a shallow ditch or canal, speckled, in the early summer-tide, with the flowers of the water ranunculus. From the stream, or, better still, from this causeway, we enjoy the beauty of the great grassy plain, and its scent as it withers in due season into hay, beneath the heat of the summer sun, and the ever-new grouping of the trees, which thus obviate the possible monotony of meadow scenery. The only building that for some time arrests the eye—if we except a pair of small stone obelisks—is the white house occupied by the Kew Observatory—a not unpleasing structure, which we shall presently notice more in detail. Not far from this, and at the same time no great distance from the old palace of Sheen, was the Carthusian convent, founded some four and a half centuries since by Henry V.

As report has it, the king was much disquieted in his conscience as to the mode in which his father had gained the crown by the deposition and death of Richard II., and as a peace-offering founded, in the year 1414, the Convent of Sheen and the Nunnery of Sion, on the opposite side of the Thames. It was incorporated under the name of the House of Jesus at Sheen, and the rules ordered that when the devotions at the one convent ceased those at the other should begin. These foundations are recorded in the speech assigned by Shakespeare to Henry prior to the Battle of Agincourt:—

> "I have built
> Two chantries, where the sad and solemn priests
> Still sing for Richard's soul."

Royally endowed, and pleasantly situated, the monastery of Sheen is not without its place in history. The Prior of Sheen was powerful enough to avert for a time, by his intercession, the due penalty of death from the pretender Perkin Warbeck, who escaped from the Tower and sought refuge within the walls of the monastery. But a second attempt to break prison brought him to the scaffold. To Sheen the body of James IV. of Scotland was brought from the field of Flodden for burial, a purpose which seems to have been unfulfilled, for the corpse, wrapt in lead, was seen lying in a lumber-room some years after the suppression of the monastery. Hard by its walls the noted Dean Colet, founder of St. Paul's School, built himself a small house, to which, for a time, the great Cardinal Wolsey retreated after his disgrace. After the suppression of the monastery its buildings became the possession of more than one noble family in succession. According to Spelman, a curse was upon it, for in less than a century and a half it went to nine owners, never once descending from father to son. It witnessed the marriage of Sir Robert Dudley and Amy Robsart, and the childhood of Lady Jane Grey. In the reign of Mary the monks came back for a brief season, but they had again to cross the seas when Elizabeth reigned. After many changes, demolitions, and additions, a part of the

THE RIVER AT KEW.

monastery, or a residence upon a portion of its site, was occupied by Sir William Temple, the well-known statesman, and its gardens became the scene of his experiments to discover "how a succession of cherries may be compassed from May to Michaelmas, and how the riches of Sheen vines may be improved by half-a-dozen sorts which are not known here."

Evelyn, who incidentally notices that the abbey precincts had become divided between "several pretty villas and fine gardens of the most excellent fruits," remarks on the excellence of Sir W. Temple's orangery, and the perfect training of the wall-fruit trees in his garden. Here the retired statesman, away from the hurly-burly of politics, meditated on horticulture and indited epistles, while King James was vainly grasping at a tyranny, and the Prince of Orange was marching eastward from Torbay. When that prince became king he visited Temple at Sheen, but after this time the latter chiefly resided at Moor Park; so that it is with this place rather than Sheen that the memory of his young secretary, afterwards the noted Dean of St. Patrick, is associated. All traces of the ancient monastery have now disappeared, the last remains having been destroyed about the year 1769.

The "Kew Observatory," which we have already noticed, stands isolated among the broad meadows, away from dust and noise, and from the reverberation of traffic. It was built for George III. by Sir William Chambers, "for the purpose of studying astronomical science with special reference to the transit of Venus." But after a time its activity declined, and for many years "Kew may be said to have quietly glided into a long winter of hibernation, being under the careful guardianship of a curator and reader." Attention was once directed towards it in a painful way, a double murder having been committed by the janitor, a man named Little, previously much respected in the neighbourhood, who was arrested in the house of his victims and duly executed in the year 1795. The observatory was suppressed by Sir Robert Peel, and the building offered to the Royal Society. That body, however, refused to undertake the charge, as it did not possess any funds applicable for the purpose, but by means of the subscriptions of various men of science, and a grant from the British Association, an observatory for meteorological and electrical observations was established under the charge of a committee, at which much important work was done. In the year 1871 the grant of the British Association was withdrawn, and a sum of £10,000 was placed in the hands of the Royal Society by Mr. Gassiot, for the maintenance of magnetic observations at Kew. It is now under the charge of a committee termed the Meteorological Council, and is the central establishment for observations relating to meteorology. Here instruments relating to this science are tested and marked, and a large amount of most valuable work is executed.

After the little town of Isleworth—which lies on a concave bend of the river, under the lee of some low islands—has been passed, the Thames is bordered on either side by a park, and, for the last time on its course to London, almost shakes itself free of the grasp of the builder. On the one hand the Royal Gardens at Kew succeed to the Old Deer Park; on the other lies the ample domain of the Duke of Northumberland, whose large but ugly house becomes a conspicuous—a too conspicuous—feature in the landscape. We may turn aside for a moment from the property of the Crown to notice that of the house of Percy, once hardly less potent. Sion House, which is separated from the river by a level expanse of meadow, interrupted only by one low mound, which supports some fine old cedars, occupies the site of the second chantry founded by Henry V., as a peace-offering for the sins of his father. The dedication-stone was laid by the king in the year 1416. It was a convent for both sexes, though the nuns predominated, these being sixty in number, while the brethren were only twenty-five. But they were entirely separated, a thick screen dividing them even in the chapel, where, indeed, both sexes were seldom present at the same time, so that all occasion of scandal was carefully avoided. The convent was endowed with the manor of Isleworth, and at a later time received many grants of property which had belonged to the alien priories. Finding the original buildings too small, the society obtained licence to raise themselves a new convent, on the site now occupied by the Duke's mansion, which is rather to the east of that on which the king built. Life glided by smoothly and uneventfully for the inmates; they became rich, and lived easily, but harmlessly, until the crash of the Reformation came. For some reason or other they had incurred the special displeasure of the king, and were accused of harbouring his enemies, and being in collusion with the Holy Maid of Kent. One of the monks, together with the Vicar of Isleworth, was executed at Tyburn. The lands were distributed, but the house and the adjoining property were retained by the Crown, the nuns retreating to Flanders. For a brief space, indeed, in the reign of Queen Mary, they returned to their old home, but on her death again became exiles. The society still continued to exist, though for a while its members suffered great poverty; but at last they settled "in a new Sion on the banks of the Tagus, at Lisbon, in the year 1594. Here they still remain, after the lapse of nearly three centuries, restricting their membership entirely to English sisters, and still retaining the keys of their old home, in the hope, never yet abandoned by them, of eventually returning to it." It is said that some half century or more ago, when they were visited at Lisbon by the then Duke of Northumberland, they told his Grace the story of having carried their keys with them through all their changes of fortune and abode, and that they were still in hopes of seeing their English home again. "But," quietly remarked his Grace, "the locks have been altered since those keys were in use"—a reply which, whether intended or not, had much significance in it. There are many good people in the world who cling tenaciously to the keys which were fabricated by the worthies of olden time, forgetful that the locks have been altered, so that their binding and loosing power is gone.

Doubtless the nuns made their own comments when in later years a gruesome story came from England to them in their new home across the sea. The coffin of Henry VIII., on its journey towards Windsor, was laid for a night within the convent walls; there the bloated corpse within burst, and the blood dropped on the pavement, so that it was licked up by the dogs, as that of the King of Israel in the streets of Samaria. A few months later the convent was granted to the Lord Protector Somerset, who began the building of the present mansion, and when he fell on the scaffold, it was given to Dudley, Duke of Northumberland. There was a curse upon it. Lord Guildford Dudley had it for his home, and from its door he led the Lady Jane, his wife, to the Tower, to claim the throne of England, and receive at last the stroke of the headsman's axe. Elizabeth granted it to Henry Percy, Earl of Northumberland, but he was no exception to the ill-luck of his house, for he was afterwards convicted of being an accomplice in the Gunpowder Plot, disgraced, heavily fined, and imprisoned in the Tower. His son, the tenth earl, repaired the house, and from beneath its roof the children of the ill-fated King Charles were conducted to St. James's Palace to bid their father a last farewell. One of them—Charles II.—held his court here during the Great Plague, and royalty has more than once in later times been a guest within the walls of Sion House. The mansion retains the general outline of the Lord Protector's building, though it has been modernised, and probably made uglier. It is a bleak-looking structure, faced with grey stone, quadrangular in plan, as we note in passing, with embattled square towers at the corner. The principal façade is relieved by an arched terrace, and over the central bay now stands the lion with outstretched tail, once so conspicuous on Northumberland House in the Strand. The gardens and grounds, laid out in the style of the last century, are fine, the plant-houses being especially noted—in fact, they "may be said to be no mean rivals to Kew."

But to these we must return, for the open meads of the Old Deer Park have now been replaced by the groves of Kew. First come the wilder portion of the royal gardens, devoted more especially to forest trees, scenes of sylvan beauty and quiet solitude, as few of the visitors find their way hither, but remain in the more highly cultivated portion, among the plant-houses and the gay parterres, in the neighbourhood of the Richmond road. Yet a more delightful spot for a ramble cannot easily be found; the great trees, feathering down to the sward, cast cool shadows in the summer heat; the long pool here glitters in the light, here lies still and dark beneath overhanging foliage. In due season many a flowering shrub adds a new and more striking diversity to the varied tints of verdure, and the water-lilies expand their cups of gold and silver among their broad floating leaves; the fowl float idly by; the birds twitter among the branches; among the scents of springing grass and of opening flowers, amid the flickering lights and shadows, and the peace of the forest, the roar of London streets dies away from the wearied ears, and the smoke of the town is forgotten in the savour of the pure air.

From the river bank we obtain glimpses from time to time of the glittering roof of the great palm-house, of the various buildings devoted to botanical science, and of the tall pagoda. The history of these gardens, now so great a boon to the dwellers in London, must be briefly sketched, for they are inseparable from the Royal River; and the site of the palace, for a time a favourite residence of kings and princes of England, is but a short distance from its bank, although the walled enclosure prevents so free a view of this as of the other parts of the gardens. The building which now bears the name of "the palace" was, in the earlier part of the last century, called the Old Dutch House. It is a red-brick structure, heavy in style, but not unpleasing, dating from the reign of James I., and probably built by Sir Hugh Portman, a wealthy merchant. Kew House, or "the palace," as it was often called, stood a little more than a hundred yards away, and was obtained on a lease by Caroline, wife of George II., and afterwards purchased by Queen Charlotte. It became the country residence of Frederick, Prince of Wales, and after his death was inhabited by his widow. Here was spent much of the early life of the young prince, afterwards George III. Brought up in the strictest seclusion, jealously guarded by his mother and her favourite, Lord Bute, he received an education which cramped his faculties and in many respects disqualified him for his future lot. "The king lamented, not without pathos, in his after-life that his education had been neglected. He was a dull lad, brought up by narrow-minded people . . . . like other dull men the king was all his life suspicious of superior people. He did not like Fox, he did not like Reynolds, he did not like Nelson, Chatham, Burke; he was testy at the idea of all innovations, and suspicious of all innovators. He loved mediocrities."

Here, then, the young princes and princesses, after the death of their weakly and insignificant father, grew up under the guardianship of their stern and unloving mother, assisted, as every one will remember, by Lord Bute, who, at one time, had a fair claim to the title of the most unpopular man in England. It was with him that Prince George was riding when the note was put into his hand which apprised him that an end had come to his grandfather's pleasuring, and that he was king, and at Kew Palace he remained till the following morning. During the first twenty or thirty years of his reign at least three or four months of every year were spent at Kew, where, as has been said, he "played Darby and Joan" with Queen Charlotte, and the young princes and princesses amused themselves like other children in the gardens. The great contrast between the mode of life of this and the preceding reigns was not altogether to the liking of the people; the rarity with which the king appeared in public, or entertained the members of the "upper ten," the infrequency of state ceremonials, gave some colour to the accusation that he affected an Oriental seclusion, and was aiming at establishing a despotism. It had also, in all probability, another ill effect, that the king and queen lost their social influence over the aristocracy, and by standing aside from their position of the leaders did not exert upon its members that influence which would naturally have resulted from the purity and simplicity of their own lives. Certainly, society at large continued hardly less corrupt under the young king, whose reputation was spotless, than it had been when his grandfather kept court with Walmoden and Yarmouth. It was in Kew Palace also that the unfortunate king was secluded during the first attack of that mental disorder which afterwards permanently darkened his life.

The original Kew House was eventually pulled down by George III., who commenced the building of a much larger palace in its neighbourhood. This, which, so far as we can judge from prints, promised to be as ugly as are most structures of that period, was an incomplete shell when the king died, and was happily destroyed by his successor. The Dutch House, however, which was inhabited by the old king, and in a room of which Queen Charlotte died, still remains, though now almost unfurnished and unused. Here, also, in the drawing-room, were celebrated the marriages of two royal dukes, Clarence and Kent, the latter the father of the present Queen.

One other memory also lingers about the precincts of Kew. On the lawn, perhaps a furlong from the old palace, we may notice a sun-dial. This marks the

THE PAGODA IN KEW GARDENS.

site of a little observatory, wherein, in the year 1725, before the house became a royal residence, James Bradley made the first observations which led to his two important discoveries, that of the aberration of light and of the nutation of the earth's axis. This sun-dial, together with a memorial tablet, was erected by William IV.

Tempting as the "Royal Gardens" appear from the river, the promise is more than fulfilled on closer inspection. Places devoted to the pursuit of science are apt, in their studious severity, to be somewhat repellent to the uninitiated; and even a botanic garden, beautiful as some of its contents must always be, is occasionally no exception to the rule. But this is not the case at Kew. There, indeed, work is not sacrificed to pleasure. Its arrangements are scientific and precise enough to satisfy the most exacting. Its museums and laboratories afford opportunity for the severest study; but yet, in many parts of the garden, nature is so happily blended with art, apparent freedom of growth and association with scientific order and exactness, that as we wander over its lawns, or linger beneath the shadow of its stately trees, we can abandon ourselves simply to the beauties of nature, and "consider the lilies of the field" without counting their stamens or their pistils. The glass-houses, open during certain hours of the day, are often bright with exotic flowers;

KEW BRIDGE.

in the great tank the huge lily from South American rivers opens its blossoms among pond-flowers from sunnier regions than our own. The palm-house enables the home-staying Briton to form some conception of the verdure of a tropical forest. For those who love the formal style of gardening, trim parterres bright with many colours, there is satisfaction on the terraces by the side of the ornamental water, while those who prefer a wilder growth need only wander away towards the outskirts of the garden. On the rockery, which, in its present form, is one of the later additions to the gardens, many an Alpine flower will be seen flourishing in the Valley of the Thames as vigorously as on the crags of the Pennine or Lepontine Alps; while in the new picture gallery—the gift of Miss North—the singular skill and enterprise of the donor enables us to wander among the floral beauties of every land, and to put a girdle of flowers around the earth in much less than forty minutes. But to see Kew Gardens in their glory we should visit them when the rhododendron and its kindred flowers are in bloom. The shrubbery of azaleas is bright with every shade of saffron, and is dappled tenderly with clusters of white and of pink. The great bushes of rhododendron are all aglow with colour, and down the long walk is a many-tinted vista of blossoming shrubs.

Formerly, the pleasaunces at Kew abounded in the absurd anachronisms which,

CAMBRIDGE COTTAGE, KEW.

in the days of our great-grandfathers, were supposed to enhance the beauties of a garden—sham ruins, stucco temples, and the like. There was even a Merlin's cave; perhaps, at times, a magician also was on view. These monstrosities have, happily, for the most part, crumbled away, or have been more promptly destroyed; fragments of them have gone to build the rockeries, and served thus some useful purpose in a district where stone is far to find. The Chinese pagoda almost alone survives, conspicuous owing to its height from many points in all the country round, and of this all that we can say is that it is a pagoda upon whose architectural merits we must leave the Chinese to pass a judgment—remarking, meanwhile, that it harms the landscape no more than a water-tower, and considerably less than a factory chimney.

Gliding along the river, past the enclosing wall of the palace, we are confronted on the opposite shore by the houses of Brentford. These we will leave for a moment to finish our say concerning Kew, whose handsome many-arched bridge of grey stone, not unlike that of Richmond, is now coming into view. This bridge was built about the year 1783, replacing an earlier structure. A short distance from it on the Surrey shore, abutting on to one side of the gardens, in the immediate vicinity of the royal palace, is the more ancient part of the village of Kew. Here is the Green, so pleasant a feature in many of these suburban townlets. On one side stands the church, with its little graveyard—the "Chapel of St. Anne," and once a true Queen Anne structure, for the ground was granted by that sovereign, and the church completed in the year 1714. It has, however, undergone many alterations, especially about the year 1838, when it was enlarged at the expense of King William IV., and another "restoration" has lately been completed, during which a chancel has been appended, and the roof of the nave raised. At the same time a mortuary chapel was added, in which is laid the body of the late Duke of Cambridge. He died at Cambridge Cottage, an unpretending mansion looking on to the Green, which remained the property of his widow until her death in 1889. Other persons of note in their day rest in this little God's acre. Aiton, the gardener; Bauer, the microscopist; Kirby, the architect; Meyer and Zoffany, the artists; and, greater far than they, Gainsborough, whom to name is enough, was, by his own desire, buried under a plain tomb in Kew Churchyard. Here also is buried Sir William Hooker, late director of Kew Gardens, to whose repute as a botanist must be added that of developing the resources of the institution, and by his influence obtaining from successive governments the means of founding a great national museum of botany. He was succeeded by his son, Sir Joseph Dalton Hooker, the present director, by whom the work thus inaugurated has been no less ably carried on.

We must now turn back to Brentford, where the little Brent, which has made its way from the uplands of Hendon, falls into the Thames. Its aspect from the river, perhaps owing to the contrast which the opposite bank has so long presented, is not attractive. Brentford is a very ancient settlement. Some have thought that this spot was the scene of Cæsar's passage of the Thames; certainly it was the chief town of the Middle Saxons. Were there not also two kings of Brentford? But when did they live? On this, history is silent; but the tradition is an old one. The town has always had a rather unsavoury reputation; "it is referred to by Thomson, Gay, Goldsmith, and others, chiefly on account of its dirt." Indeed, the remarks on it might be thus summed up:—There are three kings at Cologne, and two at Brentford, and in the matter of odours the towns are proportional.

Some of the views in the neighbourhood of Kew Bridge are very pretty; the houses by the riverside often group picturesquely; the stream is diversified by one or two wooded islands; barges floating down the Thames, or moored against its banks, combine well in foreground and middle distance; but after this there comes an uninteresting interval. The right bank is occupied largely by market gardens, the land lies low, and in places has an unkempt aspect; the passenger by the towing-

HIGH WATER AT MORTLAKE.

path sees heaps suggestive of refuse to other senses than that of sight, but as Mortlake is neared the prospect again brightens on the right bank of the river, though the left remains rather monotonous. Some attractive houses stand by the waterside. There is the well-known "Ship" Inn, and the odour which is sometimes wafted from the shore, though due to art rather than to nature, is more pleasant than that of most chemical processes, for it is suggestive of good English beer. Mortlake has lost its tapestry works and its potteries, but it has retained its brewery. Once every year Mortlake is numbered among the famous places of England, for here is a limit, generally the winning-post, of the aquatic Derby—the Oxford and Cambridge boat-race. This we shall presently mention a little more fully. Enough now to cast a glance at the townlet itself, which, like all places near London, is developing, and becoming more townlike.

The name of Mortlake appears in English records at a very early date, as it was an important manor belonging to the Archbishopric of Canterbury. The manor itself appears to have included much more than the present parish, but the house was in the village. It was a not unfrequent residence of the archbishops down to the time of Cranmer, by whom the lands were alienated in exchange to the king. Only the tower of the church is old, and this is not particularly interesting; the remainder is Hanoverian, and suitable to its period. Here are entombed Dee and Partridge, the astrologers; Philip Francis, the supposed author of "Junius"; and John Bernard, Sir Robert Walpole's "only incorruptible Member of Parliament"; while in the neighbourhood lived Colston, the philanthropist; Jesse, the naturalist; and Henry Taylor, author of "Philip van Artevelde"—a fair share of celebrities for a quiet little suburban town.

Just beyond Mortlake is the village of Barnes, with the bridge of the South-Western Railway spanning the Thames. The houses and gardens by the riverside give a bright and homelike aspect to the scene. The church incorporates fragments of an ancient building, and the rectory has been occupied by more than one clergyman of mark. Inland, stretching back across the peninsula, and thus offering a short cut from Putney, well known to the frequenters of boat-races, lies Barnes Common—a breezy spot, bright in summer with blossoming furze, which happily is still sacred from the builder, and untouched by the landscape gardener.

HOGARTH'S TOMB AT CHISWICK.

From Kew Bridge down to Putney Bridge Father Thames follows a course even more serpentine than usual. Its double loop forms an almost regular S, the axis of each fold lying nearly north and south. The general direction of its flow also has changed, from a northerly to an easterly course; thus, at Mortlake, we are again brought by the river into the vicinity, comparatively speaking, of Richmond Park, on its northern side; while, on the left bank, Chiswick may almost be said to make two appearances on the riverside. The older part, however, of this place

THE UNIVERSITY BOAT-RACE.

lies below Mortlake. It still retains traces of its ancient picturesqueness, when Chiswick House was a favourite residence of the Dukes of Devonshire, and its fêtes among the chief events of a London season. Chiswick is, however, greatly changed since then—still more from the days when it was the home of Hogarth, whose tomb is in the churchyard. New houses have sprung up, the iron-roofed sheds of a ship-building establishment—devoted especially to the construction of torpedoes—uglify, if the word be permitted, the margin of the Thames, and the incessant clang of hammers disturbs the peace of the stream. Henceforth, we find ourselves within the grasp of the metropolis. Once or twice, it is true, the river seems to slip away again into the freedom of the fields, but it is only for a brief space; it is soon prisoned again between the walls of the workshop, or doomed to remain in sight of the unsightly performances of the nineteenth-century builder.

Between Putney and Mortlake, a distance of about four miles, is the course of the annual Inter-University Boat-race—the water Derby, as it is sometimes called—which, for a brief season, diverts busy London from its daily routine, engrossing almost universal attention, and even imparting to the streets and shop-windows a tinge of blue; for, as all the world knows, dark-blue and light-blue are the respective colours of Oxford and of Cambridge. From the humblest to the highest in the land, from the crossing-sweeper to the Bond Street exquisite, from the flower-girl to the peeress, each one wears the colour of his or her favourite University; though some, it must be admitted, prudently purchase reversible ribbons, and, after the race, take care to make a change, if needful, and duly sport the winning colour. Never does the Blue-ribbon Army seem to have enlisted so many recruits as during the few days prior to Palm Sunday, the race being, by a custom which may be regarded as established, rowed on the Saturday preceding this festival. Some time, however, elapsed after the first contest took place before either time or place were finally settled. The Oxonians had a preference for the beginning of the summer vacation, a time which was not acceptable to the Cantabs; some, also, of the earlier races were rowed on other parts of the Thames—as at Henley, or between Westminster and Putney. Owing to difficulties in coming to an agreement on these points, as well as from other causes, the contest at first was of an intermittent character. The first race, rowed over a much shorter course at Henley, and won by Oxford, was in 1829. Between this date and 1845 there were only five races, all rowed from Westminster to Putney, and four of these were won by Cambridge. The first race over the present course was in 1845, and in the next year outriggers were used for the first time, all the earlier races taking place in what are now contemptuously called tubs. The race has come off regularly each year since 1856, and its direction has been, in almost every case, from Putney to Mortlake. Oxford has scored several more triumphs, on the whole, than Cambridge. The most exciting episode ever witnessed was in 1859, when the Cambridge boat sank near Barnes Bridge. The crew of that year was an exceptionally powerful one, and was looked upon as safe to win. But the builder of their "eight" had supplied them with a boat which was hardly up to their weight—at any rate, for a river liable, like the Thames, to be rough and lumpy on occasion. The ill-luck of Cambridge is almost proverbial; the water was as choppy as it could well be, and the water slowly swamped the Cambridge boat. Those who watched the race will remember how its crew struggled gallantly on, falling gradually behind their rivals; rowing magnificently, though their boat seemed held back by some invisible force, till at last it filled with water and sank under them as they bent to the stroke. Fortunately—though in those days there was little restriction on the number of steamers that were allowed to follow the race, and no means of preventing them from pressing on the losing boat in their struggle for the better point of view—no life was lost, and no one was even hurt. Now that danger — and, owing to the characteristic English recklessness and selfishness, it was rapidly becoming a very serious one—has been averted; for only four steamers are permitted to follow the competitors, these being respectively for the Umpire, the Press, and the members of each University.

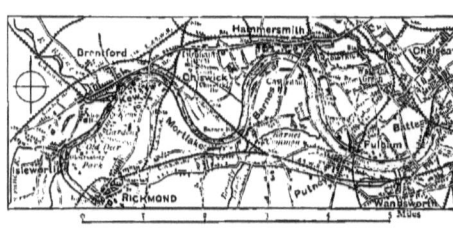

RICHMOND TO BATTERSEA.

Another source of danger has been removed of late years. In the earlier times of the race the old Hammersmith Suspension Bridge became a favourite station for spectators of the humbler rank, as it commanded a good view in both directions; and, though it was not quite half way on the course, by the time it was reached by the boats the race was often practically decided. Indeed, it was a saying, rarely falsified, that the race would be won by the boat which passed under the bridge clear of its opponent. On this bridge crowds continued to gather even when it had been condemned as unsafe; after a certain hour vehicles were stopped, and the concourse thickened; adventurous boys managed to mount the chains, to be pulled down at first, ignominiously, by the police; but at last now one, now another, as the throng gathered, contrived to elude the guardians of the law, and soon scrambled up to secure heights. The roadway became a black mass, a string of blackbeetles seemed to have taken possession of the chains, and the bridge carried a load of human beings that probably the engineer who constructed it never for a moment contemplated. There was the additional and yet graver danger from the shifting of the pressure as the crowd attempted to follow the boats when they shot beneath, or as the possible result of a panic; so that at last the bridge had to be closed alike to foot-passengers and to carriages for most of the day. The old wooden bridge at Putney used also to be crowded—as the new one still is, by those who prefer to see the start; Barnes railway bridge has another contingent brought hither by trains, and in many places the riverside is thronged. Barges are moored in the stream, stands are erected in gardens by its side; the towing-path on the Surrey side is black with people. When the boats are off a string of vehicles may be seen tearing at full speed across Barnes Common. These contain enthusiasts, who, after having witnessed the start at Putney, take the short cut across the peninsula in hopes of seeing the finish at Mortlake. A hoarse roar goes up from the crowd as the two boats, looking strangely small in the wide open space of water, are espied coming round the bend of the river; so light are they that the crews seem almost to sit upon the water. Their oar-blades flash in the sun. They dart past, perhaps in conflict, perhaps the one a length or two ahead of the other; the steamers follow close upon them, sending up a surging wave on either bank, whose arrival at the shore is signalised by conspicuous commotion among the spectators at the river brink, as the chilly water unexpectedly sweeps around their ankles; the boats pass out of sight round another corner; the race will be over in a moment, and when the news comes the brief excitement of the day is ended.

The crowd now may be reckoned by hundreds of thousands. The railways are gorged, and as the time of the race approaches, every street leading to the course is thronged by carriages and pedestrians. Even so far away as the end of Westbourne Terrace, the throng on the Bayswater Road and its steady westward progress would attract the notice of the most casual observer. Yet this intense excitement, this concentration of popular interest on the two Universities, is of much later date than the establishment of the race itself. A quarter of a century since, the attendance was by no means great; now, long as is the course, and many as are the stations which divide the attractions, only a wet day or a very early hour (the time of the race depends on the tide) keeps the concourse within any moderate limits.

Hammersmith, now united to London, offers few inducements to the tourist " in search of the picturesque," though, of course, it is hard to find any riverside place where some nook or corner may not offer a sketch to the artist. Its old suspension bridge—so noted a point, as stated above, in the history of boat-races—has now been succeeded by a more substantial structure, opened in 1887.

The reach of the river from Hammersmith to Putney is comparatively quiet, and the marshy condition of the left bank has compelled the builder to keep at a distance; so that though lines of houses may be seen inland, they are parted from the water by extensive osier beds. We turn our backs disgustedly on the cement works, and glance forward to the more open country beyond, where are houses scantily scattered among trees, and the "Old Crab Tree" Inn. On the right bank a considerable tract of meadow-land still remains unenclosed, on which occasionally there is some fair hedgerow timber, and from which, in summer, the pleasant scent of new-mown grass is wafted; willows rustle by the towing-path, and the white poplar sheds its downy seeds beneath our feet. Bushes grow freely on the river bank, and now and then, for a moment, hide the water. For the last time, if no snorting steamer or screaming steam-launch, laden with holiday-makers, chance to be in sight, or, still worse, in hearing, the Thames for a moment resumes something of its former peaceful aspect, although the fact that the tidal character of the river has now become conspicuous makes it needful to consult the almanack before paying it a visit—at least, for those who desire to appreciate the real beauty of the scene. The most tempting spot is reached as we begin to approach Putney, where we obtain from the path a view of an old-fashioned brick mansion, standing among lawns and

OLD HAMMERSMITH BRIDGE.

fields which are shadowed by some noble elms. This mansion bears the appropriate name of Barn Elms, which, as it has been remarked, seems to indicate that the trees have always been a distinctive feature of the grounds. It has long been a place of some note. Sir Francis Walsingham, minister of Elizabeth, formerly lived here, and more than once entertained his queen—too often, it is said, for the prosperity of his purse. Cowley, the poet, also was for a time an inmate of Barn Elms, and both decorous Evelyn and frolicsome Pepys came here a-pleasuring. In an adjacent building lived old Jacob Tonson, noted among the bibliophiles of the reign of Queen Anne; and here were the head-quarters of the noted Kitcat Club. In a large room erected by him was placed the famous collection of portraits of its members, painted by the hand of Sir Godfrey Kneller, which, from their being all three-quarter length, have given their name to portraits of this kind. The club-room, which was separated from the mansion by a garden, after falling into a dilapidated condition, was pulled down in the early part of the present century.

Somewhere among these trees was fought the notorious duel between two fine gentlemen of the age of the Restoration—the Duke of Buckingham and the Earl of Shrewsbury—when, as it is said, the wife of the latter, disguised as a page-boy, stood by, holding the horse of her paramour. The Earl received a fatal wound, and the lady went home to the Duke's house. It is needless to say that the Court

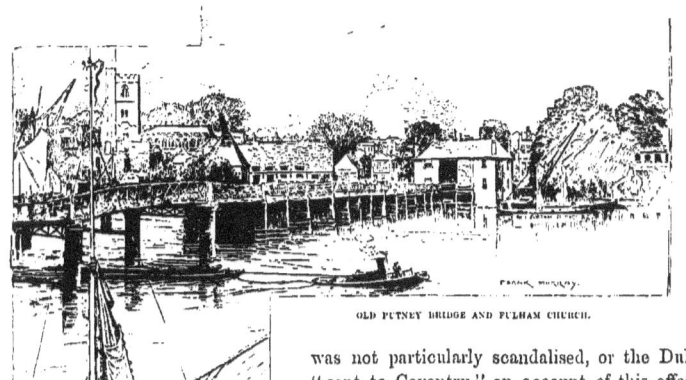

OLD PUTNEY BRIDGE AND FULHAM CHURCH.

was not particularly scandalised, or the Duke "sent to Coventry," on account of this affair. Barn Elms is now the home of the Ranelagh Club. Nearly opposite, in the immediate neighbourhood of the river, is Craven Cottage, a quasi-rustic retreat, which in its day has been frequented by various personages of note. It was built originally for the Countess of Craven, afterwards Margravine of Anspach; but subsequently has been considerably altered. Here afterwards lived Sir E. Bulwer-Lytton, and entertained Louis Napoleon, after his escape from Ham; at a later period it was the residence of an aristocratic money-lender.

Barn Elms passed, we approach the twin villages—though the term is no longer applicable, for they are now suburbs of London—of Putney and of Fulham, one on either side of the stream. Much alike in their churches, they still differ, and once differed yet more, in other characteristics. For many years Putney has been a centre of London aquatics, which have set their mark on the riverside. Except for the broader stream, an Oxonian or a Cantab might fancy himself at certain spots by Cam or Isis. There are the boat-houses on the same nondescript pattern, the sheds sheltering eights and fours and "funnies"—or whatever name be used to designate the cranky one-man racing boats—the usual flags indicating the head-quarters of the different rowing clubs, the usual specimens of the amphibious race that is peculiar to the riverside where oarsmen most do congregate; in short, the waterside at Putney is a rather odd, not wholly unpicturesque, and somewhat unique bit of Thames scenery.

The old wooden bridge, supported on piles, which formerly united Putney and Fulham was a very picturesque and decidedly inconvenient, not to say dangerous, structure. As there has already been occasion to indicate, it has now been superseded by a new one of stone, built a few yards higher up the stream.

The noise of mimic strife on the river, or at worst a holiday-maker's brawl, is all that disturbs the peace of Putney in the present day, but in olden time the town was for a time the head-quarters of an army. In the year 1642 forts were built both here and at Fulham to protect a bridge of boats which was thrown across the Thames; and again, in 1647, Cromwell encamped at Putney for some time. The memory of another Cromwell, only less noted, is connected yet more closely with the place, for here was born Thomas Cromwell, minister of Henry VIII. The old wooden bridge also must have been traversed many a time by one of our most noted men of letters; in Putney, Gibbon, the historian, was not only born, but also received his earlier education.

The ground somewhat rises from the water, on the Surrey side, towards Putney Heath, but on the Fulham side it lies low. From the stream above the bridge will be seen the trees of the domain belonging to the Bishops of London; its ample precincts are enclosed by a moat, even on the riverside, a raised causeway dividing it from the Thames. Very little of the manor-house, or "palace," is visible from the water, as its buildings are not lofty, and it is surrounded by trees. The manor has been the property of the bishopric from a very early date; even at the time of the Norman survey "in Foleham the Bishop of London held forty hides." The palace is a rambling brick structure, more like a college than a mansion, reminding us of some of the colleges at Cambridge. No part is of very great antiquity; the older forms a quadrangle, and was erected by Bishop Fitzjames, in the reign of Henry VII. Some of the earlier buildings were pulled down about the year 1715, as the palace had become in part ruinous, and was found to be needlessly large. This was done by the advice of Commissioners, among whom were Vanbrugh and Christopher Wren. The hall belongs to the older part of the palace; the chapel is new; the library was built in all probability by Bishop Sheldon, and contains a collection of books, to which Bishop Porteus was the first and an important donor. Considerable additions, increasing the comfort rather than the beauty of the house, were made in the earlier part of the present century. The library is a valuable one, and there is a fine collection of portraits of former occupants of the see, interesting to the students alike of history and of English fashions and faces—

the last subject, dealing in what we may term the natural history of the Englishman, being remarkably well illustrated by the long series of men of one profession, and approximately of one period of life. Except for its rather objectionable situation, lying so near the level of the Thames, Fulham Palace must be a most attractive residence. The grounds occupy about thirty-seven acres, and the shrubberies have long been noted, some of the rarer trees being of unusual size and beauty. Special attention was paid to the horticulture of Fulham so long since as the days of Bishop Grindall; and Bishop Compton added to its attractions by planting a large number of rare shrubs and trees—or what were in his day rare—chiefly from North America.

The church, as has been said, resembles that of Putney, but is the more handsome building, standing in a spacious and well-kept churchyard. As might be anticipated from its proximity to the home of the Bishop of the diocese, it has been carefully restored, and, though without any architectural features of special interest, is a very fair specimen of a parish church. It has evidently been much improved since the year 1816, when, in a well-known work on the "Beauties of England and Wales," it is described as "a respectable structure, destitute of uniformity," and the tower is said to be "defaced" by incongruous modern battlements, and by "a mean octagonal spire of wood, surmounted by a flagstaff and vane." Many of the Bishops of London—especially of those since the Restoration—are entombed either in the church or the churchyard—mostly in the latter. The latest to be laid there was the last occupant of the see, the amiable and judicious Bishop Jackson, who officiated in the church on the final Sunday of his life, and while walking thither suffered from a premonitory seizure of the disease which so speedily proved fatal. Some of the monuments within the church are worth a passing notice, though the more striking are seventeenth or eighteenth century work. The grave of Lowth will attract the eyes of those who honour learning. There is a not unpleasing mural monument in memory of Miss Katharine Hart, who "lived vertuouslye, and dyed godlie ye 23rd daie of Octo., 1605;" but most amusing—if we may use such an epithet—is a large monument under the tower to a certain "nobilissimus heros Johannes Mordaunt," created Viscount Aviland by Charles II. Of this worthy there is a statue, and the artist has contrived to infuse into the pose and the face such an air of infinite superiority that it must have been quite a condescension on the part of his lordship to breathe the common air.

The market gardens of Fulham, formerly so noted, are, to a large extent, covered by buildings; the once quiet village has practically become incorporated with London. Below Putney Bridge we never cease to be reminded that we are now on the very margin of the metropolis; that its growth is rapid, and its boundaries accordingly are ragged and unattractive. Immediately below the bridge on the right bank is a fine-looking terrace, and on the left some pleasant houses and gardens, survivals of more ancient days, when Putney and Fulham were country villages, and the fisheries of the latter were leased for an annual rent of "three salmon;" it is only seventy years since they were spoken of as a "source of local profit." Beyond these riverside residences lies Hurlingham, notorious for its pigeon-shooting. On the opposite shore we presently pass the houses of Wandsworth, whereof the brewery is the most prominent, if not the most attractive object.

But as we come to Battersea Reach the signs of industry and commerce thicken around us; works of various kinds line the banks; the Surrey shore, once the more lovely, is now being covered thick with unattractive buildings. On the Middlesex side some traces of older days still occasionally linger—boulders of more solid rock incorporated in the clay of modern masonry—but farewell to the characteristic scenery of the River Thames, farewell to all natural beauty; its waters have now become a great highway of commerce. They have no rest by day, and not always by night, from the fussy steamer and the laden barge; they are turbid with mud, inodorous also, sometimes, with the foulness and garbage of a huge town. The history of the Thames is the history of many a river of England, and may be summed up in Charles Kingsley's words :—

> "Clear and cool, clear and cool,
> By laughing shallow and dreaming pool;
> Cool and clear, cool and clear,
> By shining shingle and foaming wear;
> Under the crag where the ouzel sings,
> And the ivied wall where the church-bell rings;
> Undefiled, for the undefiled,
> Play by me, bathe in me, mother and child.

> "Dank and foul, dank and foul,
> By the smoky town in its murky cowl;
> Foul and dank, foul and dank,
> By wharf and sewer and slimy bank;
> Darker and darker the further I go,
> Baser and baser the richer I grow;
> Who dare sport with the sin-defiled?
> Shrink from me, turn from me, mother and child."

T. G. BONNEY.

CHAPTER X.

BATTERSEA TO LONDON BRIDGE.

The Scene Changes—A City River—Battersea—Chelsea—The Old Church—Sir T. More and Sir Hans Sloane—Cheyne Walk—Don Saltero's Coffee-house and Thomas Carlyle—The Botanical Gardens—Chelsea Hospital—The Pensioners—Battersea Park—The Suspension Bridge—Vauxhall—Lambeth—The Church and Palace—Westminster Palace and the Abbey—Its Foundation and History—Westminster Hall—Westminster Bridge—The Victoria Embankment—York Gate—Waterloo Bridge and Somerset House—The Temple—Blackfriars Bridge—St. Paul's—Southwark Bridge—The Old Theatres—Cannon Street Bridge—London Bridge and its Traffic.

IT is at Battersea and Chelsea that the Thames first acquires unmistakably the character of a metropolitan stream. Hamlets there are, higher up, which announce the proximity of a great capital; but here is the capital itself, though only the rudimentary beginnings, or, to speak more correctly, the scattered ends. Looking down the channel from this point of view, we see on both sides abundant evidence of crowded life — of industry on the one bank, and of wealth on the other. The omnibus of the river—the penny steamboat—plies to and fro on its frequent errands. On shore, the vehicles of London bring something of its noise. Yet there is plenty of quiet in both these old-fashioned suburbs; and, although innovation has been at work here as elsewhere, nooks may be found, both in Battersea and Chelsea, which have all the character of a sleepy old county town. Battersea, in particular, is the most straggling oddity in the neighbourhood of London—a grave, slow, otiose place, lulled with the lapping of waves, soothed with the murmur of trees in unsuspected gardens, troubled but little with the clamour of passing trains, and dreaming, perhaps, of eighteenth-century days, when there were mansions in the land, and my Lord Bolingbroke had his family seat near the church. The river here makes a somewhat abrupt curve, and gives a dubious outline to the whole locality. Small inlets run up between old walls, dark with the sludge of many years; and the streets and buildings have had to accommodate themselves to the caprices of the stream. Hence it is that, when walking about Battersea, you speedily lose your bearings, and, after following a devious lane which you suppose to be parallel with the river, suddenly find yourself on a bit of shingly strand, with a barge on the limits of the tide, and a general appearance as if the end of all things had been reached.

Battersea, then, is as "nook-shotten" a place as is the "isle of Albion" itself, according to Shakespeare. Gardens as old as the time of Queen Anne hide coyly behind walls that permit only the tops of the trees to be discerned. Houses, of the sedate red-and-brown brick that our ancestors loved, stand at oblique angles to the roadways, each with the silent history of vanished generations entombed beneath

OLD BATTERSEA BRIDGE, 1890.

its ponderous, red-tiled roof. Ancient taverns or inns (call them not public-houses, still less hotels or gin-palaces)—goodly hostelries of the past, broad-frontaged, deep-windowed, large-chimneyed, many-gabled—invite the most temperate passer-by to refresh himself in the cavernous gloom of the bar. The old parish church—not so old as one could wish, but having a Georgian character that is beginning to acquire the interest of all departed modes—occupies a sort of peninsula on the river, the ripple of which speaks closely in the ears of dead parishioners. On the whole, Battersea has known better days. It is now chiefly given up to factories, to the humble dwellings of factory people, and to the houses and shops of the lower middle class. But, in the National Society's Training College, it has a noble old mansion, standing in well-timbered grounds; and the free school of Sir Walter St. John (grandfather of Queen Anne's famous minister) is also interesting. The school was founded in 1700, but the building is of the modern Tudor style. To a casual visitor, however, the most noticeable thing about the suburb is the river itself, with its belongings; —the straggling banks; the rickety water-side structures; the boat-builders' yards; the heavy, black barges hauled on to the foreshore, undergoing repair, or being lazily broken up; the larger vessels, with sails of that rusty orange hue which tells of sun and breeze; and the prevalent smell of pitch, mingled with watery ooze.

Chelsea is becoming fashionable along the river frontage; but, although the stately red-brick mansions recently erected on the Embankment are sumptuous and noble, the chief interest of the locality is in the older parts. Advancing in the direction of town, historic Chelsea begins about the spot represented in our view of Cheyne Walk. The fine old house at the corner of Beaufort Street is an excellent specimen of the kind of suburban dwelling our forefathers used to build, when, the land being far less valuable than now, they spread out broadly and roomily, and were not constrained to pile storey upon storey, until the roofs seemed desirous of making acquaintance with the clouds. It is at this point that the Chelsea Embankment commences—a splendid promenade between avenues of plane-trees, which every season will make more umbrageous. Several years ago, before the late Sir Joseph Bazalgette began to reclaim the river-bank, there was no more picturesque spot in Chelsea, of the dirty, out-at-elbows order, than the bit extending eastward from Battersea Bridge to the old church. Its fantastic irregularity of roof and gable, its dormer windows, its beetling chimney-stacks, its red and brown, its look of somnolent old age and grave experience, had something of a Dutch character; but it was certainly not Dutch in point of cleanliness. Picturesque it is still; but the Embankment has swept away that side of the street which was towards the river, while the ragged tenements on the other side await the hands of the destroyer.

Old Chelsea Church is familiar to every Londoner who goes up the river in a steamboat or a wherry. Its massive square tower, its red-tiled roofs, its external monuments in the bit of green churchyard, the dusky glow of its old brick, and its general aspect of having been entirely neglected by the restorers, attract attention, and to a great extent reward it. The edifice cannot be reckoned among the most beautiful specimens of ecclesiastical architecture in London; yet its appearance is venerable and interesting, and its associations might furnish matter for a whole chapter, or even for a book. The chancel is said to have been rebuilt in the early part of the sixteenth century, and the chapel at the east end of the south aisle was erected by Sir Thomas More. The date of this chapel is about 1520; the tower of the church belongs to the reign of Charles II.; and the building generally stands on the site of one which antiquarians refer to the beginning of the fourteenth century, and of which some portions still remain. The body of More (minus the head) is stated by Aubrey to have been interred in "Chelsey Church, near the middle of the south wall;" but this is doubtful. At the place indicated, however—which is about the spot where he used to sit among the choir, and where he erected a tomb for himself during his lifetime—a tablet of black marble yet appears to his memory. More is the presiding deity of Chelsea. His house was not far from the church, in a north-westerly direction; and here he was visited by Holbein, who painted his portrait, and by Henry VIII., who on one occasion walked with him in the garden for the space of an hour, "holding his arm about his neck," as his son-in-law Roper relates—the same neck which he afterwards caused to be divided by the headsman's axe. Many persons of eminence, especially in con-

CHEYNE WALK.

nection with literature and science, lie buried in Chelsea Old Church, or in the adjoining graveyard; and the passer-by almost brushes against the urn, entwined with serpents, which marks the resting-place of Sir Hans Sloane. At the north side of the church is the grave of John Anthony Cavallier, the leader of the Camisards, a body of Protestants in the Cevennes, who, about the beginning of the eighteenth century, carried on a religious war in which Louis XIV. lost ten thousand of his best troops. Cavallier ultimately escaped to England, entered the British service, was for a time Lieutenant-Governor of Jersey, and died at Chelsea in 1740.

It was towards the close of the seventeenth century that Chelsea first became socially famous as a pleasant outlet from London; and some of the existing houses belong to that period. A few years later—in the reign of Anne—it was a place of great resort. Hither came the cits by boat, to stare at the curiosities of "Don Saltero's" coffee-house in Cheyne Walk, or to visit the Chelsea China Works, established in Justice Walk by a foreigner, the products of which manufactory (now discontinued a hundred years or thereabouts), still haunt the old shops of the suburb, and command good prices. Here also people flocked to eat buns at the "Old Chelsea Bun-house," which retained a distinguished reputation until its long existence ceased in 1839. Swift mentions these celebrated dainties in the "Journal to Stella," and seems to have had a relish for them, together with a fondness for Chelsea generally, the distance of which from town he measured not only in miles, but in steps. Cheyne Walk is the most characteristic portion of the suburb. Many of the houses are ancient; some are extremely attractive, with their substantial look of old-world liberality and thoroughness, their massive piers and wrought-iron gates, their stone globes and sculptured ornaments, their shadowy trees and draping creepers. The two most interesting of these houses, by reason of the modern associations which mingle with their antiquity, are that formerly occupied by the late Dante Gabriel Rossetti—truly a house of dream and vision—and that where "George Eliot" died, after a brief residence. But the greatest memorial figure in modern Chelsea is that of Thomas Carlyle, who lived for nearly fifty years in Great Cheyne Row, and died there in 1881. The Embankment has altered the character of Cheyne Walk, which looks scarcely so old-fashioned as it did in other days, when the river came up almost to the roadway, and boatmen lounged about on a scrap of beach, ready to take you to Putney or Hammersmith, if you disdained the steamer. There is a scene in one of Miss Thackeray's novels, which portrays, with exquisite delicacy of touch and colouring, the Cheyne Walk of a somewhat recent, yet a bygone, epoch. Still, the alterations have given an added dignity to the place, and a beauty which forbids us to regret the past. The real injury to the old row has proceeded from the bad taste of some of its inhabitants, who have faced and coloured a few of the houses in a way entirely out of keeping with the general character of the neighbourhood.

Making our way down the river, we come to the Botanical Gardens, belonging to the Apothecaries' Company of London, where all manner of simples have been cultivated since the year 1673. The ground was first enclosed in 1686, and some of the old walls remained until the alterations consequent on the making of the Embankment. An ancient look still hangs about the prim walks and orderly beds, where one seems to sniff the aromatics of departed generations. Old houses cluster round, and peer with blinking windows into the old nursery of herbs. In the centre is a statue by Rysbrack of Sir Hans Sloane, set up in 1733, in consideration of benefits conferred on the gardens by the great physician; and near the southern boundary is a rugged cedar, planted, together with another, in 1685. More interesting to the general public is Chelsea Hospital, the grounds of which should be reckoned among the parks of London. The Chelsea Pensioner, with his scarlet gabardine, flaming along the ways like a travelling fire, is a figure so peculiar to this neighbourhood that one scarcely ever sees it anywhere else. The retired soldier has a noble dwelling-house in the massive yet comely structure which Sir Christopher Wren reared for him. There is no finer specimen of brick architecture, with stone for the decorations, than the edifice which Nell Gwynne is said, by a doubtful tradition, to have assisted in founding. One might even detect a professional analogy in the style of building. The wings stretch out like troops in column; the main body is the army in mass, compact, steadfast, and impenetrable. But the battered old men have done with fighting now; they have come here to nurse their wounds and aches, and the prevailing sentiment is, as it should be, a blessed and a soothing calm. Within those iron gates, having the grounds and the river on one side, and the quiet old Queen's Road on the other, it is almost like a sanctuary. The sunlight falls asleep in the quadrangles and passages. Caught between wall and wall, detained by trees, reflected from numerous angles, it seems to double back upon itself, and fill the air with somnolent heat and glow. Here is a true place of rest; on the edge of the great city, yet sequestered; substantial, ceremonious, prescriptive; shadowed with greenery, bright with flowers and lawns, lulled with the memory of ancient days, the tender comradeship of the past. Is it not right that all wayworn men should taste a little of the lotos-eater's life ere they depart?—

> "'Courage!' he said, and pointed towards the land;
> 'This mounting wave will roll us shoreward soon.'
> In the afternoon they came unto a land
> In which it seemèd always afternoon."*

Round the precincts of Chelsea Hospital it seems as if it were always Sunday. Though frequented by Chelsea people, the grounds of the Hospital are but little known to the rest of London. Yet the east side is bordered by an avenue of pollarded Dutch elms worth going to see—an avenue dusky at mid-day, and, after dark, wanting only a ghost to make it perfect. Immediately beyond is all that remains of Ranelagh Gardens—the rival of Vauxhall in the middle of last century, when the Rotunda was the most fashionable lounge in London—now a miniature park, with

* Tennyson's "Lotos-Eaters."

trees, greensward, and flower-beds, and a large space set apart for the old pensioners, where they cultivate small plots of garden, and will sell you a nosegay of humble but odorous blooms for a few pence. It is pleasant, in the decaying light of a summer evening, to see the veterans tending their plants, watering or weeding, making up bunches of red and blue and yellow blossoms, and recollecting in their age that Adam was a gardener, not a soldier. Several of these men have faced the storm of battle, and left behind them arms or legs. Now they wait upon the gentle ways of Nature, before the setting of the sun.

With the Hospital grounds on one side, and Battersea Park on the other (the latter winning increased favour every year by its fine effects of wood and water), we come to the Chelsea Suspension Bridge, near neighbour of Battersea Bridge, which in 1890 superseded the older structure shown in our illustration on page 259, and now a thing of the past. The railway bridge from Victoria, a little beyond, is a pleasing specimen of its order. Railway viaducts are often abominations. That they *can* be otherwise is shown by some few instances. The railway bridge at Knaresborough, in Yorkshire, is really beautiful. But then it is of stone, not of iron. Iron bridges—excepting those slung on chains—can scarcely escape the reproach of ugliness. Vulcan himself must have forged the original, and infused into it something of his own deformity. But we have passed out of the stone into the iron age; the engineers have us in hand; and we must submit to a good deal of unloveliness for the sake of utility and cheapness. Stone bridges are works of architecture and art; wooden bridges have a certain rustic prettiness in the country; but the iron bridge of the railway harmonises with nothing. It so happens that just as we reach the Victoria Bridge we enter one of the most uninteresting parts of the river. With Pimlico on the one side, and the outskirts of New Battersea on the other, the eye and the mind are equally baulked of any agreeable subjects of contemplation. As regards associations, Pimlico is perhaps the most barren district in all London; and the part facing the Thames is a mere succession of commonplaces. We even hail Millbank Prison as a relief—though it would be difficult to imagine anything more dreary than that stern, gaunt structure, with a thousand heart-aches behind its walls.

Vauxhall, immediately opposite the great prison which Bentham designed as a model penitentiary, at a time when such experiments were in vogue, has some attractive memories, if only on account of the famous Gardens, which the members of the youthful generation know not, but which their elders bear in genial memory; and when we get to Lambeth, of which Vauxhall is only a precinct, we are on memorial ground indeed. Lambeth is so large a place (its circumference is said to be about sixteen miles) that in 1846 it was subdivided into four parishes; but the most interesting part is that which borders on the river. A certain indescribable quaintness—a dusky hue of tradition and romance—hangs about the neighbourhood. The very name is of unknown etymology, and has a sort of Hebrew sound, though it is probably Anglo-Saxon in some corrupted form. In earlier times, the suburb, as we read in an old account, was celebrated for "astrologers and almanack-makers"

—much the same kind of people in days when men believed in the influences of the stars. Francis Moore ("Old Moore," whose Prophetic Almanack still finds readers) was a dweller in Lambeth; and so, likewise, was Simon Forman, who was connected with the mysterious murder of Sir Thomas Overbury in 1613. Lambeth, moreover, has an ancient reputation for unusual crimes. In 1041—a sufficiently remote date for that fascinating twilight in which it is not easy to discriminate between fact and fiction—the Danish King, Hardicanute, died suddenly in Lambeth,

VAUXHALL BRIDGE, FROM NINE ELMS PIER.

at a banquet given on account of some great lord's marriage. By many it was supposed that he had been poisoned; but it is perhaps more probable that he succumbed to a stroke of apoplexy or paralysis, induced by excessive gluttony. Less open to question is the narrative of a stupendous crime committed in 1531 by a cook in the service of Dr. Fisher, Bishop of Rochester, who had a palace near the archiepiscopal residence. According to Holinshed, the cook threw some poison into a vessel of yeast, and thus not merely destroyed seventeen persons belonging to the family, but also killed some poor people who were fed at the gate. The conclusion of this horrible story is worthy of the beginning. The offender was boiled to death in Smithfield, in pursuance of a law made for that very case, but repealed in 1547. There may, however, be some doubt as to the proportions of the crime. Stow says that, out of seventeen persons poisoned, only two died.

Many parts of Lambeth still preserve a grave, quiet, thoughtful aspect, as of a locality which has had many experiences of life, and can talk to itself of ancient and shadowy days. Elias Ashmole, the founder of the Ashmolean Museum at Oxford, who is associated with the neighbourhood, and the Tradescants, father and son, whose collection of curiosities was at South Lambeth, have, so to speak, thrown a hue of antiquarianism over the whole place; while the venerable palace of the Archbishops of Canterbury gives an ecclesiastical character to the river-side. In the church very little ancient work remains, but its foundation dates back several centuries, and it has some noticeable tombs and monuments, together with the celebrated window displaying the figure of a pedlar, with his pack, his staff, and his dog. The legend connected with this pictorial representation is to the effect that some well-to-do chapman endowed the parish with an acre and nineteen poles of land (now known as "Pedlar's Acre"), on condition that his portrait, and that of his dog, should be perpetually preserved in painted glass in one of the windows of the church. Nothing, however, is known with any certainty of this ancient benefactor, and it has been suggested that the picture is nothing more than the rebus of some person whose name was Chapman, and who thus symbolically revealed himself, after a fashion very common with our ancestors. The most striking incident connected with the church belongs to the revolutionary times of 1688. We can hardly pass its walls without the mind's eye conjuring up the shivering figure of Mary of Modena, the second wife of James II., who, on a cold, rainy December night, took shelter beneath the porch, with her infant son in her arms, while she waited for a coach to convey her to Gravesend, where she was to embark for France. The infant—then only a few months old—was the future Chevalier St. George, better known to English readers as the Old Pretender. Thus the opening of his life was romantic, his early manhood was romantic, and the long remainder of his days was an ignoble commonplace.

The appearance of Lambeth Palace, whether from the river or the shore, is extremely picturesque, and London has hardly a more charming corner than that formed by the Archbishop's residence and the adjacent church. The gate-house of the Palace stands broad and square, looking up the stream, its brickwork sober with the rich red-brown of age. The grey stone-tints of the church afford a delicate contrast; and between the two are the grass and flowers of the graveyard. Behind the Palace rise the trees of the archiepiscopal gardens; and the margin of the river—formerly rugged and neglected enough—is now dignified by the Albert Embankment. The effect of the latter, as well as of the spacious buildings of St. Thomas's Hospital a little further on, is perhaps a little too modern for its surroundings; but, in the presence of Lambeth Palace, the Past is sure to overcome the Present. A large portion of English history lurks behind those ancient walls; the shades of kings and prelates haunt its chambers, its corridors, and its gardens; and the sighs of miserable prisoners might be heard within the Lollards' Tower, if the memory of bygone sufferings could find audible expression. It is believed by antiquarians that the Archbishops of Canterbury had a house on this spot in the latter part of the eleventh century; but it was not until about a century later that Archbishop Baldwin exchanged some other lands for this particular manor, which had previously belonged to the see of Rochester. The Palace dates from that period, but of course very little of the original structure now exists. If any twelfth-century work remains, it is in the chapel; the rest belongs to subsequent ages, and exhibits the influence of various styles. The Lollards' Tower was erected in the early part of the fifteenth century by Archbishop Chicheley, for the confinement (as most writers suppose) of a set of heretics who were among the forerunners of Protestantism. The dark and contracted cell at the top of the winding staircase inside the tower, with iron rings yet clinging to the walls, and the names of victims still visible in the blackened oak, is a grim memorial of the Middle Ages, not to be paralleled in London, except within the enclosure of the Tower. It is a sermon in stone and timber, preaching toleration with mute yet eloquent lips. Some modern authorities, however, deny that Lollards were ever imprisoned there; and the structure is now (officially) called the Water Tower; but the top room has obviously been used as a dungeon.

Undoubtedly, the most conspicuous figure in connection with Lambeth Palace is that of Laud. We can hardly think of the building without thinking of him. He was translated to the Province of Canterbury in September, 1633; his execution was in January, 1645; but the last four years of his life were passed in prison, so that his occupation of the archiepiscopal residence extended over little more than seven years. Into those years, however, were crowded the events and the struggles of a lifetime. The Romanising tendencies of Laud gave offence to the growing Puritanism of the middle classes, and at length he was almost a captive in his own palace, besieged by angry crowds, who would doubtless have paid little respect either to his office or his person could they have laid hands on him. He records in his Diary, under date May 11th, 1640, that a furious rabble, incited by a paper posted up at the Old Exchange two days before, attacked his house by night, and prolonged their violence for at least two hours. After that, he "fortified" the place as well as he could; but the popular resentment increased, and, in 1642 and the following year, Lambeth Palace was roughly handled by parties of soldiery. During the Commonwealth, the building was used as a prison, and the Great Hall was nearly destroyed. The latter was restored by Juxon, and is supposed to represent the original with tolerable fidelity. But it is of Laud we think, and not of Juxon, as we move from room to room; for Laud represents an era in the English Church.

Looking across the Thames, from Lambeth Palace, we get the best view of the Houses of Parliament, which gain rather than lose by the absorption of detail into the general mass. We have now passed the dull and shabby part of the river, and are surrounded by grand and august memories. The stream itself is a highway of empire; the shores are peopled with stately, with noble, or with interesting shapes. The suburbs are behind us; the ancient city of Westminster rises with its towers and steeples on the left bank. Along this channel have passed the Briton in his coracle, the Roman in his war-ship, the Anglo-Saxon and the Dane in their galleys, the Norman, the Plantagenet, the Tudor, and the Stuart, in their resplendent barges. Youth, beauty, and gallantry, genius and learning, the courtier and the soldier, the prelate and the poet, the merchant and the 'prentice, have taken their pleasure on these waters through a succession of ages which form no mean portion of the world's history. Patriots and traitors have gone this way to their death in the sullen Tower. Kings and princesses have proceeded by this silver path, amidst the flaunting of streamers and the music of clarions, to bridal

LAMBETH PALACE AND CHURCH.

pomp or festal banquet. The pride of mayors, of aldermen, of sheriffs, has glassed itself in these waves. Here, in the days of Henry II., the adventurous young men of London played at water-quintain, to the infinite delight of the spectators; here, somewhere between Westminster and London Bridge, King Richard II. met the poet Gower, and commanded him to write a book for his special reading—whence arose the "Confessio Amantis;" and here Taylor, the Water Poet, once saw the Muses sitting in a rank, who gave him a draught of Helicon, which had the unfortunate effect (not unknown in other instances) of emptying his purse.

Westminster is to the full as historical as London itself, from which, be it remembered, it is even now entirely separate, as a city with rights of its own. It might even be described as more truly the capital than London; for the Parliament, the Government Offices, and the Law Courts are situated within its bounds, and the chief palace of the Kings and Queens of England, from Edward the Confessor to Elizabeth, was at Westminster. It is curious to reflect how very unsubstantial is the claim of London, in the strictest sense of the term, to be considered the national metropolis. Of Roman Britain, York was the capital—the Eboracum of Hadrian, of Septimius Severus, and of Constantine. Winchester, as the principal city of Wessex, which subdued the other kingdoms of the Heptarchy, became the seat of government for all England in the early days of the united monarchy. Then Westminster succeeded, and it is hard to say where London

THE VICTORIA TOWER.

comes in. The very name of this city of royalty and statecraft has a grandeur about it with which its appearance corresponds. Westminster Abbey, Westminster Hall, and the Houses of Parliament are three structures not easily to be surpassed for majesty of association and picturesque dignity of aspect. It was a wise decision by which the Gothic style was selected for the new buildings rendered necessary by the disastrous fire of 1834. Anything else would have broken the continuity of the national life, and been altogether at issue with surrounding objects. Sir Charles Barry has provided London with one of its most distinctive features, and his two great towers must henceforth be landmarks of the surrounding country, even more than the dome of St. Paul's, though that, too, will always remain one of the great memorial characteristics of the vast metropolis.

Where one has so noble an edifice, it seems ungracious to repine; yet the loss of the older building was a misfortune for which nothing can compensate. It was nearly the only remaining portion of the Palace of Westminster originally founded by Edward the Confessor, and retained by our kings until Henry VIII. removed his palace to Whitehall. St. Stephen's Chapel, the Cloisters, the Painted Chamber, the Star Chamber, the Armada hangings—all these were destroyed by the great conflagration arising from the overheating of a stove in which some official had been too assiduously burning the tally-sticks whereon the Exchequer accounts were kept until the latter part of last century. The House of Commons sat within the walls of St. Stephen's Chapel, rebuilt in the reign of Edward III., and converted to the use of the national representatives in that of Edward VI. Either at that or some later period, the external walls were wainscoted; a new floor was laid above the level of the old pavement, and a new ceiling shut out the fine timber roof. The chapel, therefore, still remained, but it was almost completely hidden from view. In 1800, however, previously to the addition of the Irish members to those of England and Scotland, it was found necessary to enlarge the chamber, and, on the wainscoting being taken down, the walls erected by Edward III. shone out in all their splendour of architecture, sculpture, painting, and gilding; the whole looking as brilliant and vivid as if it had just left the hands of the workers. The alterations involved the destruction of those beautiful specimens of mediæval art; but drawings were made of most. Still, a good deal of the original palace and chapel was left, though sadly defaced by modern perversions, often in the most execrable taste. The fire carried still further what other influences had begun; and, at the present day, all that is preserved of the palatial structure which successive kings re-edified and adorned are Westminster Hall and the crypt of St. Stephen's Chapel.

With its Hall and its Abbey, Westminster can never cease to be interesting, attractive, and picturesque. Here, if anywhere, we are in the very heart of English history, and can, at our bidding, summon a long procession of sovereigns, prelates, statesmen, soldiers, wits, and scholars. Standing before the Abbey, with the river close at hand, we think of those ancient days when all the adjacent ground was a marsh, so environed with water, and beset with brambles, as to acquire the name of Thorney Island: a wild, bleak, barren spot, almost at the very gates of London, yet apart from it; inhabited only by poor and outcast people, or perchance by banditti, who levied contributions on the rich nobles and merchants, and then escaped to their fastnesses among the thickets of the fenny isle. Then—somewhere about 616—came Sebert, King of the East Saxons, who, according to tradition, founded the Benedictine monastery of which the Abbey is a noble relic. West Minster—the Minster west of St. Paul's, originally called East Minster, according to some accounts—took its rise from that time, and speedily became a place of great importance. The brambles disappeared; the land was drained; the creeks and ditches of the Thames were made to retire into their natural channel; walls and pinnacles arose out of the wet and dreary soil; and the chant of the Benedictines was heard along the river-banks, and in the neighbouring fields. After a while, houses grew up around the monastery, and population was attracted to a spot which the Monarchy and the Church were beginning to favour. The religious foundation was enlarged by King Edgar, and afterwards by Edward the Confessor; and from the time of the latter to that of Queen Victoria, all our kings and queens have been crowned within the walls of the Abbey. Many, also, are buried in the same building, which gives occasion to moralising Jeremy Taylor to observe:—"In the same Escurial where the Spanish princes live in greatness and power, and decree war or peace, they have wisely placed a cemetery, where their ashes and their glory shall sleep till time shall be no more; and where our kings have been crowned, their ancestors lie interred, and they must walk over their grandsire's head to take his crown." Whether for kings or humble men, there is no place better adapted to this vein of thought than Westminster Abbey.

The Abbey-church was dedicated to St. Peter, who, according to the mediæval tradition, appeared to a fisherman on the opposite bank of the Thames, and requested him to ferry him over to Thorney Island, where, with his own hands, he performed the ceremony of consecration. An atmosphere of legend and romance surrounds the earlier history of Westminster Abbey, and continues even as late as the days of Edward the Confessor. It is related in old chronicles that that monarch, having omitted to make a pilgrimage to Rome, which he had promised on condition that he should be restored to his throne, from which the Danes had expelled him, was enjoined by the Pope, as the necessary price of absolution, that he should expend the funds set apart for his journey on the foundation or repair of some religious house dedicated to St. Peter. The particular house was not indicated; but, just at that time, a monk of Westminster, named Wulsine, dreamed that the Apostle appeared to him, and bade him acquaint the King that he should restore the church on Thorney Island. "There is," he is reported to have said, "a place of mine in the west part of London, which I chose, and love, and which I formerly consecrated with my own hands, honoured with my presence, and made illustrious by my miracles. The name of the place is Thorney; which, having, for the sins of the people, been given to the power of the barbarians, from rich has become poor, from stately low, and from honourable is made despicable. This let the King, by my command, restore and amply endow: it shall be no less than the house of God and the gates of Heaven." This, according to the old belief, was the way in which the later Abbey arose. At any rate, Edward rebuilt the monastery and church on a larger and more sumptuous scale; and, from that time forth, Westminster Abbey became the grandest, and on the whole the most august, building in London. It was then, likewise, that the edifice first took a distinct and historical place in the annals of the English people. Until then, it is difficult to trace its history, which, indeed, is little more than a series of ecclesiastical myths. From the days of Edward the Confessor, the story of the Abbey is clear in every respect; but, in such an edifice, history itself assumes a romantic, almost a marvellous, colour. We are in the presence of eight centuries of the national life; for, although no portion of Edward the Confessor's work remains at the present day, the Abbey is so associated with the saintly monarch that it is impossible to detach his memory from the structure begun by Henry III., continued by Edward II., Edward III., and Richard II., and from time to time enlarged by later sovereigns. The building we now behold is the legacy of successive ages, which have left upon the stone itself the imprint of their thoughts, their

THE ABBEY, FROM LAMBETH BRIDGE.

aspirations, their struggles, and their hopes. In passing from chapel to chapel, from cloister to cloister, from aisle to aisle, we seem to pass through the centuries which gave them birth, and which have strewn over all the dust of their extinguished fires. But Westminster Abbey is not merely an embalmed corpse, preserving the semblance of a life which has long since vanished. It is still the shrine of England's greatest men—still the embodiment of ideas yet living in the national heart.

Westminster Hall is second only to the Abbey in historic interest. It was originally built by William Rufus, and it is probable that some of his work still exists, though the bulk of what we see is due to Richard II. The magnificent timber roof—one of the finest in Europe—belongs undoubtedly to the period of

Richard; and it is marvellous to think that this piece of wood-carving should have survived the wear of five centuries, and resisted without injury the dynamite explosion of 1885. A well-known tradition states that the roof is made of Irish oak, in which spiders cannot live; but it appears to be really constructed of chestnut. The place was intended as a banqueting-hall, and so used by King Richard; but some of our early Parliaments assembled there, and, at the very first meeting of the Houses in the new edifice, Richard himself was deposed. The Law Courts were likewise held in this building and its predecessor, from 1224 to 1882. Until a comparatively recent time, the judges sat in the main body of the Hall; and, in the seventeenth and eighteenth centuries, one side of the vast chamber was taken up by the judges, the lawyers, the juries, and the other persons concerned, while the opposite side was divided into a number of little shops or counters, where vociferous traders bawled their wares and solicited custom, until the Usher over the way commanded silence with a voice louder than their own. With one exception—the Hall of Justice at Padua—Westminster Hall is believed to be the largest chamber in the world not supported by pillars. Its aspect is indeed noble, and the recollections which crowd upon the mind on entering its walls are almost overwhelming in their historic and dramatic interest. In the Hall of Rufus, Sir William Wallace was condemned to death; while the very building that now stands has witnessed the trials of Sir Thomas More, the Protector Somerset, the Earl and Countess of Somerset, who contrived the assassination of Sir Thomas Overbury, the Earl of Strafford, King Charles I., the Seven Bishops who defied the power of James II., three of the rebel lords in 1745, Warren Hastings, and several other persons of less distinction, who still have made some mark in the political or social history of the land. Here Oliver Cromwell was inaugurated as Protector; and here, only a few years later, his head was set upon a pole, between the skulls of Ireton and Bradshaw. One could fancy ghosts flitting at night about this vast old Hall. It would be a strange gathering, drawn from the tragedies of five hundred years.

Returning to the river, we pass under new Westminster Bridge, but think rather of its predecessor, the work of Charles Labelye, a native of Switzerland, yet a naturalised British subject. This structure lasted from 1750, when it was completed, to 1853, when its destruction was commenced. Until the building of Labelye's bridge, there was actually no way over the Thames, within the metropolis, but at London Bridge; and the proposal to execute this most necessary work encountered violent opposition in the City. Old Westminster Bridge was a ponderous erection, in which, if we may accept the statement of the architect, twice as many cubic feet of stone were employed as in St. Paul's Cathedral. With its fifteen arches, diminishing in span from the centre, its lofty parapet and wide alcoves, it presented a rather handsome appearance, and many Londoners, not yet old, retain it in kindly memory. It was badly constructed, however, and several of the piers gave way in 1846. There was no alternative but to take the whole structure down; but it has an abiding place in literature, owing to the noble sonnet which Wordsworth composed there on the 3rd of September, 1803. Another literary association with the bridge is of a painful nature. When Crabbe the poet first came to London, in 1780, he was in such deep distress that, after appealing in vain to many persons of distinction, he delivered a letter at the door of Burke's house—a letter to which the great orator and statesman afterwards replied with the utmost kindness; but, pending the answer, Crabbe was in such a state of agitation that, as he told Lockhart in later days, he walked Westminster Bridge backwards and forwards until daylight. It was by such experiences as this that Crabbe acquired his realistic power of delineating the sufferings of the poor, with whom the fear of hunger or the workhouse is one of the permanent facts of life.

It is on quitting Westminster Bridge that the Victoria Embankment begins— a magnificent work, containing the finest effects of architecture, mingled with trees and shrubbery, that are to be found in the metropolis. When one recollects the unsightly mud-banks that used to stretch along the shores of the Thames in this part of its course—the grim, dilapidated buildings that approached the water's edge —the general appearance of ruin—the shiftless, disreputable air of the whole locality, save where some great building, such as Somerset House, broke the dull uniformity of dirt, decay, and neglect—it is impossible to be too grateful for what we now possess. The massive river-wall, with the bronze heads of lions starting out of every pier, the extended line of parapet, the artistic lamps reflected at night in the shining stream, the Cleopatra's Needle, with Sphinxes round its base, the avenues of planes, the green and leafy gardens, the elevated terrace of the Adelphi, the stately river-front of Somerset House, and the splendid new buildings which have been erected at various points of the route, make up, together with the broad and flowing river, a picture which it would not be easy to surpass. At Charing Cross, unfortunately, there is an irremediable contradiction to this grandeur. The railway bridge which there crosses the Thames is one of the ugliest of an ugly family; and all we can do is to comfort ourselves with a sense of the convenience afforded by such structures, and with the impression of Titanic power always accompanying the transit of vast bodies through the air above our heads. As soon as our backs are turned upon the viaduct, it is forgotten; and close by, at the bottom of Buckingham Street, we come upon a decaying relic of old London, which is worth going to see. The Water Gate, formerly belonging to York House, and built by Inigo Jones for George Villiers, Duke of Buckingham, still outlasts, in melancholy isolation, all the princely splendours that once distinguished this spot. York House was, for a short time, the London residence of the Archbishops of York, by whom it was afterwards let to the Lord Keepers of the Great Seal. It was here that no less a man than Francis Bacon was born, and he retained possession of the dwelling until his death. The next occupant was the famous Duke of Buckingham, the favourite of James I. and Charles I., who pulled down the old house, and erected a temporary mansion to supply its place. His intention was to build a more sumptuous palace on the site of Bacon's town-house; but Inigo Jones's Gate was the only portion ever erected. Of course, when originally made, it was on the absolute margin of the river, and here, at high tide, the Duke and his friends took the water in their barges, or landed after an excursion on the Thames. At the present day, owing to the formation of the Embankment, which covers the sloping shores of the river formerly left dry, or rather oozy, when the tide was out, the Water Gate of Inigo Jones is a long way inland, and looks forlornly across the intermediate gardens towards the stream from which it is permanently divorced. The edifice is a fine piece of Roman architecture, massive, rugged, yet ornamental, and admirably adapted, by the peculiarities of its structure, to serve as the approach to a mansion whose grounds came sweeping down to the edge of the waves. The house was afterwards sold by the second Duke of Buckingham, one of the profligate noblemen of Charles II.'s reign, illustrious by his own wit and spirit, and still more so by the masterly portraitures of Dryden and Pope; and a number of streets were built upon the site, some of which were called after the names and title of the Duke.

YORK GATE.

Waterloo Bridge—the grandest bridge in London, and perhaps in the world—admirably falls in with the architectural character of the Embankment and its surroundings. Nothing can exceed the magnificence of those nine broad arches, each one hundred and twenty feet in span, and thirty-five feet high; or of the columned piers from which they spring. The whole effect is colossal, yet graceful to the last degree of cultured power. Where the massive pillars meet the Embankment, they give an added grandeur to the work of Sir Joseph Bazalgette, and the triumphant arches, as they leap the channel of the river, display the happiest admixture of strength and suavity. The engineer who executed the works of Waterloo Bridge was the celebrated John Rennie; but the design was furnished by a somewhat obscure projector named George Dodd, who, in the first instance, was appointed to carry out his own conception, but who appears to have been discharged through inattention to his duties, and the lax habits which ultimately brought him to the prison where he died. The name of Rennie is so universally associated with the bridge, often to the exclusion of any other, that it seems but fair to give the credit of the plan to this forgotten and most unhappy genius.

Leaving Waterloo Bridge and Somerset House in our rear, the next object of note that we reach is the Temple, where we might linger a whole summer's day, without exhausting all the interest that attaches to that memorable spot. What one chiefly sees from the river is the green and pleasant garden, where, according to Shakespeare, the partisans of the Houses of York and Lancaster plucked the white and red roses which served as the distinctive badges of their cause. Looking northward, however, we discern some of the new buildings which border the open ground; and we know that beyond these lie the wonderful courts and alleys—the mazy lanes and avenues of old houses—which, taken altogether, make the Temple one of the most fascinating spots in London. As he passes by on the smooth waves, the man familiar with books can hardly refrain from repeating to himself the murmuring lines of Spenser, in which the poet traces back the history of that cloistral retreat to the days when it was associated with a great military and ecclesiastical Order. Spenser was a thorough Londoner, and therefore well acquainted with

> "Those bricky towers
> The which on Thames' broad, aged back doe ride,
> Where now the studious lawyers have their bowers:
> There whilom wont the Templar Knights to bide,
> Till they decayed through pride."

In the poet's time, and for nearly a hundred years after, brick edifices were very uncommon in London, and the Great Fire of 1666 would never have spread so rapidly, or extended so far, had not the majority of the houses been constructed of wood. It was the "bricky towers" of the Temple which at length stopped the westward march of the conflagration. The oldest parts of the two Inns seem almost as if they might be coeval with the days of Spenser; but the greater number of the buildings belong apparently to the latter part of the seventeenth century. Many alterations have of late taken place in the Temple, and the new work (if only for its newness) is out of harmony with the old. Could Charles Lamb revisit this beloved spot, it is to be feared that he would be much troubled by some of the recent innovations. Those who share Lamb's appreciation of old London have certainly a good deal to put up with in these days. Perhaps the alterations are necessary and unavoidable; but they are often terribly jarring, though there are persons who will scarcely tolerate even a sigh over the departed or departing relics of an interesting past. A good deal of the old Temple, however, still remains, and may perhaps survive for another decade or two. In the Temple Church we have a striking relic of the Middle Ages, elaborately, but not always judiciously, restored between 1839 and 1842; and the Middle Temple Hall is thought to contain some of the best Elizabethan architecture in London.

We are in modern times again when we come to Blackfriars Bridge; for not only is the structure one of yesterday, but that which preceded it dates back no farther than the second half of last century. The bridge erected by Robert Mylne was completed in 1769, and lasted for nearly a hundred years; but it shared the infirmity of Labelye's work at Westminster, and the subsidence of the piers became so alarming that in 1864 the whole edifice was doomed to destruction. One of the finest views of St. Paul's Cathedral, or, at any rate, of the dome, is obtainable from Blackfriars Bridge; but the appearance of the bridge itself on the eastern side is greatly marred by the railway viaduct of the London, Chatham, and Dover line.

BIT OF THE VICTORIA EMBANKMENT.

We have now passed the Thames Embankment, and the river begins to be bordered by wharves and warehouses, often black with the smoke of many years, yet not devoid of a certain rugged picturesqueness and gloomy state. Enormous cranes project from the walls; vast bales of goods dangle perilously in the air, and are lowered into the barges and other vessels which come up close to the landing-stages. Tier above tier of narrow, grimy windows rise into the sky; and gaunt openings in the walls, which seem as if they were intended for suicide, but are really meant for the reception and discharge of goods, reveal to the observant passer-by some dusky glimpses of that accumulated merchandise, the interchange of which has made London the greatest city in the world. In these sullen edifices, beetling over the water-side, you shall see nothing of beauty or of grandeur; but a man must be ignorant indeed, or grossly dull in his perceptions, if his mind do not discover, in the reaches of the Lower Thames, matter of the deepest interest, affecting not merely his own country, but her possessions in every part of the world, and to some extent the whole world itself. From this point, the wondrous city spreads around: the city with its roots in fable, and its branches in the living present; the city of commerce, of manufactures, of finance; the city of incalculable riches, and of that hopeless poverty which accompanies riches as the shadow accompanies the sun; the city which receives into its bosom the vessels and the wealth of all the globe, and which is in constant and electric sympathy with every part of Europe, with the teeming populations of the East, with the desert heart of Africa, with the young Republics of the Western Continent, and with the rising commonwealths of Australasian seas. Whence comes this marvellous power—this universality of influence? Partly from the genius and energy of the races which people Britain; but partly also from the opportunities presented by that deep and expanding stream which issues out into the German Ocean, and brings the fleets of nations to the walls of London. The greatness of England depends upon this liberal and majestic Thames—a fact so apparent, even in the time of Queen Mary, that an acute Alderman, hearing of the sovereign's intention to remove with the Parliament and the Law Courts to Oxford, observed that they should do well enough, provided her Majesty left the river behind. Even in the time of the Roman occupation, London was a great commercial city; and since then, eighteen centuries of development have reared the mighty fabric of her trade.

Though St. Paul's Cathedral is some little way from the Thames, its splendid cupola is so prominent an object from the river that it is impossible not to pause a little before Wren's masterpiece, and consider the history of this great edifice, the foundation of which takes us back to the early days of British history. By some antiquarians it has been supposed that, in the Roman times, the summit of Ludgate Hill was occupied by a temple to Diana; but this tradition was entirely discredited by Sir Christopher, who records that, in digging for the foundations of the present Cathedral, he found no evidences whatever of the existence of any such pagan structure—no fragments of cornice or capital, no remains of sacrifices. He did, however, arrive at some foundations, consisting of Kentish rubble-stone, cemented with exceedingly hard mortar, after the Roman manner. He believed these to have been the relics of an early Christian church, destroyed during the Diocletian persecution, the erection of which he considered may have been due to St. Paul himself. Whatever may be the truth of these remote traditions, it seems unquestionable that a Christian fane existed on this spot from an early period. The crown of the hill was a very likely place for such an edifice, and the proximity of the river made it easy of access from surrounding parts. The church demolished during the persecution of 302 was rebuilt in the reign of Constantine, between the years 323 and 337. In the following century it was destroyed by the Saxons, but, after the conversion of the early English, was again erected by Ethelbert and Sebert in the sixth and seventh centuries. The Cathedral which immediately preceded the present was begun about 1083, and lasted until the Great Fire of 1666. During this long period of nearly six hundred years, the edifice underwent frequent

THE RIVER AT BLACKFRIARS.

alterations, and received many additions. Some of its dimensions are thought to have exceeded those of any other church in Christendom. Its length from east to west was six hundred and ninety feet, and the spire over the central tower rose five hundred and twenty feet into the air. This spire was burned in 1561, and, from that time until 1633, the noble old pile was in a state of dilapidation, which it is surprising that so rich a city as London should have allowed to continue. But the whole condition of the Cathedral at this period was one not easy to understand at the present day. The middle aisle, usually termed Paul's Walk, was an ordinary lounging-place for the wits, gallants, and disreputable characters of the time. Under the pillars of that magnificent arcade the lawyer received his clients; the business man transacted his affairs; the idle inquired after news; servants wanting employment let themselves out for hire; and the chorister boys exacted tribute

of gentlemen who entered the Cathedral, during divine service, with spurs on.
From the period of the Reformation to the early part of the reign of Philip and
Mary, matters had been even worse; for a daily market was held in the nave,
and men would lead mules, horses, and other animals from entrance to exit.
"Paul's Walk" is one of the most frequent subjects of allusion in the works of
the Elizabethan dramatists; and there was certainly no better place in London for
an observer of manners, like Ben Jonson, to imbue himself with the humours of
men.

It need hardly be said that Old St. Paul's was a Gothic structure; but when
it was repaired in 1633, the work was put into the hands of Inigo Jones, who
was entirely a child of the Italian school. He accordingly set up a classical portico
in front of the ancient Gothic church, thus producing an effect of painful incongruity, although the portico in itself appears to have been extremely fine. The
circumstance, however, is to some degree excused by the design of Charles I. to
build an entirely new Cathedral, of which Inigo's portico was to be the frontispiece.
The Civil War put an end to this project, together with many others; and during
those tumultuous days Cromwell's soldiers stabled their horses in the metropolitan
church of London. The complete destruction of the building followed six years after
the Restoration, when the greater part of London succumbed to a disaster which
more vigorous measures might have stifled in its infancy. Another Gothic edifice
would have been more in accordance with the traditions of the place; but it is
fortunate that no attempt was made to revive an architectural style with which
all the builders of that age were entirely out of sympathy. Wren held the Gothic
forms in absolute contempt, and the towers which he added to Westminster Abbey
show how miserably he failed when trying to accommodate himself to methods
which he neither understood nor cared to understand. With the Renaissance he
was perfectly at home; and his great work, whatever objections we may make
on the score of coldness, so far as the interior is concerned, is surely characterised
by a grandeur of its own, dependent not merely upon physical size, but on vastness
of conception, and on that sense of towering magnificence, and almost infinite
dilation, which is produced by this mountain of hewn stone, extending into curved
and pillared aisles, and swelling upwards into the mimic firmament of the dome.
For nearly two hundred years Sir Christopher Wren's Cathedral has been the
central monument of London. Round its giant mass the waves of the great city
beat day by day in feverish unrest; and there is something in its ponderous bulk,
its countless reduplication of arch and column, and its soaring cupola, which seems
to image the stability of English life in the midst of constant agitation and perpetual change.

Southwark Bridge, under which we pass shortly after returning to the river,
is chiefly interesting as being the first thoroughfare which carries us over into what
is popularly called "the Borough"—certainly one of the most memorable parts of the
capital. By a kind of fiction, Southwark is accounted one of the twenty-six wards

ST. PAUL'S, FROM THE THAMES.

of London, and, considered in this relation, is entitled Bridge Ward Without. It is therefore, to some extent, a part of the City; yet it has its own government, and a distinctive character, both in general appearance and in metropolitan history. In early times, it was a sanctuary for malefactors, and in other respects possessed an evil reputation, which appears to have been not wholly undeserved. In the Bankside, Southwark, was situated the Bear Garden, of which we read so frequently in old English writers—a place where Shakespeare must have seen the bear Sackerson which he has immortalised in the *Merry Wives of Windsor*. Edward Alleyn, the actor, who founded Dulwich College, was at one time master of this objectionable place of amusement; and here Pepys went one day with his wife, and pronounced the entertainment "a very rude and nasty pleasure." A much pleasanter association with old Southwark is the fact that Shakespeare's theatre, the famous "Globe Playhouse," conspicuous in stage history, was here situated close to the river. The external shape of this illustrious edifice was hexagonal, and, though the stage was roofed over with thatch, the spectators sat in the open air, without any covering whatever. The interior was circular, and the building displayed a classic figure of Hercules supporting the globe. One would be glad to know the exact spot where Shakespeare trod the boards, submitted some of his works to public approval, and perhaps discharged the duties of a manager. But, although the theatre is commonly said to have stood in Bankside, there appears to be some doubt upon the point. Unquestionably, however, the Bankside has the best claim, and it is believed that Barclay and Perkins's brewery occupies the site, or nearly so. Originally erected in 1594, the Globe was burned down on the 29th of June, 1613, owing to some lighted paper, projected from a piece of ordnance, having found a lodgment in the thatch. This was rather less than three years before the death of Shakespeare; but the playhouse was speedily rebuilt, at the expense of James I., and of many noblemen and gentlemen. The drama that was being acted on the occasion of the fire seems to have been Shakespeare's *Henry VIII.*; and Sir Henry Wotton, who writes an amusing account of the affair to his nephew, says that the drama "was set forth with many extraordinary circumstances of pomp and majesty, even to the matting of the stage; the knights of the Order with their Georges and Garters, the guards with their embroidered coats, and the like." The new theatre was much handsomer than the old, and provided with a roof of tile, so that the discharge of ordnance should not again produce such disastrous consequences. The house was pulled down in 1644, by which time Puritanical opinions had gained so much ground amongst the London population that theatres were no longer the prosperous undertakings they had been in more careless and light-hearted days.

From the Bankside to the High Street of Southwark is no great distance; but it takes us backward from the time of Shakespeare to the time of Chaucer. The "Tabard" Inn stood in that ancient thoroughfare, and, until recently, some old, decrepit buildings flanked the back yard of this hostelry, which, though probably not coeval with Chaucer, were at any rate antique enough to suggest his period.

The Borough High Street, being the main road into the south-eastern parts of England, was from an early date celebrated for its roomy hostelries, some of which still remain in all their picturesque amplitude, with external galleries, overhanging roofs, carved timber, dusky passages, and cavernous doorways. None, however, could boast such an association as that which throws its halo round the "Tabard." We are not, of course, to suppose that Chaucer's immortal poem is an exact record of anything that happened on some given occasion; but it is more than probable that Chaucer performed the pilgrimage to Canterbury, "the holy, blissful martyr for to seek," and that, with his companions, he started from the "Tabard" in the High Street. It is also conceivable that these pious excursionists often beguiled the way by telling stories; and it is thoroughly in accordance with the manners of the time that some of the stories should be of a very questionable tendency. Pilgrimages, after a while, became a form of dissipation with which the religious sentiment was but slightly associated. As early as the fourth Christian century, Gregory, Bishop of Nyssa, dissuaded his flock from joining pilgrimages, because of the low moral tone frequently developed amongst the travellers. In the ninth century, Englishwomen had a particularly bad name for the gallantries they carried on under pretence of devotion; and in the fourteenth century, when Chaucer wrote, the matter had doubtless become still worse. One of the results of this perversion, however, was that people distinguished by every variety of character were drawn together by the common object of adoration at some famous shrine. Chaucer was thus presented with the finest possible opportunity for the exercise of those powers of observation and of portraiture in which he was hardly inferior to Shakespeare himself. Hence a poem which, notwithstanding the difficulties of its partially obsolete English, is still a living force in the literature of our race. Hence a collection of stories which touch the whole round of human nature—in its pathos, its humour, its tragedy, its devotion, its blunt and rugged realism, its high-raised phantasy, its vulgarity, and its nobleness; and hence that fascinating light of genius and human fellowship which hovers round the vicinity of the "Tabard" Inn, and will consecrate even its modern brick-and-mortar with the tenderest memories of the past.

Returning towards the river, we find on our left hand, not far from the water itself, the fine old church of St. Saviour's, Southwark (anciently called St. Mary Overies, from its position as to the bridge), which contains a handsome Gothic monument to Chaucer's contemporary, John Gower. The church has been much injured by alterations in recent times, but still presents some beautiful specimens of the Early English style. All that remains of the old church founded in 1208 is in the choir and the Lady Chapel; yet, on the whole, the effect is venerable, and the associations with the church are highly interesting. Among the persons here buried are Edmund Shakespeare, the brother of William; John Fletcher, the fellow-dramatist with Beaumont; Philip Massinger, another dramatic poet; and several persons more or less connected with the theatrical world of Shakespeare's generation.

We are now at the southern extremity of London Bridge—one of the best of Rennie's works, but a very uninteresting structure compared with that which preceded it. Still, it is impossible to pass over this granite causeway without seeing, at any hour of the day, such a spectacle of human life as penetrates both the heart and soul. All bridges are favourable to this kind of observation; for they contract and isolate the great stream of human beings, which for a brief period is

SOUTHWARK BRIDGE.

incapable of any diversion either to the right or to the left, but is brought sharply and sternly face to face with him who would take note of his fellow-creatures. Moreover, the absence of houses or other buildings at the side of the footpaths brings every figure into relief against the vast, eternal sky, and suggests, in a subtle and almost terrible way, the fragility of the individual, as compared with the infinity above his head. Beneath is the deep, dark river; above are the inscrutable heavens; and between the two are these mites and motes of a vanishing existence, suspended for a time between elements which are stronger than themselves. On London Bridge one sees all the chief varieties of human character, passing on from morn to eve, and often far into the night, with that look of patient endurance, or of half-suppressed suffering, which comes out so strangely when large multitudes of men and women are brought together, without any community of interest, or knowledge of each other's cares. The City man, the lawyer, the clerk, the rugged labourer, the railway servant, the desperately poor, who are evidently on the tramp, either from London to the country, or from the country to London; the lurking thief, the flashy swindler, the Jew with his bag, poor women with their heavy bundles, and heavier faces, and perhaps still heavier hearts; the street Arab on

CANNON STREET STATION.

the look-out for stray halfpence, the girls who sell cigar-lights, the meagre seamstresses going to and fro with their work, and, at one season of the year the vast emigration of hop-pickers, making for the fields of Kent—all these are here, together with many other types of character that demand recognition from thoughtful minds. Under certain climatic conditions, the effect is almost phantasmal in its reduplication and variety, its familiar aspects and its mysterious depth of life.

The complete demolition of the old bridge in 1832 was a matter of necessity, since the decrepitude of the former had at length gone beyond all hope of further patching, and the growing traffic of London required a broader and more convenient way from the City to the Borough. But no more interesting structure was ever devoted to the labourer's pickaxe. A bridge appears to have existed as early as 978; another, built of wood, in 1014, was partly burned in 1136; and this was succeeded, some years later, by the edifice which was destroyed within the memory of some still living. The design was given by Peter of Colechurch, chaplain of St. Mary Colechurch in the Poultry. The construction occupied thirty-three years, from 1176 to 1209, which, considering the breadth of river to be spanned, the massiveness of the work, and the primitive nature of engineering science at that time, does not seem excessive. Peter's bridge was of stone, not of timber, and consisted of nineteen arches, a drawbridge for large vessels, a gate-house at each end, and a chapel in the centre, dedicated to St. Thomas of Canterbury. According to an old tradition, the course of the river was diverted into a trench while the works were proceeding—a trench which, commencing about Battersea, ended at Redriffe. Traces of this vast ditch were remaining about Lambeth Marsh in the middle of the seventeenth century, when small lakes of water appeared here and there, with intervals of fenny ground between. The bridge was built on piles, and these masses of timber, driven into the bed of the stream, must have lasted until the destruction of the bridge itself. On the outside of the timber foundations other piles were fixed, which rose up to low-water-mark, and formed projections into the river, having somewhat the character of open boats or barges. The object of the external masses, which were called "starlings," was to break the rush of water as it dashed towards the bridge itself; but the narrow arches and their timber defences constituted a peril in the navigation of the river, and were the occasion of several accidents to boatmen not thoroughly masters of their calling. The operation of "shooting the bridge" was an exceedingly awkward one, and many persons were afraid to undertake it. The water formed a little cascade in these menacing straits, and the strength and rapidity of the current would sweep away small boats, and leave their occupants little chance of their lives.

In many ways, London Bridge was perhaps the most characteristic structure of its kind in the world. The chapel of St. Thomas, erected on the eastern side of the bridge, over the tenth or central pier (which was carried a considerable distance eastward along the channel of the river), appears to have been a very beautiful Gothic building, reared upon a massive and graceful crypt, which could be approached not only from the bridge, but by a flight of steps leading from the starling of the pier. A tower, often grimly adorned with the heads of distinguished traitors, stood near the centre of the bridge, and the sides were covered with substantial houses, which were not taken down until 1757-8. The tower in the middle part of the bridge was removed towards the end of the sixteenth century, when its place was occupied by a wooden edifice called Nonsuch House, constructed in Holland, brought over to England in pieces, and put together with wooden pegs, to the exclusion of all iron. It crossed the bridge on an arch, and presented a singularly picturesque appearance, with its timber carvings, its four square towers, its domes, its spires, and its gilded vanes. The heads of the traitors—or of those who were described as such—were transferred from the demolished tower to the gate at the Southwark end, which was henceforth known as "Traitors' Gate." Such was the singular aspect of Old London Bridge, which, whether viewed from the river or from the roadway, must have looked like some fantastic vision. Its history is no less full of variety and of strange experiences. Terrific fires occurred from time to time, by which, on some occasions, large numbers of lives were lost. Arches and piers were carried away by high tides, or rendered frail by the incessant action of the water, so that large structural repairs were frequently needed. Here, in 1263, Eleanor of Provence, the queen of Henry III., was attacked by the Londoners, when, during the De·Montfort troubles, she was endeavouring to escape to Windsor. Eleanor was proceeding up the river in a boat, and the exasperated citizens, assembling on the bridge, assailed her, not merely with insulting words, but with dirt and stones, so that she was obliged to return to the Tower. It should

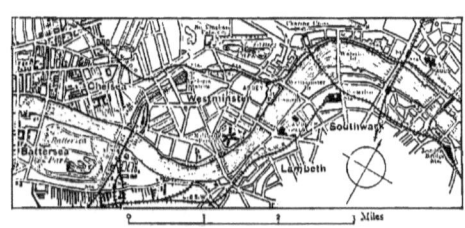

BATTERSEA TO LONDON BRIDGE.

be observed that, although the bridge was for the most part flanked by houses, there were open spaces every here and there, very convenient for pelting a queen who happened to be unpopular. By this way, Wat Tyler obtained an entrance into the City at the head of his Kentish men. Single combats; desperate faction-fights, attended by much slaughter; triumphal processions of conquering kings; splendid pageantries of the great and noble; the mournful pomp of royal funerals; the sumptuous entry of foreign princesses; Wolsey in his grandeur, Wyatt and his insurgents, Charles II. on his return from the Continent, when he at length succeeded to the throne; knights, citizens, men-at-arms, priests, 'prentices, beggars, ruffians, fugitives; the rich, the poor, the mighty, the humble, the downcast, and the prosperous—all this wealth of human action, suffering, despair, and hope, gives an enduring charm to the memory of Peter of Colechurch's structure, and furnishes such a record as few other buildings can parallel. The story of London Bridge is a romance of the deepest interest, of the most gorgeous and the most gloomy colours. But we touch only on its more salient points, and, passing on along the eternal river, leave the shadow of this English *Ponte Vecchio* behind us like a dream.

EDMUND OLLIER.

IN THE POOL.

CHAPTER XI.

LONDON BRIDGE TO GRAVESEND.

Hogarth's Water Frolic—Billingsgate—Salesmen's Cries—The Custom House - Queen Elizabeth and the Customs—The Tower, and Tower Hill—The Pool—The Docks—Ratcliff Highway—The Thames Tunnel—In Rotherhithe—The Isle of Dogs—The Dock Labourer—Deptford and Greenwich—Woolwich Reach and Dockyard—The *Warspite*.

HERE is a mighty change in the river after it has passed Fishmonger's Hall. When the tide is running out it races through the arches of London Bridge, and swirls round the buttresses, and eddies to right and to left in such manner that the Thames waterman, having remembrance of many disasters brought about by absence of knowledge or want of care, amazes his passenger by his singular method of progression, rowing round a clump of barges, getting under the hull of a steamer, shooting across the river with the current, creeping slowly along by the wharves, and otherwise manœuvring as if he were a general preparing to take a town. It requires a long pull and a strong pull to shoot the bridge against the tide, and often has it happened to the idler, leaning over the buttresses, to behold an upturned boat floating below him, and behind it the boatman and his passenger sustaining themselves above water by clinging to the oars.

The Thames waterman of the present year of grace is by no means such a picturesque person as the oarsman of former days. There is no salt water flavour about him; he wears no indication of his calling; he is, to all appearances, merely

ST. MAGNUS' CHURCH AND THE MONUMENT.

a landsman in a boat. It was otherwise in jolly William Hogarth's time. That great humorist drew, as the tailpiece to an eccentric book, a queer little design of a grinning Thames waterman, stout and jolly, seated on crossed oars, his legs drawn up to his chin, a drinking-glass and an earthen pipkin dangling from his gigantic heels. He was a creature all hat, boots, and broad grin; whereas the waterman of to-day is rather a solemn sort of person, very indifferently clad, who takes your shilling or your half-crown as if you were doing him an injury.

The tailpiece aforesaid adorns the last page of an entertaining account of how Hogarth, and three friends of his, set off on a holiday excursion to Gravesend, Rochester, and Sheerness, sailing over just that space of water which we are about to explore. The four boon companions set off from the Bedford Coffee House, in Covent Garden, on the 27th of May, 1732. They spent a day in the neighbourhood of Billingsgate, drinking, apparently, and Hogarth drew a caricature of a long-shore humorist who was known as "the Duke of Puddledock," which said caricature, the rhyming chronicler of the expedition records, in execrable verse, "was pasted on the cellar door." Thackeray calls these four, "a jolly party of tradesmen at high jinks," and high jinks they certainly had, Hogarth and one of his companions playing at hopscotch in Rochester Town Hall. They went down the river in a "tilt boat," laughing and shouting and drinking, exchanging jokes with the watermen, singing each other to sleep with jolly choruses, and behaving generally in a manner that was highly indecorous and reprehensible. It was six o'clock in the morning when they reached Gravesend, something like twenty-four hours after their start. With the tide in your favour, on the steamers which leave London Bridge twice a day, you may now make the same bright and agreeable journey in two hours and a half.

And such a journey should every one make who wishes to realise, however faintly, the picturesque magnificence, the prodigious commerce, the splendid importance of the Thames. The crowded shipping of the Pool, the steamers coming and going, the vessels lying at anchor here and there, as if the river were a huge dock, only feebly represent the vast tonnage which is borne on our grand and historic river every day of every year. Behind the great piles of warehouses—towering over the housetops, ornamenting the sky with a curious fretwork of masts and spars and cordage—lie scores and hundreds of the vessels of all nations, crowded into dock beyond dock, making a line of rigging, of glittering yards and masts, of furled sails and flaunting canvas, on either side of the Thames for mile on mile.

It is on the Tower side that the line is least broken. London Bridge is scarcely left behind ere St. Katharine's and London Docks come in sight; then follow the enormous acreage of the East and West India Docks, then come the docks at Millwall, and the Albert and Victoria Docks, stretching onward to North Woolwich—a vast contiguity of dock property, basin beyond basin occupied by some of the finest shipping that roams the seas.

Earlier in the century, when the screw-steamer was as yet undreamed of, and there had been no vision of the steam-tug which is so vast a convenience to-day, this portion of the river presented at certain seasons a much more stirring sight than now. Fleets of vessels, with their sails spread, came in at every tide; hundreds of ships lay crowding in the Thames at the mercy of the wind; it was a long panorama of seafaring life, with no bellying smoke to impede the view. All that has been changed by the wand of Science and the genius of Discovery. If a vessel lies in the stream instead of in the docks, it is for purposes peculiarly its own; and the dock gates, instead of opening to whole fleets driven up by a prosperous wind, swing open to solitary, but more gigantic, vessels propelled by steam.

Not that sailing-ships are no longer numerous in the Thames! The old East Indiaman has departed, the ships of John Company are broken to pieces; but the tall three-master is by no means an unfamiliar object, and on the Thames waters below London Bridge one may encounter schooners and brigs and brigantines galore. Nor has the number of lighters and wherries and dumb-barges diminished. On most shipping rivers these auxiliaries of trade have almost or wholly disappeared. The uncouth keelmen of the Tyne are a race with few survivors, and when the universally popular "Keel Row" is sung or played, it is almost invariably without reference to its former signification. On the Mersey no long line of coal-barges blocks the stream. The barge and the keel have, indeed, had their day, and are now little more than encumbrances; yet it is probable that they will be familiar objects for years to come on the Thames. When the docks were made, the watermen rose up in revolt against a threatened invasion of their privileges, and were fortunate enough to secure for themselves new rights which ensured their continuance and prosperity. So it happens that in addition to the sailing-ships and schooners which may be seen at anchor along either side of the Thames, there is a great number of smaller craft, inconvenient but full of interest, greatly in the way, but very delightful to the artist and the heedful possessor of a "quiet eye."

No effective justice has ever yet been done to the lower portion of the Thames. You will find it stated in most books on the subject that the river ceases to be picturesque when it has passed St. Paul's. A French poet calls it "an infected sea, rolling its black waters in sinuous detours"; and that is the despondent view that has been taken by the majority of English writers. Yet in the eyes of those who have roamed about this section of the river, and have loved it, only at London Bridge does the Thames become really interesting. In the higher reaches it is an idyllic river, swooning along through pleasant landscapes; after St. Paul's it takes on a new and more sombre sort of glory, assumes a mightier interest, and is infinitely more majestic in the lifting of its waters. Above London Bridge, even when the wind is blowing, the waves are small and broken, like those of a mountain lake; in the Pool the water surges and heaves in broad masses, the light seems to deal with it more nobly, and the Thames assumes such majesty as becomes a stream which flows through the grandest city, and bears so great a portion of the commerce, of the world.

As for picturesqueness, one may behold a score of the finest possible pictures from London Bridge itself. The grey tower of St. Magnus' Church, smitten by a passing ray of sunlight, stands out bright and shining behind the dark mass of buildings over Freshwater Wharf; beyond it, more dimly seen, the Monument lifts its flaming crown; the Pool is alive with hurrying steamers and clustering sails; Billingsgate is in the midst of its traffic; the white face of the Custom House looks down into the dun waters; and yonder rise the more sombre walls of our most ancient fortress, the venerable quadragon of the White Tower, with its four dark cupolas, dominating them all.

Round the spot on which Freshwater Wharf now stands clustered Roman London. There are still some half-hidden relics of it under the recent and handsome Coal Exchange in Thames Street. There, descending to the foundations, one may find a hypocaust full of fair spring water, a pavement floor, an ancient and austere seat built of Roman tiles, and some pieces of ruined wall. It is the lower portion of a Roman house, the most interesting and complete bit of evidence still remaining in London of the Roman occupation of Britain.

The front of Billingsgate has altered its aspect of late. A wharf has arisen where, heretofore, a couple of narrow gangways descended sheer to the foreshore of the Thames, when it was exposed, and to the water, when the tide was in. Many a Billingsgate porter has lost his life hurrying up those gangways, yet, so conservative is the City of London in its habits, that it is only a few years since the conclusion was reached that the market would be no worse, and human life would be all the safer, for a pier. With that very modern improvement one of our London "sights" has changed its aspect. No longer may we behold the four lines of white-jacketed figures, two bustling up from and two hurrying down to the boats. Yet the white-jacketed figures are there, and they bustle about as of old, though the work has become indescribably easier, and is carried on by men in less constant peril of their lives.

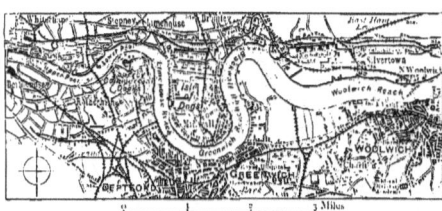

LONDON BRIDGE TO WOOLWICH.

To see Billingsgate in the full tide of its work—and England has no other sight to compare with it—one must rise with the sun in summer, and long before the dawn in winter, when heavily-laden market-carts from Kent are rumbling over London Bridge, whilst the homeless tramp is still composing himself to slumber, and while still the mists cling to the surface of the river so heavily as to seem beyond the power of any mere London sunshine to raise or dispel.

At five in the morning, summer or winter, rain or shine, Billingsgate seems to shake itself and start on a sudden into active and turbulent life. In the night a series of long, low, snake-like steamers have crept up the river, bearing freight from the fishing-smacks which are pursuing their dangerous fortune in the North Sea. Just below where they have dropped anchor cluster several broad-beamed, highly-polished, Dutch schuyts, bringing oysters or eels to market, and reminding you, by their bulk and build, of the stout, prosperous, slow-moving citizens of Amsterdam. Little

BILLINGSGATE EARLY MORNING.

panting steam-tugs are hurrying here and there, and amid a confused glare of lights, and a tempest of smoke and steam, the Billingsgate porters, having waited for the five o'clock bell, rush out in streams to schuyt and smack and steamer, pushing, shouting, swearing, surging to and fro in the mist and steam and glare, working with the energy of gnomes doomed to perform an allotted task ere the first beams of morning surprise them at their toil.

Thames Street, and Fish Street Hill, and Pudding Lane, and many a street and alley roundabout, are crowded, packed, jammed, with vans and carts and trollies. The stranger wanders bewildered and afraid among all these, in danger of being knocked down by laden porters, run over by market-carts, hustled out of all self-possession by feverish buyers, or lost amongst such a wild and interminable confusion of vehicles as no other place in the world can show.

For all that is known to the contrary, Billingsgate has been a fish-market from the time when the ancient British inhabitants of the proud hill on which the City of London stands put off in their coracles to seek the means of livelihood in the broad waters which dock and warehouse and wharf now confine in the comparatively narrow channel of the Thames. There was a toll on fishing at Billingsgate when the Saxon Æthelstan reigned. William III. made the market open and free for all sorts of fish in 1699. Since that day many attempts have been made to establish fish-markets elsewhere in London, but up to this time with uniform non-success. It is not yet quite a score of years since the present Billingsgate Market was completed. You may still read, in even recent books, of "the elegant Italian structure" of Mr. Bunning, with its towering campanile, its fine arcades, and its picturesque blending of brick and stone. Mr. Bunning's market, however, was too small for its purposes; and in 1874 the present building was begun, and, in spite of vast difficulties, was finished without disturbance of the business of the day. It preserves much of the old "elegance" of structure, and is partly Italian in style, but the smoke of the steamers clings to it, and has blackened it so that, between the grey buildings above Freshwater Wharf and the shining walls of the Custom House, it looks like a patch of shadow in a field of light.

Fish was once indifferently delivered at Billingsgate or at Queenhithe, on the other side of London Bridge. Henry III., at a loss how to furnish pin-money for his wife, gave to her a tax on the fish landed at Queenhithe Pier. It was a tax, too, which the fishmongers were very reluctant to pay, and many were the fines inflicted on shipmasters who tarried at Billingsgate instead of making their way to the royal quay. Billingsgate fought that hard battle against royalty with great resolution, and ultimately won. Since then it has become obstructive on its own account, and has, in turn, successfully resisted any invasion of its own exceptional privileges. The dealers at Billingsgate must in those early days have been as rich, and quite as exclusive and privileged, as are their successors of this latter part of the nineteenth century, for it is recorded how, when the news was brought to London of the victory which Edward I. had obtained over the Scots, they paraded the city with over a thousand horsemen, accompanied by the sound of trumpets, and the streaming of banners, and all the fine pageantry of a picturesque time.

The daily supply of fish to Billingsgate amounts, on an average, to 500 tons. It is difficult to realise how prodigious a quantity is this; but the imagination is assisted by reflecting that one ton of fish is equal in weight to twenty-eight sheep, so that the day's supply of 500 tons is equivalent to a woolly herd of not less than 14,000. London in this manner draws to itself the great bulk of the fish that are caught around our coasts; but, it must be understood, Billingsgate does not exist for the advantage of metropolitan consumers alone. Most of the large provincial towns draw upon the great fish-market of the Thames, and almost as soon as the day's supply is landed and sold much of it is speeding off in fast trains to the great centres of industry, where it is again distributed, it may be, to less important communities, and to small hamlets nestling amid ancestral trees.

At Billingsgate you may make your purchases by the ton or the single fish. There are fish-salesmen of varying degrees, some selling, in large quantities, the fish as it is landed from the boats, others selling over again to shopkeepers and to costermongers what they have only themselves purchased some half-an-hour before. The more respected and prosperous dealers, coming early, with long purses, have the pick of the market, and are speeding off home again before the bell of St. Paul's has tolled the hour of nine. Then the costermongers come crowding in, shouting, pushing, swearing, exchanging jokes, impugning the freshness of the fish, boiling into anger at the prices asked from them, and filling the market-hall with an amazing clatter of Cockney tongues.

The attendance of the London coster is regulated by the supply of fish. Sometimes only a few scores of these itinerant dealers are to be encountered in Old Thames Street; sometimes they are present by hundreds and thousands. It has never yet been discovered how the intelligence of a profuse and cheap fish supply is diffused over London; but it invariably occurs that when the market is overstocked every costermonger in town has knowledge of the fact long before noon. It is much as if the street-dealers were connected with Billingsgate by electric wires. "Barrows" come racing by dozens over London Bridge; Covent Garden Market is suddenly deserted by the most numerous class of its customers; from Shadwell, from Kentish Town, from more remote Hammersmith, the costermonger rushes off to Billingsgate as if for bare life, and by mid-day. cheap fish is being cried all through the London streets and far off at the doors of "Villadom" in the suburbs.

The late Henry Mayhew has striven to give an idea of the confused cries of Billingsgate in his wonderful and painstaking work on "London Labour and the London Poor," where the sounds heard above the general din are represented thus:— "Ha—a—ansome cod! best in the market! All alive! alive! alive, O!" "Yeo, ye—e—o! Here's your fine Yarmouth bloaters! Who's the buyer?" "Here you are, guv'ner; splendid whiting." "Turbot, turbot! All alive, turbot!" "Glass of nice peppermint this cold morning, a ha'penny a glass, a ha'penny a glass!" "Fine soles, oy, oy, oy!" "Hullo, hullo, here! beautiful lobsters, good and cheap!" "Hot soup, nice pea-soup! a—all hot, hot!" "Who'll buy brill, O, brill, O?" "Fine flounders, a shilling a lot! O ho! O ho! this way—this way—this way! Fish alive! alive! alive O!" And in such fashion is business carried on at Billingsgate every morning, amid a turbulence not to be described.

It is a prosaic, evil-smelling business, this of fish dealing, relieved by no such spectacles as were to be witnessed in the time of Stowe, when, "on St. Magnus' Day the fishmongers, with solemn procession, paraded through the streets, having, among other pageants and shows, four sturgeons, gilt, carried on four horses; and after, six-and-forty knights armed, riding on horses, made 'like luces of the sea;' and then St. Magnus, the patron saint of the day, with a thousand horsemen." The salesmen reserve their solemn rites in these days for the dinners in Fishmongers' Hall, and the only "Knights" they can boast of are those ludicrous "men in armour" who make a part of the Lord Mayor's Show.

Close by Billingsgate lies the long frontage of the Custom House, conspicuous no less by reason of its bulk and position than for that leprous whiteness which, on certain kinds of stone, is one of the effects of the biting and crumbling atmosphere of London. The site is one that should be dear to lovers of English poetry. Here Geoffrey Chaucer officiated as Controller of Customs, the stipulation being that he should write the rolls of his office with his own hand, and perform his duties personally, and not by deputy. It may be that whilst his pen was thus unpoetically employed, his mind wandered off to the "Tabard" Inn, by the end of London Bridge, to its jolly landlord, "bold of his speech, and wise and well-taught," and to the curiously compounded bands of pilgrims who gathered there on their way to the shrine of St. Thomas at Canterbury. Here, also, came William Cowper, in one of his fits of insanity, intent on suicide. The water was low, exposing the foreshore, and there was a careless porter sitting on a bale of goods. It seemed to the poor stricken poet as if the man were waiting there to prevent the execution of his purpose, "and so," he says, "this passage to the bottomless pit being mercifully closed against me, I returned to the coach," which was really the only sensible thing he could do.

The present Custom House, built in 1825, contains one of the longest and the most dingy-looking rooms in England. Here may be encountered strings of British merchants and rough ship-captains waiting to transact business relating to their cargoes. At one counter is kept a record of vessels and their owners, at another the clearance of ships outward is the subject of concern; at a third the skipper must hand in a list of every article on board his vessel, and thence proceed from counter to counter until he has satisfied all the requirements of the law. In one corner of the building there is a Custom House Museum, containing many quaint official documents, detailing how John Doe, being a Papist, did not receive his quarter's salary, and how some other servant of the Customs has been docked of his wages because of the indiscretion of somebody else's wife; containing, also, curious articles which have been employed in small acts of smuggling —a stewardess's crinoline that has been puffed out with a bottle of right good Hollands, a book which has been made to do duty as a brandy-flask, quantities of snuff that have been shipped as oilcake, and many other curious examples of unexpected failure to evade the law. Those whose business it is to detect cheats of this description love to retain some memorial of their prowess, and in this manner it happens that the Custom House Museum is valuable chiefly to those who care to study human ingenuity in connection with dishonest purposes.

There is in existence a curious record concerning the Custom House and Queen Bess. "About this time (1590)," writes the quaint author of "The Historie of the Life and Reigne of that famous Princesse Elizabeth," "the commodity of the Custom House amounted to an unexpected value; for the Queen, being made acquainted by the means of a subtle fellow, named Caermardine, with the mystery of their gaines, so enhanced the rate that Sir Thomas Smith, master of the Custom House, who heretofore farmed it of the Queen for £14,000 yearly, was now mounted to £42,000, and afterwards to £50,000, which, notwithstanding, was valued but as an ordinary sum for such oppressing gaine. The Lord Treasurer, the Earls of Leicester and Walsingham, much opposed themselves against this Caermardine . . . but the Queen answered them that all princes ought to be, if not as favourable, yet as just, to the lowest as to the highest, desiring that they who falsely accused her Privy Council of sloth or indiscretion should be severely punished; but they who justly accused them should be heard. That she was Queen as well to the poorest as to the proudest, and that therefore she would never be deaf to their just complaints. Likewise that she would not suffer that these toll-takers, like horse-leeches, should glut themselves with the riches of the realm, and starve her Exchequer; which, as she will not bear it to be docked, so hateth she to enrich it with the poverty of her people." From which lion-like speech it appears that Queen Elizabeth more than suspected her Privy Councillors of having intercepted moneys which should have found their way to the Exchequer of the Crown.

After the Billingsgate fever is over, everything round about the Custom House seems quiet and sleepy and still; yet an almost inconceivable amount of business is transacted within its walls. Every merchant receiving a cargo, every shipmaster going out or coming in, has unavoidable business here. There is a series of counters, distinguished by the various letters of the alphabet, and from one to another the visitors to the Custom House continue to circulate, engaging in one sort of transaction at one counter, and in some other sort at a second, third, and fourth. It is a long and wearisome process, the discharge of the various duties appertaining to the entry and the clearing out of ships—a process which, be it said, seems much less trying to the clerks than to those on whom they are called upon to attend.

In front of the Custom House there is a broad quay, used as a public promenade, a true haven of rest to him who has lost heart and energy in the almost vain attempt to escape from the crowd and the bustle of Thames Street. At this spot, on New Year's Morning, the Jews of London were wont to assemble to offer up prayers in remembrance of that sad captivity when their people sat down by the waters of Babylon and wept. The custom has been discouraged of late years, but there are still some professors of the ancient faith who follow the rule of their forefathers, and offer up the time-worn prayers on the spot which was consecrated by them in "the days that are no more."

It is difficult to break away from that portion of the river on which we seem already to have lingered too long. The Thames is here full of interest and of crowding associations. Over the water, behind the great, grim warehouses, slopes downward into Bermondsey that Tooley Street which the three tailors—"we, the people of England"—have made famous throughout the world. From amid grimy roofs and grey-brown walls rises the tower of St. Olave's Church, half-buried and lost amid a London of which its builders never dreamed. Down here, in narrow street and dim entry, the bewildered stranger begins to feel that, after all, man is too small for the planet on which he lives. Great walls—of granary, and store, and manufactory—reach over and above him, and dwarf him into extreme littleness. He seems to be walking beneath high cliffs by the sea. The whole air trembles and throbs with noise and travail. Here and there, through some unexpected narrow opening, may be discerned a thin strip of river, with ships and boats. At intervals of every two or three hundred yards these openings occur, and they lead down to old-fashioned Thames stairs, where the waterman plies his trade. Lingering about the landing-places, or the streets and alleys adjacent thereto, one meets occasional blear-eyed, evil-countenanced, ill-clad men, who approach with a sinuous stoop of the shoulders, a deferential ducking of the head, and a dirty thumb raised to the brim of a greasy hat. These men will do anything for money except work. If you employ one of them to conduct you to the stairs and to call a boat he will pretend to hurry forward, but, without progressing much, will look furtively behind, seem to measure your size and estimate your running powers, and then proceed slowly in front, his evil-looking thumb continually beckoning, and his croaking voice ejaculating, "This way, sir; this way; this way, if *you* please." He means no mischief probably, but as you walk through those parts of London in such company you are thankful it is daylight, and that even the alleys and courts of the Surrey side are not absolutely impervious to the sun.

Some of these strange places have equally strange names. Pickle Herring Street, and Shad Street, and other cramped thoroughfares with ancient and fish-like designations, suggest that here, also, almost directly opposite to Billingsgate, there must have been a market once. There is scarcely even a shop or a public-house now. This is the London that really works with a will. To the right are tanneries and tallow-chandleries—their odour loads the atmosphere as if it were a thick fog, incapable of any effort to rise—to the left are vast granaries and wharves; and between them the narrow spaces are filled up with hurrying vehicles and toiling men.

From the south to the north side of the river there is a continual stream of labourers, some making their way under the river, like moles, by means of the subway, some streaming down to the boat-landings and casting off in batches into the tide. The subway is an iron barrel, some six feet in diameter, which has been driven underground far below the bed of the Thames. Walking through it, one hears, as a series of dull, only half-audible, thuds, the lashing of a paddle-steamer overhead. No other sound reaches that cramped, underground chamber, in which one seems to be walking as in a coal mine, from the dark into the dark. After this dreary journey, we ascend a flight of stairs that is wearying, and that seems to be endless, and emerge on Tower Hill, into the sunshine, and the presence of green trees, and the sight of what is most venerable in the whole English realm.

Tower Hill is a sort of oasis in a desert filled with the whirling sands of traffic—the terminus to the great lines of warehouses which fill Thames Street. Surrounded by shops and offices and public buildings, it is, but for the country cousin newly arrived to behold "the sights," almost as quiet as some retired corner of the parks. Standing here, where so many historic heads have fallen, one may behold the river streaming by, and watch the sun lighting up the polished masts of a hundred vessels slumbering in the Pool.

On Tower Hill stands Trinity House, which claims notice here because of its close connection with the river and with ships. Queen Elizabeth made the Masters of Trinity the guardians of our sea-marks, and they have now the sole management of our lighthouses and our buoys. Part of their business is to mark out the locality of wrecks, and to announce to the shipmasters of all nations any changes in the entrances to English ports. At Trinity House is one of those numerous London museums which are seldom seen—a museum of models of lifeboats, buoys, light-houses, life-saving apparatus, and other objects connected with the safety of ships and voyagers at sea. Here the curious visitor may spend an hour or two with advantage, and it will be matter for wonder if he does not come away oddly instructed in many intricate matters connected with the sea.

To all fairly informed Englishmen, the history of the Tower of London is so familiar that it would be an impertinence to recount any portion of it here. The "towers of Julius, London's lasting shame"—not that Cæsar really had anything to do with them—have the peculiarity of being known, through some sort of representation, to most, even, of those stay-at-home people who are said to have country wits. And let it be said at once that at the first glance they are not nearly so imposing as they are usually made to appear. "And that is the Tower?" an American observed to me lately; "and that is the Tower? Well, then, I guess the Tower was not worth crossing the Atlantic to see." Yet, even this unfavourable critic saw reason to change his views. It is from the river, and not from Tower Hill, that the first inspection of this venerable edifice should be made. Seated

THE TOWER, FROM THE RIVER.

on an idle barge, one may contemplate it at leisure; and it is only after leisurely contemplation that its fine grouping, its richly varied colour, and its compact massiveness force themselves on one's slow appreciation. From just behind where we are supposed to be seated, the adherents of the Earl of Salisbury poured stone shot into the Tower precincts when Henry VI. was king. Facing us, the lower portion now hidden by a quay wall, is the round arch of Traitors' Gate—

> " Through which before
> Went Essex, Raleigh, Sidney, Cranmer, More."

with those steps still intact on which the Princess Elizabeth seated herself, petulantly declining to make such an entrance to the Tower as would declare her to be a traitor to the realm.

Up to quite recently, to the time of Mr. Shaw-Lefevre's occupancy of the office of Board of Works, indeed, the Tower, as seen from the river, was much disfigured by modern buildings of exceeding ugliness, which public feeling had long since condemned. Most of these have now disappeared, but one such building, bearing the appearance of a granary, still remains to break the face of the White Tower with its dull red-brown. Beyond it one catches glimpses of quaint gabled roofs, characteristic of periods as widely separated as those of Elizabeth and Queen Anne. To the left are more buildings of old red-brick, with ivy clustering over them, and beyond, home of many sad memorials, rise the walls of the Beauchamp Tower, with, beside them, a curious lumber of quaint, many-windowed, square turrets, jumbled together in different ages for diverse purposes, and now used as lodgings for the Beefeaters and the guard.

In the church of St. Peter ad Vincula, the situation of which one may guess from the river, were interred the headless bodies of Queen Catherine Howard, of Anne Boleyn, of the Countess of Shrewsbury, and of Lady Jane Grey; of Sir Thomas More, of the first Cromwell, of Seymour, Lord High Admiral, of his brother, the Protector Somerset, and of many others whose illustrious positions were the occasions of their own misfortunes. "There is no sadder spot on earth than this little cemetery," says Macaulay. "Death is there associated, not, as in Westminster Abbey and St. Paul's, with genius and virtue, with public veneration, and with imperishable renown; not, as in our humblest churches and churchyards, with everything that is most endearing in social and domestic charities, but with whatever is darkest in human nature and in human destiny; with the savage triumph of implacable enemies, with the inconstancy, the ingratitude, the cowardice of friends, with all the miseries of fallen greatness and of blighted fame." The most ancient and illustrious building that is mirrored in the waters of the Thames is, indeed, also the home of the grimmest memories. The Tower is a sad, depressing place to visit, the concrete representative of all the darker events of our history.

The character of the Thames below London Bridge is best expressed by the normal appearance of the Pool. And let me at once explain that the Pool is the wide, curving stretch of river which extends from just above the Tower to the neighbourhood of "Wapping Old Stairs." Here, in most abundance, you find "toil, glitter, grime, and wealth on a flowing tide." Mr. W. L. Wyllie's picture, purchased out of the funds of the Chantrey bequest, is a wonderfully characteristic description of the aspect which the Thames presents in this busy portion of its course. In the foreground a couple of coal-laden boats, with a little hasty steam-tug beside them, are making slow headway against the tide; beyond these, a great iron steamship rears up its vast bulk; a couple of heavily-laden Thames barges are flying along under full sail; on either side of the river there are confused masses of rigging, with here and there the hull of a ship, half visible through whirling clouds of smoke and steam. On the waterway, kept clear of all vessels at anchor for a breadth of 200 feet, the strong white sunlight gleams, making clear to the spectator what the poet Spenser had in his mind when, in his rich vocabulary, he spoke of "the silver-streaming Thames."

Spenser's phrase is one which has been greatly misunderstood. The Thames, even in its quietest and least corrupted days, can never have been a very pellucid stream. When Taylor, the water poet, plied his craft upon it he must have found almost as much difficulty in looking into its mysterious depths as we find to-day. Certainly, to the local colour of a swift-flowing river which brings down continuous deposits of mud from far-off meadow-lands, such a word as "silvery" could never properly be applied. Only when the sunlight struck the river, and its rippled surface tremblingly gleamed back to the sky with a reflection of its own brightness, could Spenser have been delighted by the aspect of the " silver-streaming" waters, ebbing and flowing through London's heart, and bearing onward their heavy burden to the sea. Leaning over the rail of a steamer outward bound, one is apt to forget everything else in the contemplation of these brilliant and rapidly changing effects of light, which seem to chase one another as if in mere wantonness, and which, in no mood of wantonness, at every capricious curve of the stream cast on the thick dusky waters some new and strange glory.

The Pool is full of such life and movement as is to be encountered on no other English river, for here the crowded ships do not merely lie at anchor, waiting on wind and tide; they are busy loading and unloading freights. One hears the grating of cranes and the shouts of men; the peculiar " dumb-barges " of the Thames cluster round the hulls of screw-colliers from Newcastle, and receive from them their separate loads of coal; and excitable little steamers are running in and out as if they had lost their way among the crowded shipping. In midstream the traffic is almost as busy and confusing as that of a London street. Vessels are coming up with the tide; barges are slowly floating onward, their brown sails spread, tacking to the wind, their decks washed now and again by some arrowy wave from a paddle-steamer. There is, as Mr. Jefferies says, "a hum, a haste, almost a whirl," for on the river work proceeds at a more rapid pace than in the docks, and the Thames, it must be remembered, is the busiest port on the surface of the globe.

It is hard to say whether the Pool is most beautiful and striking at moonlight or in the dawn. Turner loved it best at the hour before twilight, when the sky was robed in gold and crimson and purple, and the Thames was ablaze with the light of the setting sun. At such seasons it is indeed very glorious; yet to me it has always seemed most beautiful in the morning, when the light is slowly diffusing itself from behind a bank of purple cloud, and the face of the White Tower is touched into pale gold, and there is a glittering radiance on turret and roof, and the craft anchored in the stream are reflected to every mast and spar and half-furled sail, and the river trembles in the new radiance as if it were divided between delight and fear. Everything is very still and soft and shadowy. It is such a scene as seems appropriate to happy dreams. In another hour or two the river will be awake, the twitter of birds flitting across the waters will be drowned in the shouting of labourers and the shrieking of cranes; the stream, its brief glory departed, will be churned up by paddle-wheel and screw; the swarthy steam-colliers will come, hard and clear, out of the soft haze, and the Thames will become a workaday river again, wonderful still, but, after such a vision, too grimly prosaic and real. Yet it is well to have seen it once with the dawn upon it, if only to learn how those have libelled it who deny that it is "picturesque."

In the seventeenth century there was, in the Upper, Lower, and Middle Pools, space for 900 vessels. Nearly that number might now be packed into the London and St. Katharine's Docks, which lie just below the Tower, hemmed in by what was once fashionable London, now fashionable no more, but famous the world over as the accustomed haunt of the seaman on shore. Before docks were constructed along Thames-side, vessels were unloaded into barges and wherries, and river-robbery was a thriving trade. Numbers of men lived, and grew rich, on what they had contrived to steal from cargoes that were waiting to be discharged at the wharves about London Bridge. Ships were sometimes as much as six weeks in unloading, and a whole host of lightermen, carmen, porters, and nondescripts thrived on the unconscionable delay. There were good pickings in those times, and it is wonderful, when we consider with how much rascality and obstruction our commerce had to contend, that England ever became a great nation of carriers and traders.

The docks nearest to London Bridge cover the site of a church, a hospital, and a graveyard.

More than 700 years ago, or, to be precise, in 1148, Matilda, the wife of King Stephen, founded on a site just below the Tower a hospital which was dedicated to St. Katharine. It endured, in one form or another, to 1827, when the building was pulled down, and the hospital was removed to Regent's Park. In that same year was commenced the construction of the St. Katharine's Docks, which, by the employment of 2,500 workmen, were completed in the brief space of eighteen months. They cover an area of twenty-three acres, ten of water and thirteen of land. The docks are the most prosaic of all those which are to be found along Thames shore. To the river they present a dull heavy frontage, which suggests no connection with ships; on the land side they are shut off from observation by a prodigiously high wall. Entering through the gates you find three great basins, with ships lying close to the wharves, and you have towering above you gigantic warehouses, dull and dismal, but capable, you would suppose, of finding storage room for half the commerce of a city. The cellars of St. Katharine's Dock are complex and amazing, but the docks themselves are going out of vogue, for many of the ships which used to frequent them are now intercepted before the lights of London come in sight, the Victoria and Albert Docks, much lower down the river, absorbing a great proportion of the traffic which was wont to make its way into the Pool.

The London Docks, much larger than those of St. Katharine, are beginning to share in the same neglect. They are of more ancient date than their neighbours, having been designed by Rennie, the architect of London Bridge, in 1805. As many as three hundred vessels can find a comfortable haven here. The warehouses will contain 220,000 tons of goods; there is storage for 130,000 bales of wool; the wine-cellars are among the marvels and attractions of London. "Here," Mr. Sala has remarked, "in a vast succession of vaults, roofed with cobwebs many years old, are stored in pipes and hogsheads the wines that thirsty London—thirsty England, Ireland, and Scotland—must needs drink." Curious persons come here with tasting orders, and are shown round by brawny coopers, who seem marvellously wasteful of good wine, and are more generous to their visitors than the most prosperous of City merchants, with the best plenished of wine-cellars, is to his friends. Many a visitor to the wine-cellars of the London Docks has found occasion to regret, when he has reached the open air, that he has been so easily tempted to pass too frequent opinions on too many varieties of wine. In the cellars an amateur wine-taster is apt to overrate the strength of his head; above ground once more, the breath of the river brings him to a sense of his own incapacity for frequent and varied potations, and he shamefacedly

LIMEHOUSE CHURCH.

betakes himself to a cab, to escape as quickly as may be from the scenes of his bibulous indiscretion.

On the dockside one encounters men of all nationalities—the swart Lascar, the dusky Sulioto, the quiet pigtailed Chinaman, the grizzled negro; Germans, Swedes, stout little Dutchmen; Americans, Fins, Malays, Greeks, and Russians. Nowadays an English ship is a polyglot institution. In the Sailors' Home, near to the gates of St. Katharine's Docks, one may hear men conversing in all the European languages; in the Asiatic Home, close to the India Docks, there is such a confusion of tongues as dismayed the builders of Babel. Entering and departing, the owners of all these voices, and the thousands of dock labourers,

THE RIVER BELOW WAPPING.

lightermen, loafers, visitors, must pass the inspection of the police, who stand at the dock gates always on the watch, and who do not scruple to submit to close examination the garments of all those whose pockets may happen to bulge unduly, or who, having entered in the morning with a perfectly erect spine, stoop inexplainably at the shoulders when they have completed their business at night. In the docks there is a perfect system of espionage, and "the Queen's Tobacco Pipe," until recently located at the London, and now at the Victoria Docks, has smoked many thousands of little presents of tobacco that large-hearted sailors had intended for the gratification of their friends.

The long, narrow, grimy, and dissolute lane known to Englishmen everywhere as Ratcliff Highway, and now disguised under the name of St. George's Street East, begins its career near the gates of St. Katharine's Docks, and winds along like a great slimy snake towards Limehouse and Blackwall. It is unvisited of all those who have not business to transact with men who go down to the sea in ships; for this is peculiarly the sailors' quarter of London. Jack is to be encountered at every step, not infrequently reeling somewhat, and with a lady of loose manners on his arm. The shop-fronts are hidden behind strange collections of oil-skins, sea-boots, mattresses, blankets, and the miscellaneous assortment of articles specially provided for emigrants and sailors. The public-houses, of which there are many, resound with the noise of mechanical organs and string bands. The language one hears is of a strictly nautical description; and every third house or so is a lodging-house for sailors.

Of late years Ratcliff Highway has improved in character somewhat, and many of the men who were wont to be fleeced and robbed in it have been rescued from the crimps and sharpers by the Sailors' Home; but it has still much of its old disrepute left, and discreet persons do not perambulate it after nightfall without the escort of the police. Here the seafaring men of various nationalities separate themselves into groups, and form little colonies of their own. The public-house is their forum, to whatever nation they may belong. In one of these, English is spoken, in another German, in a third Norwegian, in a fourth Greek. Even the negroes have a special house of call of their own. As for the Chinamen, they prefer to smoke opium in quietness, and so they divide themselves between various Chinese lodging-houses, where they can eat, in the properly orthodox manner, with chop-sticks, and assemble round a table at night to gamble with their friends. It is a strange, stirring, disordered place, Ratcliff Highway. Its population is changing with the arrival or departure of every ship, yet its aspect and its frequenters always seem to be the same, similar in manners, bent on the same amusements, afflicted by the same vices, reeling into or out of the same doors. There is nothing here except an occasional piece of nautical slang to suggest the jolly British tar. To a great extent, indeed, the tar has ceased to be either jolly or British. The majority of the sailors to be met with in Ratcliff Highway are visibly and distinctly foreign. There are no white "ducks," or raking straw hats; nobody publicly "shivers his timbers," or speaks in that mixed and technical language which helps to make the characters of Captain Marryat so delightful. Only on the stage is it nowadays possible to encounter the sailor of tradition. The seaman who frequents Ratcliff Highway outwardly resembles the stoker of a railway train, attired in his second best suit. There is nothing romantic about him, nothing picturesque; and if the river and the docks were not so near, and the shops were not so nautical-looking, and one's ears were not occasionally saluted with "How goes it, Captain?" and "Hallo, mate!" there would be nothing to suggest his connection with the sea.

All this was very different in Ned Ward's time, when that lively writer was collecting materials for his *London Spy;* very different, indeed, when men who are only now middle-aged were in the bloom of their youth. "Sometimes we met in the streets with a boat's-crew," says Ward, "just come on shore in search of those land debaucheries which the sea denies them; looking like such wild, staring, gamesome, uncouth animals, that a litter of squab rhinoceroses, dressed up in human apparel, could not have made a more ungainly appearance. . . . Every post they came to was in danger of having its head broken. . . The very dogs in the street shunned them. . . . I could not forbear reflecting on the 'prudence' of those persons who send their unlucky children away to sea to tame and reform them." And well he might wonder at that same prudence now, if he saw how miserable and forlorn the British tar can look when his money is spent, and how little his appearance is suggestive of those high spirits which a life on the ocean wave is supposed to engender.

From Wapping, to which Ratcliff Highway will bring us, you may pass, through the famous Thames Tunnel, under the river to Rotherhithe. Not, however, as formerly, when the tunnel was reached by sets of circular stairs, and toyshop keepers drove a meagre business under a dripping and gigantic arch. At that period, the tunnel contained a central arcade lighted by gas; nowadays, it is so dark that no man can discern when he enters and when he leaves; for it has been absorbed into the great railway system, and instead of traversing it on foot one is whirled through it in a train, so that the traveller might be carried underneath the Thames, at a depth of more than seventy feet below the surface, without knowing that he had been on anything else but an ordinary underground railway. The Tunnel cost nearly half a million of money to construct, and twenty years elapsed—from 1823 to 1843—from the time when it was designed by Brunel and the day when it was opened to the public. As a place of resort for sight-seers it proved a gigantic failure; as a railway tunnel, it is a means of communication between the two most populous and busy districts of London.

Not that there are many signs of business to be encountered when one leaves the tunnel by means of the railway station at Rotherhithe. At the first glance the district round about seems quiet and sleepy and secluded. Mr. Walter Besant came upon it unexpectedly and with great joy, for here he found a world altogether in contrast with that which he had left a little higher up the Thames—houses of quiet old sailors, little churches and chapels, rows of small dwellings with flowers blooming on the window-sills, timber-yards and lagoons and canals, and a general air of retirement and repose. It is a narrow strip of shore, Rotherhithe. On one side it is washed by the Thames; on the other, it is hemmed in by the Surrey Commercial Docks. Sailor life in its better aspects is to be encountered here, for the neighbourhood has been haunted by seamen from Saxon times downwards, and the influence of the quaint older world has not yet passed away. It was through being a "sailor's haven," say the antiquaries, that Rotherhithe came by its name. Here Canute cut deep trenches, which, according to one of the friends of Samuel Pepys, who saw the remains of them in the course of a walk from Rotherhithe to Lambeth, were intended to divert the course of the Thames. At Rotherhithe Edward III. fitted out one of his fleets, and close upon its borders, in Bermondsey, lived some of our early kings.

The signs of the Rotherhithe inns—the "Swallow Galley," and the "Ship Argo"—seem to carry us back to "the stately times of great Elizabeth;" and though the place itself must have altered greatly since then, the manner of life of some of its inhabitants is much like that of their predecessors must have been when stout, high-decked ships sailed by on their way to the Spanish main, and Rotherhithe sent out its contingent of vessels and men to fight against the Invincible Armada.

All but completely cut off from the rest of the world for many generations, Rotherhithe has naturally made the river its highway, and so, leading off from its quiet, old-world streets, there are everywhere passages which end in boat-landings and stairs. The names of many of these latter recall memories of a bygone time. There are King and Queen Stairs, Globe Stairs, Shepherd and Dog Stairs, Redriff Stairs—Redriff being the name under which, at one time, Rotherhithe was known—and others that must have received their designations when most of the land beyond Rotherhithe was marsh and wilderness. When the tide is out the stairs are left high and dry, and the river becomes a narrow channel between muddy flats, on which barges lie grounded, and the ribs of old wrecks are to be seen, and a steamer heels carelessly over, one side of its keel washed by the lapping tide.

On the opposite side of the river lies Wapping; the unique spire of Limehouse Church is visible, rising high above masts, and roofs, and chimneys, a landmark for miles; Stepney stands proudly dominant on its elevated banks; and Ratcliff, half enveloped in thick atmosphere, proclaims itself by the gleaming sunlight on its multitudinous roofs. Of Wapping one cannot think without recalling one of the tenderest of English popular songs :—

> "Your Molly has never been false, she declares,
> Since the last time we parted at Wapping Old Stairs;
> When I swore that I still would continue the same,
> And gave you the 'bacca-box, marked with my name."

Wapping Old Stairs are still discernible from the river, but are grievously difficult of identification, for, as at most other places along these shores, great warehouses have taken the place of most of the quaint-timbered old houses of former days. Yet at Wapping something of the old appearance of things is visible still. Leaning forward on to the shore, supported on mossy piles, green as the herbage of spring or brown as the weeds of the sea, are groups of strange old houses, with bay-windows, and overleaning balconies, and wooden walls, "clouted over" with planks until they look like a suit of mended clothes. The seafaring man's love of vivid colour is everywhere visible. The half-ruinous, time-worn buildings are painted after the manner of a Dutch barge — green contending with red, and raw yellow striving to hold its own against the imperial blue. In some of those curious accidental lights which are so frequent on the Thames, the low bank of Wapping assumes a peculiar glory of its own, heightened by the brown sails of barges sweeping past, and as full of colour as any picture which even Turner ventured to paint.

But from Wapping we must return once more to Rotherhithe, and to the Surrey Docks. They are ensconced in a graceful bend of the river, ere it curves back again round the far-projecting Isle of Dogs. There are, it is said, no older public docks in Great Britain, the Act by which the docks on the Surrey side were created bearing the date of 1696. Even before that period, indeed, there had been docks in the same situation "of considerable importance and benefit to the shipping." But docks which seemed large and important in the time of Queen Anne would be ridiculously small and inefficient with our present trade. The Howland Dock occupied ten acres when it was made; the Surrey Commercial Docks cover 330 acres now. They derive an historic and romantic interest from the fact that here the prize ships were brought to be delivered of their cargoes, when the jolly Jack Tar got his share of the prize-money, and, leave being granted, incontinently went off to squander it among his friends. There is a story of one such British sailor who entered the Bank of England with a warrant for twenty pounds, and exclaimed to the amused and amazed clerk:—" That will bother you, I reckon, mate; but never mind, if you haven't got the whole of the money in hand I'll take half of it now, and call for the rest another time, when it suits you."

From the lower portion of the Pool the river assumes wayward and eccentric

ENTRANCE TO THE EAST INDIA DOCKS.

habits, broadening and curving, and looking at every turn, when the tide is in, like a long chain of lakes. There is abundance of motion, and the freest wash of water, for the wind has large surfaces on which to play; and, also, the Thames is perpetually churned into long sweeping billows by the wheels of steamers passing to and fro. Henceforward every moderately straight portion of the river assumes the name of Reach, the meaning of which is obvious enough. There is first Limehouse Reach, then Greenwich Reach, and Blackwall Reach, and Bugsby Reach, and Woolwich Reach, and so onward to the magnificent reach at Gravesend. It is a devious course that is pursued by the vessels making their way down the Thames, but one which is full of perpetually varying interest, of ever-changing effects, of keen delight and breezy sensation to him who has the faculty of observing other things besides the muddiness of the Thames water.

Writing of this part of the river, Defoe describes grim sights in his "Journal of the Plague." He says of some of those who were terrified for their lives, that "they had recourse to ships for their retreat. . . . Where they did so, they had certainly the safest retreat of any people whatsoever; but the distress was such that people ran on board without bread to eat, and some into ships that had no men on board to remove them farther off, or to take the boat and go down the river to buy provisions where it might be safely done; and these often suffered and were infected as much on board as on shore. As the richer sort got into ships, so the lower rank got into hoys, smacks, lighters, and fishing-boats; and many, especially watermen, lay in their boats; but those made sad work of it, especially the latter, for, going about for provision, and perhaps to get their subsistence, the infection got in among them and made fearful havock. Many of the watermen died alone in their wherries, as they rid at their roads, and were not found till they were in no condition for anybody to touch, or come near them." A grim picture of a weird imagination! Is it possible to form a conception of anything more awful than these boats, floating up and down with the tide, unnoticed, masterless, unowned, with their dreadful burden of bodies dead of the Plague?

The Isle of Dogs is only an island because it is cut across by the entrances to the East and West India Docks. It is a vast space of dock property, hidden behind devious streets and towering wharves. Originally it was "the Isle of Ducks," the ducks to which allusion was made having a vast swamp to wade and flounder in, and a solitude peculiarly their own. Considerably less than a century, however, has sufficed to change the whole aspect of the place. Where the ducks disported themselves are now situated the East and West India and the Millwall Docks. Originally an attempt was made to construct a shorter course for vessels passing up and down the Thames. A new passage was made straight through the peninsula, where the West India Dock is now situated, but this, like the Thames Tunnel, proved to be a sad failure, vessels maintaining their course round "the unlucky Isle of Dogges," just as they did in Pepys' time. Round the long curve, engineering and ship-building yards have arisen, with houses of workmen attached thereto. The isle is populous, and dismal; and he would be a shrewd observer who should guess that it had not been built on till shortly after the century began.

One of the illustrations accompanying this narrative (page 313) represents a bend of the river at Millwall. It conveys, with much completeness, an idea of the character of the banks. Near by, a little further up the river, is the entrance to Millwall Docks. The name arises from the fact that, in former days, the only buildings on the Isle of Dogs were windmills. One of them was left till quite recent years, a quaint Dutch-looking structure, built very solid, to resist the high winds that blow unimpeded over the dismal peninsula, which even the ducks had abandoned.

A little lower down the river there were erstwhile landmarks of another sort. Gaunt gibbet-posts stood along the shore, with bones of pirates bleaching upon

THE WEST INDIA DOCKS.

them, and music of creaking chains. A reminiscence of this variety of ancient Thames scenery survives in the name of Execution Dock, which designation is only less repellant than another favourite place-name of the same period—Hanging Ditch, to wit.

The docks at Millwall are chiefly employed by steamers of large tonnage trading between London and the various European and American ports. Great bands of emigrants set out from here to the New World, and as their ships swing into the

MILLWALL DOCKS.

river there is much signalling to friends on shore and much pathetic leave-taking on the docks.

The docks are two, joined by a bridge, and a tonnage of more than 1,000,000, "gross register," passes in and out annually. Ready access to the railways is to be found at Millwall, but many of the vessels unload into dumb-barges, which swarm all over the Millwall waters, one man on board each barge, propelling his craft with a pole, and seeming to take his labour like a light recreation, and as if there were not the slightest need for hurry in all the world. These dumb-barges, sluggish and unwieldy, it is the common habit to denounce as one of the nuisances of the Thames. They float upward or downward with the tide; they are now "end on" across the river, floating sideways, and now lazily making a tolerably straight course; they get in the way of passing steamers, and are indescribably slow in getting out again. The single man on board seems to be influenced by the habit of the craft which he controls. He steadfastly declines to regard himself as an inconvenience, and if the tide drifts him into the middle of the stream he makes no haste to leave a clear course again, but prods slowly away with his long pole, utterly careless of mankind, and with an indifference to oath and objurgation which is positively sublime.

Compared to those at Millwall, the East and West India Docks, stretching

MILLWALL.

right across the neck of the peninsula, and making an island of it in fact as well as in name, are of really gigantic dimensions. A few years ago it was possible to declare, with a fair amount of truth, that the West India Docks were the largest in the world. From the land side they are approached by Commercial Road, the smaller of the two great highways which are the main arteries of East London. At the first glance the unsuspecting stranger might easily be led into supposing that he had come suddenly upon an important series of fortifications. The stone archway, crowned by a stumpy tower, which forms the entrance, is impressively massive, and even forbidding; the surrounding walls are very high, and seem to frown down a not unnatural curiosity to penetrate their secret; there is also a ditch which suggests a moat, and which looks as if the constructors had contemplated the possibility of having some day to cut off all communication with the docks otherwise than by water. Altogether, the West India Docks convey the impression that they are carefully guarded and somewhat mysterious, so that the nervous stranger within their gates traverses them not without fear and trembling, and is apt to become alarmed lest he should inadvertently trespass beyond his scanty privileges.

And from the Thames, also, the West India Docks look important and imposing, the tall warehouses rising as high as the masts of the vessels which break the regularity of their fronts, thus forming one of the most striking objects of the north shore. Of all the docks on the river, none is so likely to convey a concrete idea of the vastness of our trade, of the manner in which British intelligence and enterprise draw to the heart of London the spoils of the whole world. "I would say to the intelligent foreigner," exclaims one lively writer on the subject, "look around and see the glory of England! Not in huge armies bristling with bayonets, and followed by monstrous guns; not in granite forts, grinning from the waters like ghouls from graves; not in lines of circumvallation, miles and miles in extent; not in earthworks, counterscarps, bastions, ravelins, mamelons, casemates, and gunpowder magazines, lies our pride and our strength. Behold them in yonder forests of masts, in the flags of every nation that fly from those tapering spars on the ships, in the great argosies of commerce that from every port in the world have congregated to do honour to the monarch of marts, London, and pour out the riches of the universe at her proud feet."

It is impossible in any brief description to give more than a general idea of the extent of accommodation for commerce provided by the warehouses, sheds, and cellars of the West India Docks. The rum shed alone, with cellars corresponding in size, covers a space of 200,000 square feet. In one building vast quantities of tea are stored, in another, innumerable bags of fragrant coffee; here are sheds full of solid blocks of mahogany, yonder are bags of indigo, boxes of fruit, bales of cotton, bundles of hides, and sacks of tallow. The average number of vessels lying at one time in the East and West India Docks is 215, all ships of large tonnage. The Dock Company keeps 2,500 persons in regular employment for the loading and unloading of ships, and employs as casual labourers nearly 3,000 men besides.

The London dock labourer, to whom of late much public sympathy has been extended, is of two classes. The regular labourer, since the great strike, has been fairly well paid for an unskilled artisan, and is assured of constant employment. The casual labourer, with whom in many cases hard work is a new experience, is in a very different case. The starving and the outcast of all classes, to whom the whole of London holds out no other prospect of employment, stream down as a last resort to the Isle of Dogs, a sort of "going to the dogs" which is very grimly real. In some cases, when additional labourers are wanted, handfuls of tickets are thrown out among the eager and struggling crowd, and he who is fortunate enough in the scramble to secure one of these is provided with half a day's labour; in others, a foreman stationed at the gates secures the men he desires by scanning the throng and pointing to one and to another who seems most capable of hard labour, or in most need. It is a sad, a distressingly pathetic sight, this almost fiendish struggle for a few hours' work; and amongst those who engage in it are men of fair birth, of good education, of proved ability, but of irretrievably fallen fortunes, and with characters irrecoverably lost.

When the India merchants created the West India Docks they had a capital of half a million pounds sterling. Enormous is the capital that has now been sunk on the estate. At one time, when the owners had made a larger profit than they were permitted by Act of Parliament to divide, they bought a quantity of copper and roofed their warehouses with that expensive material. The wharves, warehouses, and quays have now a storage capacity of over 170,000 tons; in the cellars 14,000 sheep can be stored; the weekly wages paid at the West India Docks alone amount to £5,000; the revenue of the company owning them is close on £400,000 a year. Then there are the East India Docks beside—the docks that are sung about in fo'castle songs, that are haunted by the wives and sweethearts of sailors on the look-out for the good ship that is homeward bound, that are dreamed of by Jack Tar when he is thousands of miles away on the sea.

"Why?" asked Mr. W. Clark Russell, "are the East India Docks the most popular of all docks among sailors?" "There are two reasons," was the reply. "Until the Victoria Dock was opened these docks were the lowest down the river. They were consequently the first at which a ship arrived on her return home. The East India Dock was always so popular, owing to its convenience, compactness, and management, that, whenever there was room, and arrangements would admit, ships entered it. The advantage was great to the sailor. Once on shore, he had nothing to do but jump into the train on the pier-head and be off. Another reason was, the East India Dock was the home of the emigrant ship; and as it was the first place where Jack met his Polly, so it was the last place in which he bade her farewell and took his glass of grog."

Along the bank of the Thames, opposite to the Isle of Dogs, lie, making a long semicircle of streets, the twin towns of Deptford and Greenwich. Behind them rise the Kentish hills, dark with trees, among which the shadows seem continually to sleep. Deptford is redolent of historic memories. Its old church, with embattled tower, easily perceived from the river, contains the bones of Captain Edward Fenton, one of Frobisher's companions; Drake was knighted here, on board his own ship, by that unmarried queen who so appropriately ruled our country in the most adventurous period of English history. Here, Peter the Great came to learn ship-building, residing at Sayes Court, the house of the precise John Evelyn, who complains grievously of the semi-barbarian monarch who broke down his hedges, and filled his house with "people right nasty." Master Samuel Pepys, Clerk of the Acts of the Navy, was necessarily a more frequent visitor to Deptford, and he records, very early in his famous Diary, how he repaired to Deptford after sermon, "where," he says, "at the Commissioner's and the Globe we staid long; but no sooner in bed but we had an alarm, and so we rose; and the Comptroller comes into the Yard to us; and seamen of all the present ships repair to us, and there we armed with every one a handspike, with which they were as fierce as could be. At last we hear that it was five or six men who did ride through the guard in the towne, without stopping to the guard that was there, and, some say, shot at them; but all being quiet there, we caused the seamen to go on board again."

Up to quite recent years Deptford was famous for its dockyard, established by Henry VIII., and employed in the construction of vessels of war through the greater part of three centuries. The site is occupied by a dockyard no longer; convicts are no more brought to labour in gangs on the construction of men-of-war; there is no sound of hammers, nor any shouting of overseers; the keel of no mighty ship is being laid; the greatness of Deptford has departed with the era of our wooden walls.

But as Deptford has lost its importance in our naval system it has become one of the centres of our trade. Facing the river, like the vanished dockyard, and occupying a portion of its former site, is the great collection of buildings known as the Foreign Cattle Market. All cattle landed in London from abroad are brought here to be slaughtered, and in the vast shambles, which no person of nice tastes should think of visiting, beasts are being killed and dressed and quartered from morning to night, with an expedition which strikes the beholder as something unnatural and amazing. From Deptford, it is probable, proceeds the greater portion of London's meat supply, and even Smithfield gives a less striking idea of the vast capacity of Englishmen for the consumption of animal food than do the Deptford shambles.

Near the point to which we have now come the Ravensbourne enters the Thames, and with it the first black instalment of the sewage of London. The little river rises on Keston Heath, and flows sweetly through a lovely country, wandering, as a poet has sung—

"In Hayes and Bromley, Beckingham Vale
And straggling Lewisham, to where Deptford Bridge
Uprises in obedience to the flood."

On the bridge stands the boundary stone which marks the extremity of Surrey and the beginning of Kent, and just beyond it, a little nearer to the Thames, are stationed, above ground, one of the sewage pumping stations of the London County Council, and, below, the point of meeting of some of the principal sewers of south-east London.

Before us when we return to the river lies Greenwich Reach, broad and beautiful, uncrowded by shipping, with curious half-wooden houses on our right, with shoals of boats drawn up on the shore with the magnificent front of Greenwich Hospital reflecting itself in the waters. On the high bank above the landing-stage there is an obelisk erected to Belot, the Arctic explorer, unimpressive and meagre enough in itself, but beautiful as a tribute of praise from Englishmen to a daring sailor of another, and, for many centuries, a hostile nation.

At Greenwich we are enjoined to

"Kneel and kiss the consecrated earth."

It was the hospital which was thus alluded to as the means of consecration, but the glorious building is not a hospital any more. The fine old sea-dogs who used to find shelter here, and narrate to each other the story of their adventures an all the seas of the world, preferred fourteen shillings a week and their pension outside the walls to the liberal rations and the small allowance of money to which they were entitled by their residence indoors; so the place famous to all Englishmen through many generations has become a Royal Naval College, where young officers and engineers are trained in the technical and scientific branches of their work.

The site is one of the most illustrious to be found within the sea-washed borders of the British Isles. Here did the first of those who "dined with Duke Humphrey" come to carouse, for here Humphrey, Duke of Gloucester, had a manor-house, which he rebuilt and embattled, enclosing what is now known as Greenwich Park. Humphrey's choice of a site for his residence was approved by many an English king, for Edward IV. finished and beautified the Duke of Gloucester's palace; Henry VII. made it his favourite residence; Henry VIII., his brother, the Duke of Somerset, the Queens Mary and Elizabeth, were born within its walls; there the young King Edward died, and, a few days before his death, was lifted up to the windows by his courtiers, that his clamorous people might perceive him to be still living. Greenwich Palace was to Queen Elizabeth what Osborne is to Queen Victoria; James I. was wont to escape to it from London; the unfortunate Charles made it his home; and when his son, who "never said a foolish thing and never did a wise one," came to his throne, he determined to build at Greenwich the finest royal palace England had ever had. "To Greenwich by water," writes Pepys, "and there landed at the King's House, which goes on slow, but is very pretty. . . . Away to the king and back again with him to the barge, hearing him and the duke talk, and seeing and observing their manner of discourse. And good Lord forgive me, though I admire them with all the duty possible, yet the more a man considers and observes them, the less he finds of difference between them and other men."

The building which Pepys had seen in course of erection, occupying the ancient site of "the Manor of Pleasaunce," as the palace at Greenwich was wont to be called, owes its present magnificence to the genius of Wren, and its dedication to the purposes of a naval hospital to the humanity of the Consort of William III. "Had the king's life been prolonged till the works were completed," writes Macaulay, "a statue of her who was the real foundress of the institution would have had a conspicuous place in that court which presents two lofty domes and two graceful colonnades to the multitudes who are perpetually passing up and down the imperial river. But that part of the plan was never carried into effect; and few of those who now gaze on the noblest of European hospitals are aware that it is a memorial

GREENWICH HOSPITAL.

of the virtues of the good Queen Mary, of the love and sorrow of William, and of the great victory of La Hogue."

The battle of La Hogue was fought on the 24th of May, 1692. It concluded

VIEW FROM GREENWICH PARK.

a "great conflict which had raged during five days over a wide extent of sea and shore." The English had gained no such victory over the French for centuries, and England, in spite of much popular sympathy with James, in whose interest our ancient enemies had planned the invasion of our country, went wild with enthusiasm. Many of the wounded were brought to London and lodged in the hospitals of St. Thomas and St. Bartholomew, and it was shortly afterwards announced by the queen, in her husband's name, that the building commenced by Charles should be completed as a retreat for seamen disabled in the service of their country. However, as Mr. Ruskin observes in reference to the crown of wild olive, "Jupiter was poor." Little progress was made with the new hospital during Queen Mary's life, but on her decease her husband resolved to make it her monument. The inscription on the frieze of the hall gives to the queen all the honour of the great design; and though the hospital has now been diverted to other uses, the memory of what it once was can never perish, and the grand edifice will remain to Englishmen for ever

> "The noblest structure imaged on the wave,
> A nation's grateful tribute to the brave."

Wren's subject seems to have inspired his higher genius, and none of his works, not even St. Paul's, is a worthier memorial of his powers.

It was Charles II. who planted the trees of Greenwich Park, now remarkable for their great size and nobleness. From the crown of the steep ascent on which the Observatory stands, once the site of "Duke Humphrey's Tower," there spreads before the observer one of the broadest and most impressive prospects to be encountered anywhere round about London. Far down below lie, first, the Naval School, and then the two great wings of the hospital, each lifting its beautiful dome aloft into the blue of the sky; in front, the eye wanders, over the Albert and Victoria Docks, to the valleys of the Lea and the Roding; to the left, the river, broader than at any previous portion of its course, bends suddenly round the Isle of Dogs, beyond which lies London, dim and distant, its white towers and spires gleaming out of the haze, its great cross of St. Paul's glittering in the sunlight; to the right, the Thames—laden with ships, alive with barges—flows on, a wide shining space of water, past ship-building yards, and warehouses, and dry docks, until it loses itself in the grey distance of the Kent and Essex marshes.

Where the Lea—Walton's river—after flowing through Bedfordshire, by pleasant Hertford, on to Enfield, and Edmonton, and Bow, ends in an estuary of unfathomable mud, and joins the Thames at Blackwall, we are near to the entrance of the Victoria Docks.

At Blackwall, docks were being constructed in Pepys' day, and he makes this curious entry in his Diary:—"1665, Sept. 22nd. At Blackwall, there is observable what Johnson tells us, that, in digging a late Docke, they did twelve feet underground find perfect trees over-covered with earth. Nut trees, with the branches and the very nuts upon them; some of whose nuts he showed us. Their shells black with age, and their kernell, upon opening, decayed, but their shell perfectly hard as ever. And a yew tree (upon which the very ivy was taken up whole about it), which, upon cutting with an addes, we found to be rather harder than the living tree usually is." Similar curiosities, it is probable, lie waiting for discovery all along the Thames shore; and at the "New Falcon" at Gravesend there is a perfect specimen of moss, with still just a tint of green remaining in its fronds, which has been dug up from many feet below the surface at Tilbury.

Far down the river the docks are spreading—growing longer, and deeper, and roomier with the necessities of our trade. From the entrance to Victoria Dock at Blackwall to that of the Albert Dock at North Woolwich is a distance of more than three miles. The Albert Dock itself is a long, straight expanse of two miles of water, lined

THE ALBERT DOCKS.

on either side by great ocean steamers lying stem to stern. It is always resounding with the "Yo, heave oh!" of sailors, the shouts of bargemen, the cries of dock labourers, the screaming and panting of steam-cranes, the exclamations of bewildered passengers on the look-out for the vessel which is to bear them over seas. Up the River Thames every year there makes its way a vast fleet of 6,000 steamers and 5,000 sailing vessels, with an aggregate of 6,000,000 tons burden. To one who desires to understand clearly what life, and excitement, and perpetual going and coming this entails, there could be no more stirring or instructive sight than the Victoria and Albert Docks. Some of the great steamers are like floating streets, almost as populous, with rooms like palaces, and decks as clean as village hearthstones. From gigantic port-holes strange wild faces and turbaned heads look out; the quays swarm with coolies in blue and white tunics, with negroes in cast-off garments from Wapping, with Chinamen in curious pointed shoes, and pigtails neatly tied up for convenience. Above decks the officers may be heard giving their orders in Hindostanee; the red-turbaned sailors speak to their mates in unknown tongues; the howl with which a rope is hauled in or a bale is lowered is not unlike the cry of tigers in the jungle.

The Victoria Dock is very roomy, and comparatively quiet. It is a series of great basins, with surrounding quays and projecting jetties. Here are vast tobacco warehouses, and coal-sidings, and cellars for frozen meat. Not many passengers

WOOLWICH REACH.

come to or depart from Victoria Dock, which is used chiefly by cargo steamers, bringing for the consumption of Englishmen every variety of foreign produce. The tobacco warehouses are one of the sights of dock-side London. They contain as much tobacco, in bales of raw leaf, as would seem to be a sufficient supply, not for England alone, but for the whole world, for many years to come. The refrigerating chambers, spreading far underground, and designed for the reception of frozen meat from Australia, New Zealand, the River Plate, and Russia, provide accommodation for no less than 60,000 carcases. At the Victoria Dock are now located the furnace and chimney which jointly make up "the Queen's Pipe." Here, also, are landed many of the cattle which are slaughtered on the other side of the river, at Deptford.

The Royal Albert Dock was opened for traffic no longer ago than the year 1880. It is used by the great lines of passenger steamers—the Peninsular and Oriental, the British India, the Orient, the Star, and a score of others. Immense sheds run alongside the quay, capable of storing a prodigious number of cargoes, and a vessel may be unloaded and loaded in the course of a few hours. In the centre of the basin there is a movable crane, which will take up a waggon containing twenty tons of coal, and empty it in a few seconds into the hold of a ship. The Royal Albert is at once the most pleasant and the most exciting of all the docks of London. From the quays, it looks like part of a great river unusually busy with ships. There is no cessation of activity from the dawn of the day until dark. By one tide a great steamer is departing for Australia, by another for Calcutta or Bombay. It is no unusual thing to find that five or six great ocean steamships are timed to leave the dock on a single day, to sail for ports so widely divided as Sydney, Calcutta, Hong Kong, Port Natal, Japan, and the River Plate. So important, indeed, has become the traffic of the Albert Dock that it has become necessary to make for it a new inlet from the Thames.

Between the river and these gigantic docks lies the little colony of Silvertown, still looking new and clean, so recently has it been founded on the verge of the Essex marshes. Originally the Messrs. Silver commenced a rubber manufactory here, and, finding how far they were from the centres of population, had to build rows of cottages for their workpeople. Silvertown is now renowned for its electrical engineers, and has become quite a busy and prosperous centre of industry.

But in reaching Silvertown we have almost missed the fine sweep of Woolwich Reach, which is just as long as the Albert Dock, and is one of the most beautiful stretches of the Thames. When night has settled down upon the river, and the moon makes "a lane of beams" along the slowly heaving water, and the lights burning on the misty banks tremulously reflect themselves in broken pillars of flame, Woolwich Reach, with its level shores, and its indications of great activities in temporary repose, is in itself sufficient to relieve the Lower Thames of the common and vulgar reproach that it lacks beauty. There is a quiet, solemn, lapping of the waters; barges at rest, sailing ships at anchor, a yacht lying here and there, break the line of the sky with their tapering masts and their sails partially furled; a belated steam-tug pants upward, with asthmatic breath, and from either shore comes the dull regular throb of half-suspended life. The Lower Thames is never so imposing as in the night-time, when the moon is pouring down long streaks of light on the throbbing waters, and even the brown piles of the river bank seem tinged with gold.

The three prominent objects on the Woolwich side are the barracks, the dockyard, and the arsenal. Not that the arsenal can be said to be very prominent, either, for it lies by the side of the river like a low line of sheds, very bare and poor-looking, very disappointing, very unlike what one would expect the chief arsenal of England to be. The barracks alone relieve Woolwich from monotony. They rise high above the town, with their great central quadrangle, and its four spires, looking not unlike an enlargement of the Tower of London. Not far away rises the square tower of Woolwich Church, with a populous graveyard beneath, climbing over the summit of a hill. The houses of Woolwich rise above each other like irregular terraces, for here the land is more abrupt and uneven than elsewhere on Thames-side, as if it were asserting itself before it came to the dead level of the neighbouring marshes.

The once-famous dockyard, closed in 1869, is represented by great, empty, stone spaces, sloping to the river, and a pair of large, singular-looking sheds, stored full of gun-carriages and implements of war. Looking on it nowadays, it is hard to believe that up to comparatively recent years it was employed in the construction

WOOLWICH ARSENAL.

of our navy. No hammers resound there now, and the dockyard, silent and sleeping, might well be the type of an age of national amity and absolute peace.

But not so with the arsenal, which is busy night and day in forging the bolts of war. A while ago a shower of military rockets burst upwards from this busy centre of martial industry, spreading some ruin and much consternation throughout the towns of North and South Woolwich, scattering to right and to left, and penetrating the walls of houses situated a mile or so from the opposite bank of the Thames. Such accidents are always possible, despite the extremest care, and Woolwich sleeps, like Naples, in more or less constant fear of eruption. The choice of the place as a site for the Royal Arsenal was brought about by the discovery there of a kind of sand peculiarly adapted for fine castings, a fact which may help to explain the derivation of the name from Wule-wich, "the village in the bay." On the opposite side of the river, under the shadow of the trees which line the banks below North Woolwich Pier, elephants may occasionally be seen wandering, as calmly as if this were their natural habitat, for here are the North Woolwich Gardens, where, as at Rosherville, lower down the river, the folk of East London come now and then to "spend a happy day."

Off Woolwich, lies the *Warspite*, a noble example of those English frigates which did good service when England was still defended by its wooden walls. And

WOOLWICH.

the *Warspite*, which was formerly known as the *Conqueror*, is doing extremely good service now, for it is the training-ship of the Marine Society, which, at the suggestion of Jonas Hanway, the first Englishman who had the courage to carry an umbrella, was formed in 1772 for the purpose of equipping wretched and neglected boys for the sea. Since that date 60,000 boys, none of them criminals, but many in great danger of falling into crime, have passed through the Society's hands, and have started life with honest purposes. A finer looking lot of lads than those who swarm about the decks and the rigging of the *Warspite* it would be difficult to find even in a public school, and it is a proud day for the Marine Society when, once a year, a *fête* is celebrated on board the noble old war vessel, and the boys go through their evolutions in the presence of Royal and distinguished strangers.

Admiral Luard, who commanded the *Warspite* whilst it was still called the *Conqueror*, and carried a thousand sailors and marines, related a few years ago how narrow an escape it had of going down with all hands. Overtaken in a typhoon off Sumatra, it lay for many hours on its beam ends, its hold fast filling with water, and altogether in a condition so hopeless that all on board gave themselves up for lost. However, good seamanship and excellent behaviour on the part of the men saved the vessel to perform its present humane duty, and to endure as a type and example of the sort of ship which once maintained our supremacy on all the seas of the world.

PLUMSTEAD.

Below Woolwich the Thames flows through low-lying lands, flat and marshy, bounded at the distance of a mile or more by thickly-wooded hills, at the feet of which nestle here and there grey church towers, and little red villages, and occasional small towns. Looking down over Plumstead, which is a singularly prosaic place in a remarkable fine situation, the river is a mere thin streak, running between artificial banks, like those of a Dutch canal. Over the green marshland below us the river was once wont to spread itself like a great inland sea; and at various periods, since stout walls were built to confine it to a reasonable course, it has burst open its barriers and flooded the country for mile on mile. In this manner was created Dagenham Breach, where the river wall now encloses Dagenham Lake, famous for its bream fishing. On the Plumstead side the river wall was broken down in Queen Elizabeth's reign, and repaired at tremendous cost. Dagenham Breach, on the opposite shore, between where the River Roding and Raynham Creek open on the Thames, was made so late as 1707, when the swollen river, breaking down its barriers, rushed over 1,000 acres of land, and carried 120 acres into the stream. The land swept away made a sand-bank a mile in length, and stretching half-way across the river. The damage was afterwards repaired by Captain Perry, who had been engineer to Peter the Great, and who was voted £15,000 for an undertaking which had cost him £40,472 18s. 8¾d.

The land enclosed by the Thames' walls is mainly waste, but has a quiet, singular beauty, which would be more appreciated, doubtless, but for the fact that here, on either bank, London pours its two immense streams of sewage into the river. Where the Plumstead and the Erith marshes join each other, there may be seen at low tide a couple of culverts, from which issue, twice a day, two thick, black, poisonous streams. Just above them there is a substantial pier, and further back, a large white building with a tall chimney, beside which the Nelson column would seem to be dwarfed. Further back still, surrounding a covered reservoir, there is a quadrangle of small, neat houses, occupied by some of the workpeople of the London County Council. These are the sewage works at Cross Ness. They are surrounded by gardens, inside which the ground rises abruptly to the height of the dykes. All around seems clean and pleasant, but underneath, built on arches of the Roman aqueduct pattern, there is a huge reservoir, which receives most of the sewage of the south side of the Thames. The large white building which was first discernible is the pumping station, where there are four great engines capable of lifting 120,000,000 gallons of sewage in the twenty-four hours. For sixteen hours each day the sewage is being pumped up from low-lying culverts into the reservoir; for four hours at each tide it is being liberated into the Thames, which thereafter, for some miles, becomes a pestilential river, bearing its dark and unwholesome burden up and down and round about with every tide.

One might stand on the quiet Plumstead marshes and suspect nothing whatever of all this. From thence the river is made invisible by its dyke; but one observes, with an interest not unmixed with wonder, the funnel of a steamer skirting along the level landscape, or the rich brown sail of a Thames barge, or the bellying canvas of one of those sailing-vessels which, to the number of 5,000 annually, still make use of the port of the Thames.

The larger sewer works of what is called the Northern Outfall are situated on the opposite side of the river, at Barking, where there still remains a relic of the once famous Barking Abbey—the ancient curfew tower, from which the inhabitants were wont to be warned to extinguish their fires. Barking Abbey, which was a foundation of the Benedictine order, dates back to the year 670, and was the first convent for women in England. It originated with the Saxon saint, Erkenwald, Bishop of London, whose sister, Ethelburgha, was its first abbess. This lady made the convent so renowned that two queens—the wives of Henry I. and of King Stephen

—thought it an honour to be appointed to the office which so distinguished a woman had held. All the abbesses of Barking were baronesses in their own right, and took precedence of all abbesses in England. The last of the long line was Dorothy Barley, who was compelled to surrender the abbey to "Bluff King Hal" in 1539. The abbey church stood just outside the present churchyard, and was 170 feet long, with a transept of 150 feet. The curfew tower is the old gate of the outer court, and the room, of which the window is shown in the engraving on the

DAGENHAM MARSHES.

next page, was anciently the chapel of the Holy Rood. In the near neighbourhood is the house from which Lord Monteagle carried to the king a warning not to attend the Houses of Parliament on the day fixed for the carrying out of the Gunpowder Plot.

In Barking—sometimes called Tripcock—Reach we are afloat on a tide of sewage. It discolours the water all around; it is sometimes churned up by the wheels of the paddle-steamers; the odour of it assails the nostrils at every turn; and yet Barking Reach is, with this exception, an altogether delightful place on a spring or summer day, all the more delightful if the day is one which follows upon or precedes a day of rain; for the sky should be full of grey clouds and capricious light to do justice to the landscape below Barking Reach. Fortunately, even a vast burden of sewage, the refuse of the mightiest city in the world, cannot destroy the natural beauty of the river. By Erith and at Greenhithe it beslimes the low, muddy flats left exposed by the receding tide; but out in the centre of the Thames how can it avail against the influence of wind, and cloud, and sunlight? The river smiles and sparkles, and reflects grey cloud and blue sky, just as if it had no secrets to hide; and over the flat meadow-lands the shadows chase each other like happy children at play. Steamers, barges, sailing-vessels, coming and going, are almost as frequent here as in the higher reaches. It is the peculiarity of the Thames that it is never forsaken, or solitary, or at rest.

On either bank, unsuspected by the chance excursionist, are frequent powder magazines, which are a sort of introduction to Purfleet, where there is such a store

BARKING ABBEY.

of explosives as, if they were fired, would shake London to its centre, and possibly to its foundations. At Purfleet, by the way, the river banks vary their monotony by rising up sheer and white, in modest imitation of the chalk cliffs of Folkestone and Dover. As we proceed further down the river the smell of chalk-burning will taint the air somewhat disagreeably, and great white clouds of smoke will fly in our faces and almost hide the sky.

Purfleet is a pretty and interesting town, notwithstanding the uses to which it has been put, and the danger there must always be in living there. The chalk hills are crowned with pleasant woods, and over the river one looks across Greenhithe to the Kentish hills. Of the country in that direction Cobbett, writing his "Rural Rides," had only a disparaging account to give. "The surface is ugly by nature," he said, "to which ugliness there has just been made a considerable addition by the enclosure of a common, and by the sticking up of some shabby-genteel houses, surrounded with dead fences and things called gardens, in all manner of ridiculous forms, making, all together, the bricks, hurdle-gates, and earth say, as plainly as they can speak, 'Here dwell Vanity and Poverty.'" But Cobbett was by preference unjust, and the little grey houses, each with its own circle of trees, are an essential portion of the charm of these riverside landscapes, which, else, would look dead and solitary.

Off Purfleet, on one of whose chalky cliffs the standard of England was unfolded when the Spanish Armada threatened our liberties, lies the reformatory training-ship *Cornwall*, once known as the *Wellesley*, the flagship of the brave and

BARKING BEACH.

adventurous Lord Dundonald. These handsome old hulks, some of them used as reformatories, some of them as training-ships for boys who have been rescued from poverty, and one large group as a fever and small-pox hospital, are very frequent between Erith and Northfleet, and greatly increase the interest of a voyage down the Thames. Off Greenhithe, a famous yachting centre, the *Arethusa* and the *Chichester* lie moored; at Gray's Thurrock, on the opposite side of the river, lie the *Exmouth* and the *Shaftesbury*, the latter being the vessel which has been found so costly by the London School Board.

The good-looking town of Erith faces the river just above Purfleet, half-surrounded in the summer months by a fleet of small yachts at anchor; and, just below, the Rivers Cray and Darent, making a clear fork of shining water, meet together and flow as one stream into the Thames. "Long Reach" Tavern, a quaint, solitary place, once much frequented in the old prize-fighting, cockfighting days, by persons who are usually spoken of as belonging to "the sporting fraternity," stands on the flat muddy ground of this estuary of the conjoined rivers; and from this point the River

Thames bends inland towards Dartford, again taking a new direction at Ingress Abbey, where Alderman Harmer once lived, in a house built out of the stones of old London Bridge.

Around Ingress Abbey lies the village of Greenhithe, another yachting station,

AT PURFLEET.

with forty feet of water at the end of the pier at low tide. Stone Church, said to have been designed by the architect of Westminster Abbey, and beautiful and elaborate enough in some parts of it to suggest close kinship with that great edifice, stands on a proud eminence above the village, and is visible for miles around.

At Greenhithe the cement works commence, and extend themselves to Northfleet, which is a town perpetually enveloped in a cloud of white smoke, floating over the river in great wreaths, so that Tilbury and Gravesend, lying only a brief distance away, are in some states of weather completely hidden from sight until Northfleet has been passed.

To Tilbury is now to fall the often forfeited glory of containing the largest docks in the world. The heavy traffic of the Thames is gradually being arrested at a lower portion of the river. "One thing hangs upon another," remarks a recent writer, "and just as Tenterden Steeple is accountable for Goodwin Sands, so the Suez Canal is responsible for the Albert Dock, and for those that are being made at Tilbury. The long, weight-carrying iron screws that are built to run through the canal are not adapted for the turnings and windings of Father Thames in the higher reaches, and so, after the fashion of Mahomet, the docks now are sliding down the river to the ships instead of the ships coming up the river to the docks." Thus it happened that some years ago the population of Gravesend began to be increased by immense gangs of navvies, builders, and masons, who during the day-time were engaged on the Tilbury side of the river in digging vast trenches, building huge walls, and scooping out of the peat and clay accommodation for the merchant navy of England.

The new docks at Tilbury are the property of the East and West India Dock Company, which is forestalling competition by thus competing with itself. They are being dug out of what has for centuries been a great muddy waste. An army of nearly 3,000 labourers has been employed on the excavations. When the docks are completed eight large steamers will be able to take in coal at one time; the largest vessels built will be able to enter the gates with ease; there will be wharves and warehouses capable of accommodating no inconsiderable portion of the entire trade of the Thames. Branch lines of railway will run along the wharves, and be connected with each warehouse. The main dock occupies fifty-three acres of ground. The jetties surrounding the basins will be forty-five feet wide. At Tilbury, it is probable, the great work of furnishing dock accommodation for the shipping using the Thames will be finally brought to an end. It is all but impossible to

ERITH PIER.

imagine that the time will ever arrive when the Albert, and Victoria, and Tilbury, and East and West India Docks, will be too small for the demands of a trade almost inconceivably greater than that which passes through the Port of London now.

The proximity of this prodigious undertaking has driven away much of the solitariness which, for some centuries past, has hung around Tilbury Fort. That renowned but practically valueless fortification is best known through the popular engraving after Clarkson Stanfield's picture. That artist, however, has used a

TILBURY FORT.

painter's licence to the full. He has given to Tilbury Fort a massiveness and a dignity to which it can by no means lay claim. It is, on the contrary, rather mean-looking, and is only saved from insignificance by its great stone gateway, which is a sort of loftier Temple Bar.

It was when, in 1539, three strange ships appeared in the Downs, "none knowing what they were, nor what they intended to do," that the idea of building a fort at Tilbury arose. Henry VIII., alarmed at possibilities, built bulwarks and block-houses both at Tilbury and Gravesend. It is stated, on authority which is somewhat doubtful, that Queen Elizabeth reviewed her troops at this place when the realm was threatened by the Spanish Armada, and that here she declared that she thought it "foul scorn that the Pope or any other foreign prince should dare to interfere with her." On authority that is still more than doubtful, she is said to have slept in the one room over the great gateway. The statement that an Irish regiment, stationed here just before the abdication of James II., crossed the river and burnt and pillaged

Gravesend, but was afterwards defeated with great slaughter, is more authentic. At Tilbury Fort, Sheridan has laid the scene of the burlesque tragedy embodied in *The Critic*, the heroine of that piece being the Governor's daughter, who went mad in white satin, to the accompaniment of her faithful friend and companion, who considered it to be part of her duty to go out of her senses in white linen, as became the meaner condition of one who was paid to serve. A great mystery is preserved concerning Tilbury Fort by the military authorities, and any stray artist found sketching in its neighbourhood is usually treated as if he were making drawings for the advantage of the enemies of his country.

And now, having passed the Gardens at Rosherville, ingeniously constructed out of old chalk-pits, having seen Tilbury old and new, and having come to the end of this portion of our journey by water, it is time for us to land at Gravesend, where Hogarth and his merry companions put up at "Mrs. Bramble's," and, it is probable, took shrimps and tea. Gravesend, let it be said at once, is rapidly losing some of its most pleasant features. Coal-staithes and wharves have invaded the picturesque foreshore. Very dull and depressing is the entrance to the town, after passing those wonderfully grotesque baths which were built according to the sham Oriental taste popularised by George IV., who, as Praed says, was renowned

> "For building carriages and boats,
> And streets, and chapels, and pavilions,
> And regulating all the votes,
> And all the principles, of millions."

"The first gentleman of Europe," it may be confidently stated, never built a boat half so neat, smart-looking, and handsome as the yachts which, at the proper season, lie in the river in front of those sham Oriental baths mentioned above. At Gravesend are to be seen assembled the finest yachts which frequent the Thames, vessels, some of them with twelve or fourteen stout sailors to man them, and as clean and smart-looking as anything to be seen within the whole compass of the seas.

As becomes one of the oldest ports in the kingdom, Gravesend—it was called Gravesham in Domesday Book—is a town of narrow streets, of quaint shops and houses, of old-fashioned inns and close courts and alleys. The face which it turns to the river is like that of a battered old sailor—scarred, sun-beaten, weather-worn, but pleasant and honest withal. As in most seafaring towns, there is one long, cramped street, in which the houses seem to elbow each other, running, a little back from the river, almost from end to end! Far as it is removed from the sea, there is a fine salt-water savour about Gravesend, and it has also the recommendation of being situated in a pleasant country, for, after ascending its steep streets and threading here and there a leafy lane, there bursts upon the sight a glorious stretch of agricultural land, beautifully uneven, with hills of gentle slope, and occasional patches of woodland and garden and copse.

Of the history of Gravesend there is little that need be said. James II. lived here, as Lord High Admiral, when he was Duke of York, and escaped hence in a girl's clothes when he was flying from his enemies. On a hill behind the town there stands an old windmill, which is also a landmark, and which occupies the site of a beacon, the lighting of which was a call to arms. Aymer de Valence, one of the heroes of Thackeray's boyhood, and of many thousands of other boys of his period, founded and endowed a church just outside Gravesend when Edward II. was king. In 1780 five thousand soldiers were marched here to make a sham attack on Tilbury Fort, and were handsomely refreshed, *at the expense of the General*, when they had energetically stormed that fortification with blank cartridge. About

GRAVESEND.

that time, or a little later, there was a great scheme to make a tunnel under the Thames, between the town and the fort, which scheme ended in nothing but the formation of a company which appears to have spent fifteen or sixteen thousand pounds to no purpose.

At the present day Gravesend is much resorted to, first for the sake of Rosherville Gardens, and then for tea and shrimps, for which it has a reputation quite unique. Sam Weller's pieman could make a beef, mutton, or "a weal-and-hammer" out of the same festive kitten. The good folk of Gravesend can serve up shrimps in ways so various, and so tempting, that it is possible to dine off shrimps alone. At Gravesend, too, whitebait may be eaten with as much pleasure as at Greenwich, and the visitor to one of the inns of the place may watch the boatmen fishing for the whitebait which is shortly to be served up to him hot from the kitchen.

It is at Gravesend, indeed, that whitebait is now caught in most profusion. The boatmen pursue the dainty little fish in small open boats, and take it in long, peak-shaped nets, very small of mesh and delicate of workmanship. Whitebait first became celebrated in connection with the British Parliament towards the end of the last century, when Sir Robert Preston, member for Dover, was in the habit of asking his friend, Mr. Rose, Secretary of the Treasury, to dine with him at Dagenham when the session closed. Whitebait must have been had at Dagenham in plenty, and Mr. Rose made favourable report of it to Mr. Pitt; so it came about that the Premier was invited to try the whitebait for himself. Then it was that an annual Ministerial dinner was organised, the scene of the whitebait banquets changing from Dagenham to Greenwich, with an occasional dinner at Blackwall. "Yesterday," says the *Morning Post* of September 10th, 1835, "the Cabinet Ministers went down the river in

AT GRAVESEND.

the Ordnance Barges to Blackwall, to the 'West India' Tavern, to partake of their annual fish dinner. Covers were laid for thirty-five." And for something like that number covers still continue to be laid, though the Ministerial whitebait dinner now depends on the taste of Premiers, and is no longer *de rigueur*.

The whitebait itself has been almost as much the subject of discussion as the origin of salmon. Is it the young of herring, or of sprats, or of fish of many varieties? The question would seem easy enough to answer, though it can scarcely be said to have been finally answered even now. The one thing really certain about whitebait is, that it is a very dainty fish, equally good whether white or "devilled," as grateful to the palate whether fried in flour or broiled with a little cayenne. Scientific opinion, after once appearing to be convinced that whitebait is young shad, now inclines to the conviction that it is the young of a variety of species. The whitebait itself, however, seems to conspire in the concealment of its identity. Kept in captivity on one occasion, it will turn into herring, kept in captivity on another, it becomes the common sprat. Some specimens, indeed, have been known to assert themselves as pipe-fish, gobies, and stickleback, so that, though the whitebait fishermen resolutely assert the individuality of the species, it will perhaps be on the whole more safe to take sides with the men of science—and the accomplished cook.

There is, from some points of view, no more interesting spot on the Thames than Gravesend Reach. Here, after narrowing for a portion of its distance, the river spreads out again, and proceeds on a perfectly straight course to Cliff Creek. Gravesend Reach is three miles and a half in length, and is usually more populous with shipping than any other point between the Nore Light Ship and the Pool. All outward bound ships must take their pilots on board at Gravesend, and so it frequently occurs that here the last farewells are said and the last kisses are given. In the Reach, vessels wait for the changing of the tide, so that at one period of the day it is full of ships with their sails furled, and, at another, of vessels newly spreading their canvas to the wind. A breezy, stirring place is Gravesend Reach, enthralling at all hours and in all weathers, stormy sometimes, sometimes as calm as a lake on a windless night, but most beautiful on grey, uncertain days, when the light shivers downward through flying clouds, and breaks and sparkles on tumbling crests of wave; when the ships at anchor sway hither and thither on the turbulent waters, and make with their masts and cordage a continuous and confused movement against the sky; when the barges coming up from the Medway tear and strain under their canvas like horses impatient of the bit; when the half-furled sail flaps and battles in the wind, and the sea-birds, now darting to the water, now leaping towards the flying clouds, seem to be driven about against their will. Gravesend Reach, where David Copperfield said adieu to Mr. Peggotty and Mrs. Gummidge, where little Em'ly waved her last farewell, where we lost sight of Mr. Micawber and the twins, where so many tears have been shed, and so many hearts have seemed to be broken! What a ceaseless current of commerce flows through it, inward to the mightiest of European cities, outward to every country that the sun shines on. Whither is bound the vessel that is unfurling its sails yonder? Whither! To far Cathay, it may be; to obscure ports on the furthermost verges of the world.

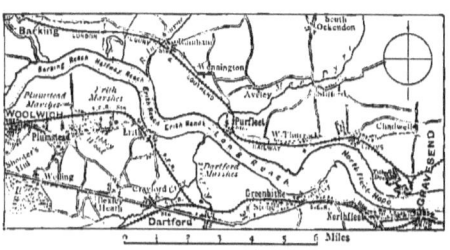

WOOLWICH TO GRAVESEND.

AARON WATSON.

CHAPTER XII.

GRAVESEND TO THE NORE.

Morning on the Lower Thames—Gravesend—Pilots and Watermen—A Severe Code—Tilbury and its Memories—The Marshes—Wild-fowl Shooting—Eel Boats—Canvey Island—Hadleigh Castle—Leigh, and the Shrimpers—Southend and the Pier—Sailing—Sheerness—The Mouth of the Medway—The Dockyard—The Town and its Divisions—The Nore—A Vision of Wonder—Shoeburyness—Outward Bound.

THE beautiful stretches of the Upper River must always offer an attraction to men who have an eye for colour, and to whom the curious spectacle of cultured wildness is pleasant.

But there are some who, while they remember the long reaches where the willow herbs shine and the glassy river rolls, think kindly of the other reaches where the signs of toil begin, and where the great stream pours on between banks that have nothing to redeem them save strangeness of form and infinite varieties of bizarre tints.

A voyage in a small boat from the hill where the Greenwich Observatory cuts sharp against the sky, down to the rushing channels where the black flood flows past the Woolwich Piers, is always unpleasant to those whose senses are delicate, but as soon as we reach Gravesend we come to another region, and there those who care little for brilliance of colour, those who care little for softness of effect, those who care only for stern suggestion, find themselves at home.

One of the pleasantest experiences in life is to wake in the early dawn, put sail on a fast yacht, and run on the tide from Gravesend, past the grim end of the Lower Hope. The colliers weigh anchor, the apple-bowed brigs curtsy slowly on the long rush of swelling water, and as you look up from your cabin you receive sudden and poignant suggestions that tell of far-off regions, and that take you away from the grim world that you have just left behind.

Here is a clumsy black brig bustling the water before her! The ripples fly in creamy rings from her bows; her black topsails, with their queer patches, flap a little as the wind comes and goes, and you hear the hoarse orders given by the man who stands near the helm, and who is in authority for the time. Then a great four-master spreads her wings, and while the little tug puffs and frets around her as though there were important business to do, which did not allow of a moment's consideration, the big ship slowly slides away, and gathering power under her

43

canvas, surges into the brown deep, and takes the melancholy emigrants away towards the Nore.

Then the "tramps" of the ocean—the ugly colliers—are not without interest. One of them foams up to you, and you know that the man in command of her has perhaps not slept for seventy-two hours. He has made his wallowing rush from the North country; he has risked all the dangers of fog and darkness and storm, and he has brought his vessel up to the derrick with satisfaction. Then in a few hours the swarm of "whippers" have cleared her; the rattle of the great cranes has rung through the night, and the vessel has been emptied in a time that would seem astonishing to those who manage sanitary corporation business on shore, and who condemn us to endure the presence of ghastly stenches and unspeakable sights for hour after hour. The anchors are whipped up and the ocean "tramp" tears away on her trip to the Tyne.

There is not a single sight or sound that does not convey its own interest. If it is autumn time, the racing yachts are clearing for action, dapper men are bounding hither and thither, as though there were nothing in life to be cared for excepting success in the race that must shortly be begun. The gun fires, and the lazy breeze of the morning strikes the huge spinnakers, while the razor-bowed craft move slowly out, and gradually gather speed until the troubled water foams in crisp whirls and rolls away aft in long creamy trails.

The upper reaches of the river are lightsome, and given over wholly to pleasure. Every turn conveys the sensation of wealth and comfort; every delicate shallop that floats luxuriously past the locks hints of money acquired in the crush of the great city; but in the Lower River any day the story of stress, and struggle, and coarse labour may be read on the spot, and perhaps nowhere in the world—not even in the huge docks of Liverpool—can so vivid an idea be gained of the mercantile greatness of England. No attempt is made to disguise the natural ugliness and coarseness of every feature in the scene; steamers surge up at half speed, and the vast waves that they throw curl against the bank and bring away masses of mud; the barges glide lazily, the black shrimpers troop down the current with their ragged sails, and everything speaks of a life given over wholly to rough toil.

It is true that many parties come from the City in steamboats, and in the summer evenings the air is full of music, and shrill sounds of laughter ring from the splashing boats as they pass you; but these are only stray visitations, and no one who knows the Lower River, no one who has felt the sentiment of the locality keenly, can ever associate it with light-heartedness.

Gravesend is a pretty town that straggles around the base of a bluff hill. From the summit of this hill you can look far over the plains of Kent; you may see the waves coiling and whitening round the Nore; you can see the towers of Rochester; you can see the great desolate stretches of marsh-lands that lie between Malden and Wallend. The town is wholly given over to shipping business, and although smart villas display their finery on the outskirts, yet somehow we feel these to be merely excrescences. They are very gaudy, the gardens are oppressively handsome, and the wealth of the owners is undoubted; but the lover of Gravesend cares only for the narrow streets that straggle down to the river; for the odd little shops where all requirements of seafarers may be satisfied; for the narrow wharves, past which the tide rushes from Northfleet Hope. For all who have read nautical literature the place is peopled with memories. Here the great Indiamen lay, in the times when the long six-months' voyage round the Cape had to be taken by officers and civilians. In these narrow, sloping streets the women stood and watched the passing of those they loved as the monster ship slid down on the tide. The very name of Gravesend brings up memories that can hardly be put into words; for in old yellow letters, in old books, in old newspapers, the word is always associated with meetings and partings, with great changes of fortune, with the keenest moments in the drama of life.

The town has the reputation of being the Cockney's Watering Place, but to those who know it intimately the normal life goes on unaffected by the incursion of the chattering crowds brought down by the steamers. The Whitechapel tripper at once betakes himself to the public-house, or to the tea-gardens, or to the dancing-rooms; while the watermen, the seamen, and the shrimpers go on composedly with their old-fashioned tasks. The pilot goes out with his smart cutter. He is a comfortable man, with a healthy air of authority, and there is something in the very roll of his voice that speaks of riches and monopoly. The Guild of Pilots keep their business very much to themselves. It would be hard to find one of them who is not exceedingly well to do, and any accidents that may happen do nothing to diminish a pilot's means. If on some dark and foggy night he makes a mistake, as some great ship gropes her way down the misty reach, it matters little to him; for even if he cuts down the ship, and drowns the whole crew, he can make himself perfectly easy. His money is settled on his wife, and the cleverest lawyer in the world could not wring anything in the way of composition out of him. The watermen still ply in their flitting wherries, but the glory of their trade is departed. Long ago the tilt-boats left for London Bridge with every tide bearing their loads of passengers.

The sternest rules were made for the guidance of the watermen. There is one curious order made by the Court of Rulers, Auditors, and Assistants of the Company of Watermen, forbidding any indecent behaviour or expression towards their fare, or whilst plying or rowing on the river. It runs thus:—

"Whereas, several watermen, lightermen, and the apprentices of such, whilst they are rowing, working upon the River Thames, and at their several respective places of resort, or plying places, between Gravesend and Windsor, do often use immoderate, obscene, and lewd expressions towards passengers, and to each other, as are offensive to all sober persons, and tend extremely to the corruption and debauchery of youth. For prevention therefore of such ill-practices for the future, it is hereby declared and ordained by the Court aforesaid, That if any waterman or lighterman, after the 16th day of October, 1701, shall upon the said river, or at any place of their resort, as aforesaid, be guilty of using any such lewd expressions, and be thereof duly convicted by one or more witness or witnesses, or by confession of the offender before the

Rulers of this Company, he shall forfeit and pay for every such offence the sum of 2s. 6d. And if any waterman's or lighterman's apprentice shall herein offend, the master or mistress of every such offender (the offender being duly convicted as aforesaid) shall forfeit and pay the like sum of 2s. 6d., and in case of refusal the offender shall suffer correction, as the Rulers of this Company shall in their discretion think fit and necessary; which said forfeitures (when paid) shall be applied to the use of the poor, aged, decayed, and maimed members of the Company, their widows and children."

This enactment is two hundred years old, and lasts up to this day. The wherries were regulated with equal strictness. No boat was allowed to take more than seven passengers at a time, and the sum of 2s. 6d. was charged on each passenger embarked over the number.

Everybody who knows the build of the wherries knows that on the Thames it is extremely difficult to turn to windward in a small boat. When the tide is running out it is, of course, impossible to turn at all; but even when the flood is running up it is extremely hard to "beat;" and the wisdom of the old masters is very prettily shown in one enactment, which declares that "If any master carrying passengers to and fro from London to Gravesend shall at any time hereafter turn to windward in any of the said boats wherein are any of her Majesty's subjects, he shall forfeit and pay for every such offence the sum of 10s." This severity of regulation made the river as safe for the ordinary travellers as it now is for those who use the large steamboats. Many persons were drowned first and last, but the number of deaths due to the upsetting of watermen's boats in the whole of the last century did not in sum equal the number destroyed in the massacre which took place when the *Bywell Castle* ran into the *Princess Alice*.

The regulations as to fares and fines are all very curious, and a glance at the droll bye-laws of the Watermen's Company seems to lift a curious veil between us and a dead society. Here is one terrible code of punishments:—

	£	s.	d.
Private watermen reviling passenger ...	0	2	0
Swearing or cursing	0	2	0
Towing a boat while carrying a passenger	0	2	6
Plying when his boat is not at the stairs	0	5	0
Working with a wrong number... ...	0	10	0
Marrying in apprenticeship	10	0	0
Refusing to carry a fare	0	2	6

Bum-boats selling goods before sunrising and after sunsetting are very hardly dealt with. For the first offence the boat forfeits 40s., and for every succeeding offence £4.

The fares made for the year 1785 were easy enough. From London to Gravesend the figure required was six shillings, and the other fares were proportionately reasonable. Thus quiet City men ran down from London Bridge on one tide and returned on the next, but the tilt-boat is now as extinct as the caravel. A few smart wherries dodge about the lower reaches waiting for inward bound vessels; but the watermen is no longer jolly, and in a few years it will be found impossible to find a youthful member of the craft, for no parent would apprentice his son to a trade in which few men can earn enough to keep body and soul together. A righteous retribution seems to have doomed a race of harpies to extinction. In the old days, when a towering East Indiaman came up the river, and the tanned soldiers and the weary civilians crowded joyously to the side, the watermen pounced on their prey, and each eager passenger had to run the gauntlet of a band of marauders. Times have altered, and the keen, ragged men who ply the wherries are only too glad to take a passenger to the Nore and back for a sovereign.

Across the water Tilbury Fort frowns over the bulge of the reach. The guns command the Lower Hope, and it would be impossible for an enemy's vessel to sail so far as Northfleet without being badly mauled or sunk. The place is associated with the names of a great queen and our greatest soldier. There the fierce Amazon mustered her troops and spoke rough words of encouragement to them; there General Gordon walked, with his quick, quiet movements, and his curt, low speech. Gordon planned the fortifications at the south of the river, but he travelled from bank to bank with that eager activity which marked his every action. His work is masterly in conception and execution, and if the torpedo service is properly organised it is hardly likely that the roar of a foreigner's cannon will ever be heard in London again. A roistering multitude once fluttered the people of the infant village of Tilbury, and rioted through the quiet place on the southern shore. The old historian grows quite haughty in his malice as he tells how "A rude rout of Rascals, under the leading of Wat Tyler, a taylor, who commanded in chief, with their grave ministers, John Ball, Jack Straw, a thresher, Jack Sheppard, of the Council of War, under the title of King's Men, and the servants of the Commonwealth of England, after ransacking and demolishing all the fair structures of the nobility and gentry of the Essex side, summoned K. Richard II. to give them a meeting, who accordingly, accompanied with most of his best counsellors, took his barge and went to Gravesend, but seeing the rabble so ragged and rogue-like, a company of swabs, composed of the scum of the people; it was held no discretion for the King to venture his person among them, and so returned to the Tower from whence he came."

Poor Richard let the "swabs" pass up the north side of the river, and he met them there with a gallantry which is a little unlike the conduct of the driveller who long afterwards fell from the throne which he had covered with dishonour. All the scene is dull and peaceful now. Gordon is gone from us, and his name will pass, like that of the "swab" Tyler, into the quietude of the history-books.

"So much carry the winds away."

North of Tilbury, and away to the eastward along the Essex shore, stretches a strange, level country netted with winding streams. As the tide runs out, the little ditches send down runnels of clear water. Charles Dickens was always fascinated by this region; but, strangely enough, his works have given everybody a false impression of the whole marsh country. People think of slime, and darkness, and poisonous exhalations, and an atmosphere of horror and crime. They think of the faces of hunted convicts and the grim night-scenes in which Joe Gargery and his pet took part; but at certain times of the year the marshes are really cheerful —the clear streams glitter in the morning sun, and the larks sing their hearts out high up in the air. The multitudinous notes fall around you from the shining heights like a shower of pearls, and for miles the eye is met by a blaze of colour and dazzling glitter. The ragworts spread in blinding sheets of yellow; the purple stars of the mallow peep modestly out from the coarse grass; and amid all the riot of sound and colour the peaceful cattle stand, and give a sense of homely companionship to the scene. When the tide flows, the river slips into the channels, and the tiny runnels of spring water are driven back to their sources; the ditches fill and overflow; the fishes, in many cases, catch their prey within a yard of where the cattle were feeding; and the grass becomes impregnated by the tide. It is this daily advance of the brackish flood that makes the marshes so valuable as grazing grounds. The cattle eagerly tear at the salty grass, and its nutritive qualities are so great, that it is sometimes found that a whole herd turned out on the marshes within a week or two weigh on the average half a stone more than they did when they first fed on the saltings. In winter, truly, the marshes are bleak and inhospitable; but in the soft, rich mud of the ditches the wild-fowl swarm, and the sportsmen have good times when the weather is frosty. A man who is not greedy, and who will be content with a very moderate bag, can hardly find a better place for exciting sport than within these northern saltings. Sometimes a redshank starts up, whistling desperately, and goes off down wind until the charge stops him; the ringed plovers cower low in the ditch, and shoot along under the bank with steady, level flight, until they are forced to sweep out over the grass and give the gunner his chance. At the fall of the evening the wild note of the curlew sounds with piercing cadence. There be many men in London who count Benfleet Station as the entrance to Paradise; and it is a very pleasant sight to see the smart shooters dispersing on a brisk, frosty morning. Below Canvey Island, and over the immense flats of that dismal place, the heaviest bags can be made with a big duck-gun. Most of the yachts on the river have a punt with the orthodox engine of destruction attached. There is something murderous and commercial about the duck-gun. To get up to a flock of birds needs a certain cunning and skill, which almost rise to the dignity of a fine art, and the excitement is amongst the keenest forms of pleasure that sport can give; but when the black, screaming flock has risen, and the boom of the huge gun has sent the echoes flying, then the sight of destruction, struggling, and suffering is apt to pain the sentimentalist. An hour's wander with a small gun—an hour that will bring its couple of brace of birds—offers the more artistic form of sport.

The shooting country is hardly broken between Benfleet and the Blackwater. Everywhere the eye travels over dark ditches, speckled flats, and stray groups of birds. At times the ground seems to be covered by a struggling army, whereof the squadrons perform strange evolutions. Then the wary gunner, watching with his glass from afar off, knows that the troops are on the feed, and takes his measures accordingly.

In choosing a boat for the river work, you can hardly do better than follow the model adopted by the waterside folk. Right round the coast, the action of years of experience has enabled the inhabitants of every place to choose the exact kind of craft best suited for their locality. In the North the delicate "cobles," with their light draught astern, are adapted for the long, shelving beaches. No fisherman ever thinks of running into the cove without preparing to make for the beach with his craft stern-foremost. The Yarmouth men have their stiff "hookers," which draw a good deal of water, and are without the dangerous "crankness" that characterises a "coble." The Suffolk men, on the stony stretches of coast between Southwold and Aldborough, have broad, clumsy, longshore boats, which stand a great amount of knocking about. In this way the unconscious process of adaptation, involving the transmission of hints from one generation to another, has made the Thames boats all that can be desired for their work. For all practical purposes a Gravesend wherry will see you safely to the mouth of the river, and beyond the Nore; while the average Thames "hooker," or "bowler" boat, as it is called, will stand up well to any weather that she is likely to encounter.

Take an ordinary wherry, and an hour's sail from Gravesend brings you into a foreign colony. Clustered thick in the sheltered haven of the river lies a fleet of vessels, strange in build, startling in colour, outlandish in rig. Their bulging bows are like the breasts of some Titanic women. The low sweep of their bulwark makes it astonishing that they can ever go to sea without being swept, even when the enormous boards are hung in position to keep out the rush of water and to stiffen the vessel. Quiet, good-humoured men lounge on the spotless decks of these ships, and address you in broken English or in a strange tongue. As you walk, you hear the sound of wallowing, and when you look into the gulf of the hold you see a strange, weltering mass of snaky-brown things of which the aspect makes an unaccustomed man shudder. Tons of eels welter in these watery caverns, and the landsman sees with astonishment that the sides of the vessel are thickly perforated to allow the rush of the sea, and that each ship is neither more or less than a huge floating sieve. In quiet ponds in Holland this harvest of eels is raised, and the vessels go to this point in all weathers. If they sailed past Gravesend, not one fish of their cargo would survive; so they remain at the bend where the water is salt, and the Thames flows through and through their holds until the last consignment has gone to Billingsgate. Then the quaint vessels warp themselves out of the haven. With their slow, blundering appearance they always seem as if they must come to mischief, yet somehow or other the quiet, phlegmatic Dutchmen make their queer craft do exactly what they wish. These fellows are not fond of the English fishermen, and a fight between the nationalities sometimes enlivens the dreary monotony of the haven; but to any one who boards their ship in a polite manner, and shows signs of good breeding, they are most complacent, and one learns to like their grim simplicity.

The river widens sharply out to the eastward of Thames Haven. On the south the Kentish Marshes stretch from the bluff of the Lower Hope to St. James's, and deep creeks run away southward towards Cooling, Halston, and Hoo St. Mary. It is very difficult to traverse this huge flat without a guide who knows the place pretty well. Men who have shot over the country winter after winter sometimes miss the exact spot at which a ditch may be crossed, and are kept wandering for an hour

AT CANVEY ISLAND.

at a time before they can extricate themselves from the labyrinth of deep, muddy channels. Like the Essex Marshes, the Cliff Marsh, the Halston Marsh, the St. Mary Marsh, and the rest, are the delight of wild-fowl shooters. A dingey can traverse most of the creeks for some distance, and birds may be got in hard weather without adventuring amongst the swamps, where a slip would produce the most unpleasant consequences. Like the Essex Marshes, too, this peninsula, which lies between the Medway and the Thames, is very beautiful in the summer for those who have learned the true sentiment of the country. Rank and luxuriant life spreads everywhere, and although sauntering is not a very pleasant employment, owing to the difficulty of negotiating the ridges between the ditches, yet the blaze of colour, and the jargon of song go on, and very pleasing thoughts come over the mind. The tide has a strong sweep, but a yacht will lie very comfortably clear of the foreshore. There are particular places, which the yachtsmen and bargemen know well, where no possible force of the tide would tear the anchor out of the ground. The present writer has again and again been caught at nightfall by the ebb, but there never was any danger, though the rush of the river went by like a mill-race. On one occasion the steering-gear of a steamer gave way as she was passing down at nightfall, and she plunged in amongst the stray vessels which were anchored alongside of the dreary flats, cutting one ship down, and bringing herself hard on the mud; but a catastrophe of this kind is hardly likely to occur once in twenty years.

A small boat soon shoots round the Lower Hope and into the westerly channel that flows around Canvey Island. At high tide the boat will travel easily up to the sea-wall, which rears itself like a strong fortification at the innermost edge of the saltings. The wall is overgrown with sea-weed, and the very steps by which one gains the Coastguard Station are slippery with sea-grass. Inside the wall the stretch of the island lies, as it were, in a great basin. Corn waves, bright meadows shine in the summer, and marshy streams creep slowly into the channels that cut the weird place away from the mainland. A wild and forbidding place is Canvey Island. The strong sea-wall is gruesome with its shaggy wreaths of trailing weed. The inner side is well covered with coarse grass, and from thence away to the northward a flat of somewhat repulsive aspect runs as far as Benfleet. The island has a peculiar population. The coastguards' hamlet lies close to the wall, and the men are ordinary sailors; but in the villages of Canvey, Knightswick, Panhole, and Lovis, there is a scant population of people who have their own ways, their own traditions, and their own methods of regarding a stranger. They are singularly hospitable, for free-handed sportsmen find the island a happy hunting-ground, and the people expect and give kindness. The one little inn by the Coastguard Station is, perhaps, the quaintest in all Essex. Memories of smugglers, of desperate water thieves, of old collier sailors seem to hang about its low walls. No one need expect comfort there, but the keeper purveys for all comers with a rude hospitality which is amusing. On the Fobbing side of the island the ditches are very deep, and the sides soft and treacherous. Once a bird is shot there it is very difficult to recover it. All the dogs kept on the island have a singularly business-like air, but no one would care to let a valuable dog follow his game down these steep, gluey, ramparts. To the east, however, the saltings stretch far towards Canvey Point; and it is not only safe, but absolutely pleasant to walk over them before the tide creeps through the rough herbage.

Hardly a shore-bird known in the British Islands fails to visit Canvey. Looking through a telescope from Benfleet Station, it is easy to pick out the flocks as they consort in their different communities, and squat among the mud, or pick their way carefully through the twining grass. At one time, on a frosty morning, it is possible to see dotterels, plovers, redshanks, gulls, and pipers, all busy on the eastern flats; while to the west the cunning curlews dodge on the slippery banks of the Fobbing ditches. The foreshore is perfectly free to strangers; although one proprietor in the island has ventured to dispute the fact. A private grant of the shore was made two hundred years ago, and below the sea-wall no visitor can be considered as a trespasser, while a boat may bring up anywhere in the channel. Canvey is not an inviting spot for camping out. On a gusty night, when the rushes moan and shiver, and the great river sounds hoarsely, it is hardly possible to look out into the darkness without feeling a sense of strangeness and even of fear. The island seems to have no salient points; the hill, topped by the house known as the Hall, rises a little, but it is more like a cloud than like a solid mound. A shadowy figure from the coastguard's hut sometimes paces up and down, but even this gives none of the refreshment of human companionship. The writer once took refuge in the Channel at midnight during very bad weather. The boatmen did not care

THE FRINGE OF THE MARSHES.

to land, and we sheltered ourselves as best we could from the storm. The island then showed in all its mystery through the drift of rain and the flying haze. It was an experience never to be forgotten; but no one is recommended to try it. It is better to seek the hospitable shelter of an inn, and put up with rough fare, or any fare, rather than remain in the open amid that abomination of desolation.

The sea-wind comes with sharp, stirring breath after we pass the long spit that shoots out from the weird island; the river is still yellow, but when the breezes set the foam dancing the crests of the waves are of pure white. In the reach at Erith there is sometimes a heavy roll that travels as swiftly and as high as the jumping seas of the Channel, but the curling crests of the waves are yellow, and they hint of foulness beneath. All changes when the estuary fairly breaks open to receive the unchecked wash of the tide, and it is exhilarating to sweep over the full-bosomed river that swells as though it would fain topple across the low rampart of the Kentish marshes and flood all the sluggish runlets. We take it for granted that any one who cares to enjoy the sights of the Lower Thames fully will use a sailing-boat.

HADLEIGH CASTLE.

The discreet navigator may then explore to his heart's content. On the southerly shore there are few buildings which have any interest, but on the Essex bluffs there are many places worth going ashore to see. The low hills command a fine outlook to the southward, and every salient point has been selected at one time or another for building purposes. Looking northward from the dull level of Canvey Island, one sees a strong tower that forms a central mark in a pretty landscape. At first sight the building looks firm and uninjured, but when you climb the bosky hillocks upon which it stands, and approach within a hundred yards, you find that the imposing shell is but a ruin after all. This is Hadleigh Castle, which is said to have been built by the proud favourite Hubert de Burgh. Six centuries, with frost, and fire, and snow have spent their wearing influence on the stately ruin. Where once the mad Earl of Kent held high revel the owl makes her nest and the garrulous jackdaws flutter and babble.

> " 'Tis said the lion and the lizard keep
> Their court where Famshyd gloried and drank deep."

The old story holds true alike on the Essex hills and on the plains of Persia. Where Hubert de Burgh gloried and drank deep the wild birds harbour and the moaning winds pour unchecked through the desolate towers. Hadleigh Castle could only have been built by a man who took long views of life, and who felt his hold on his place in life very secure. Even now, though the towers are hollow, and the grass makes the battlements shaggy, the castle has an air of grim strength, of steadfast power, that give pause to the mind. All round the grey walls the birds flutter in changing flocks. Far down the slope the river rolls and the ships glide without ceasing, while the trees rustle and the grass gleams as the breeze flies over. There is movement and colour everywhere, the trains rush along the embankment just below us, and amid all, scorning change, fronting, incurious of night-time or day, the centuries' enormous weariness, stands the structure that was built in the dark ages. Dark ages! Can we equal this nobility of outline, this triumphant strength, nowadays? When all the rickety streets of modern London shall have sunk in decay, when perchance the great city is but a fading memory, the rugged Castle of Hadleigh will remain in disdainful steadfastness—a monument of human pride and skill, and alas! of human folly and failure. Elizabeth came here, as did her savage father before her. Generations of ladies, gay and courtly knights, met in their turn within those tremendous walls, and now the curious traveller may wander unchecked amid the remnants of magnificence. Let no one who sails on the Lower River miss seeing Hadleigh Castle, for it is a worthy example, all mutilated and imperfect as it stands, of a noble school of architecture; and there are no ruins of a finer and grander type even on the storied banks of the many-memoried Rhine.

The view from a steamer is very well in its way, but the quaint glimpses of mysterious creeks, the chance views of forlorn waterside cottages, the flashes of colour from red-tiled roofs and glowing gardens can only be seen at their best from a stiff boat that can either creep inshore or bowl over the solemn flow of the outer current. Leave the chilly stillness of a channel like that which bubbles around Canvey Island; spread the boat's wings, and in a few minutes you may have the whitening ripples purling clearly along under the quarter, and you see the fleet waves coiling and plashing at the Nore.

To the north of Canvey Point lies the village of Leigh, which may be called the Yarmouth of shrimpers. The bulk of the village lies close to the water's edge, but the church, with its picturesque tower, crowns the top of the hill, and forms a conspicuous landmark. The black boats bustle out of the haven in swarms,

and settle like ungainly sea-fowl as their trawls go down. It appears as though nothing were being done—as though the boats were merely anchored in a clumsy fashion, but, all the time, the brailed-up mainsail is imperceptibly dragging each vessel along, and the nets are gathering their prey from the muddy bottom. Solemn, grimy men move listlessly about, or sit amidships, as if they were burdened with misanthropy; the rudder takes its own way, for the drag of the net usually serves to keep the boat on her course; the sail flaps mournfully, and the jar of a shaken block cuts the air like the report of a pistol. Yet the lazy-looking craft are busy, and the bubbling boiler amidships is kept always ready. When the haul is made, and the wriggling myriads of shrimps are sorted out, then the boiling-nets come into requisition; the crustaceans are swiftly dipped into hot water, and the impassive fishermen prepare deliberately for another haul. No one who goes down Thames should miss landing at Leigh, and, if possible, he should contrive to spend a Saturday evening with the men. They are a civil race, and they take a stranger's presence as a compliment. Many of them are yachtsmen, and the admirable semi-naval discipline of the yachts has leavened the manners of the place. The rough fellows sing their silly songs, and exchange wise remarks about fishing and yachts (which are the only subjects of worldly interest to them), and they are always ready to take a visitor into their confidence. Barring the slight polish acquired from mariners who have seen the strange regions of Cowes and Dartmouth, these villagers are like survivals of a dim past. In fact, so thoroughly marine is the general atmosphere, that shore-going costume seems incongruous in Leigh, the presence of a dealer is painful, and one feels as if it were a sin against propriety to wear anything but old-fashioned garments. It is worth while to pay a visit to the station in the evening when the last up-train is about to start; the platform is crowded with hampers of all shapes and sizes. They contain shrimps ready for transmission to the all-devouring Metropolis.

It is best to run well out to the southward after leaving Leigh, for then the pleasant slope of the hills that fringe the northern shore is well seen. Stray copses straggle here and there; lines of fir-trees strike against the sky like regiments with arms at the carry, and pleasant houses peep from their pretty perches. Southend is already feeling its way toward the west. The central ganglion of the town is perched in its little basin in thick clusters of houses, that seem to climb over the rounded wold; but the stray villas are planted like pioneers, and by-and-by the lines will be completed, and Southend will perhaps come to be in touch of London.

The magnetic attraction of the great city is felt everywhere. We are so secure now that bodies of men no longer huddle themselves within the solid safety of stone walls. Every modern English town has a tendency to sprawl. Only cross this river and run southward to the Foreland, and you are within sight of quaint old towns that had a serene, corporate existence, and nestled inside their defences like discreet swarms of bees. Rye, Sandwich, and all the rest, resemble the cyries of seabirds planted safely in snug coves, but this Southend sprawls like its own wriggling pier. Carlyle foretold the junction of London and Reading, and there is a sad probability that this will come about. In the same way Southend at last will blend with London, and we may have the jingling horror of a tramway from London Bridge to the low bluff that fronts the Nore.

As we move eastward a strange serpentine shape rises out of the water. At first it is like a cloud, then it takes on the appearance of a huge centipede with an abnormal number of feelers and a blunt, horned head. That is Southend Pier, which strikes for a mile and more over the mud-flats. The lighthouse rounds off the end of this odd structure with a somewhat dignified suggestion of solidity, but the long, straggling chain, alas! looks as if it were all unfit to stand the fierce rush of the North Sea. It is quite easy to land on the hulk that creaks and sways below the lighthouse, but the present writer never cares to trust a small boat against the outer edge when the river is running hard. There are strong steps at intervals all the way along, and it is best to go round the pier-head and place the boat according to the wind. When once the upper pathway is gained, it seems as though the town were within easy reach. But let no one try fast walking along that treacherous road; it is meant for men who care for gentle pedestrianism, for meditation, for quiet glimpses to seaward, for lazy criticism of passing vessels. Indeed, there is enough of interest to take away all desire for hurry. Around the piles the grey water laps and swirls, scooping out round holes in which black colonies of mussels nestle. Little fish pursue their nervous activities in the clear pools; the scream of sea-birds comes

SOUTHEND AND THE PIER.

faintly from far away, and the keen breeze makes hoarse noises in the labyrinth of the piles. At low water the flat seems interminable, and it must be owned that it does not look very pleasant. Glossy hillocks of mud thrust their shoulders out of stray ponds of salt water, and every hillock seems to be composed of a rather nasty kind of gruel. Lumps of sea-weed lie about the greasy surface: they are like currants in a monstrous, uneatable custard. The gulls settle and chatter around the bitter lakelets, and they are the only beings that find the flats easy to walk on. It is hard to say how far one would sink if he were daring enough to adventure himself among the wreathing mazes of mud. Perhaps the footing is more solid than it seems, but we never cared to try. Slowly and warily the traveller moves over the puzzling planks, and as each new landmark shows itself, the length of the pier impresses itself on the imagination of tired humanity. The men below who wallow in their enormous boots among the oyster-beds take matters easily, and tend their precious charges with deliberate care. They are like wild denizens of the gruesome, glistening waste; and they are as much at their ease as the sea-birds. But the stranger only longs to be rid of the jolting monotony of the cross-planks, and as the town comes sharply into view one is tempted to leave off contemplating the green piles, and the busy fishes, and the long melancholy of the sea-marsh, and the most phlegmatic of new-comers is inclined to break into a trot. The leisurely persons who stroll out to inhale the wind from the Nore may take their ease as they will, but, after the first minutes of interested observation, the foreigner longs for human companionship; he longs to be rid of the dominion of this intolerable roadway. The town straggles down a brief, steep bank of clay, and spreads itself over a fine level. It has all the outward appearance of a southern watering-place; the bathing-machines stand along the low sea-wall, the boats repose on the beach, and the strollers wander listlessly over the very narrow border of sand. The old town is quaint and pretty, and the new town is flashily handsome. London has set its mark deeply everywhere, and, from the smart cabmen, who salute with demure shrewdness, to the imposing platform where the band plays, everything tells of city influence. Southend is a lesser Ramsgate, and, in its way, it is a very fair imitation of that other dependency of Cockaigne.

When the tide flows, the scene is really pretty. The suggestive flat is so very, very level that the first rush of the tidal wave sends foaming streams careering among the winding hollows and pools. Like magic the vanguard of the sea gains the limit, and soon the wide sweep from Southend to Canvey becomes a shallow dimpled lake. The sense of depth is wanting, but if you only look at the surface, then you may take for granted that you are on the border of a very noble bay. As the tide gains, the little yachts rise from their bed of mud and curtsy at their moorings, the fishing-boats glide in, and the curve of the beach is full of animation. We know nothing of the bathing, but we should incline to think that there may possibly be a good deal of suspended matter in the water. Be that as it may, the bathers enjoy themselves mightily, and, even were there no bathing, the compensation offered by the sailing-boats that shoot over the wide bight is worth reckoning. To sail on a water where is depth enough to float you, but hardly enough to drown you, must be pleasing to the non-adventurous mind.

Southend is very modern, and has not yet gathered any great population; but it is so cheery, and the powers that rule municipal affairs are so firmly resolved on making it "attractive," that it has a promising future. When the Thames no longer discharges filth to the sea, and the sands regain their purity, it will be delightful to walk over that noble level; but our generation will hardly see such a blessed transformation. From much experience we can say that, in winter-time, the pier offers very inspiriting views. The waves fly hard over the sands in heavy weather, and their eager rush breaks them into short combers, that strike the piles, and set the timbers quivering. Sometimes the spray drives high, and at night the roaring darkness is as wild as the clamorous mystery that meets you as you gaze seaward from the cliffs of Bamborough or the wind-swept marshes of Southwold. So far as creature-comforts are concerned, the traveller is practically in London. The people have been too wise, so far, to set up as plunderers, and tired brain-workers who wish to escape easily from London for a short time may get a breath of sea air without paying too heavily for the medicine.

On the Upper River a certain amount of enjoyment may be had by sailing a small centre-board boat; but precisely the same quality of enjoyment may be derived by using the same boat from Southend. It is not at all whose business will allow them to run to Southampton, or Brighton, or Margate, but every one can easily get to Southend, run at intervals into the very midst of the fresh sea-breezes, and return with very little more trouble than is needed to travel from Uxbridge Road to Charing Cross. As we have so often insisted, the great blessing of the royal river is, that its pleasures are so easily accessible to the poor man. A sound longboat may always be had at a moderate price in Victoria Docks, and a fresh-built boat, on the longboat pattern, need not cost more than £30 when the most minute articles employed in fitting are paid for. The exhalations from the Kentish and Essex marshes, which become unspeakably horrible when mixed with the suspended carbon that floats above the City, are never felt at sea, and the priceless boon of health may thus be had at a less price than that paid by many middle-class families for the ministration of the physician.

A splendid run from Southend to Sheerness may be had in any state of the tide. A yacht must go through the passage called the Swashway, where the soundings are deep, but a wherry will easily pass the sands. There is nearly always a good breeze, and when the wind is strong enough to set the scuppers awash, the sensation of skimming from land to land with the speed of a bird is something to be remembered. At first, Sheerness is like a low-lying cloud, but gradually the pouring mouth of the Medway becomes distinct, and soon the front of the forts is seen, and we realise the full strength of the place of arms, which has been created on an island that once was a dismal swamp. England paid dearly before the value of Sheerness as a strong position was recognised. Twelve guns were mounted there after the Restoration, but the bold Ruyter minded the puny armament very little, and destroyed our fleet after passing under the very nose of the batteries. It must have been a wild time when the apple-bowed Dutch men-of-war cleared the Swashway, and held on straight up the Medway. Well might the people "think of Oliver, and what brave things he did, and how he made all foreign princes fear him." The Admiralty showed vigour when the dreaded Ruyter was out of sight, and from that day until the present scarcely a year has passed that has not seen some addition made to the colossal works which were begun in the time of Pepys.

At the latter end of the last century lines of old war ships were formed into breakwaters, and each vessel was utilised as a barracks. Chimneys of brick were

SHEERNESS DOCKYARD, LOOKING UP THE MEDWAY.

built on the hulks, and the lines of ships looked like floating streets. Under the shelter of these queer barriers the most extensive works were carried on in safety, and there is hardly a spot in the world where the victory of man over dumb obstacles is more triumphantly made apparent than in the monster basins where the war ships rest. A right instinct told our engineers that Sheerness protects the heart of the nation, and the energy displayed in building the stone wall, which runs for a quarter of a mile parallel with the pier, was worthy of Stephenson himself. After the great dock had been completed, which was to accommodate a dozen first-rates, it was found that, in order to make room for the huge structures, enough soil had been excavated to raise the level of the whole swamp more than fifteen feet. The history of other engineering achievements has been written at a mighty great length, but this—perhaps the most extraordinary feat on record—has met with scant notice.

The age of iron has come in, but memories of the old times hang round the town. Here is a burly hulk, moored in the swinging tide. Long ago she carried her two tiers of guns; those slovenly sides were polished like a violin, and there was not a reef-point out of place. She could not sail much better than a floating haystack, and her mode of getting through the water consisted of going three miles ahead and two to leeward. But she was good enough to fight anything that she met on blue water, and she took her share of hard knocks in her time. The remnants of the men-of-war meet us everywhere, and whispers of boyish romance come to the mind as we think of their clumsy majesty.

But there was not much romance in the life that went on in the ships that made our boast, and no glamour of poetry or rollicking fiction affects the minds of those who know the facts. When towering liners lay in this anchorage, and their strength was the wonder of the foreigner, it was too often true that the life of the men on board was one round of sordid slavery, starvation, and hopeless suffering. The men who fought our battles were fed worse than dogs, and flogged worse than convicts. Think of all that happened when the ships were running toward the sea, down this very brave river that we have traced so far. The water-casks were filled from the befouled flood, and in a few weeks the horrible stuff was so putrid that it had to be strained through linen before it could be used, and men turned sick at the smell of the nauseous draught, which was all that they had for drinking and cooking. This unspeakable nastiness was of a piece with the rest of the life on shipboard. The work of the fighting-machine went on smoothly under iron discipline, but in most cases each ship was an abode of vice and random tyranny. We hear ridiculous talk of the great days of the Navy. In those great days the men between decks lived in squalor to which paupers from the slums would object; many of them were stolen away from home and from love to go and dwell in that dim quarter among the odious hammocks; they endured shameful stripes, they drank poisonous water, they ate meat that a kennel of hounds would have refused, and they were regarded as having forfeited their manhood. Then in time of need they had to stand to their guns and run the chance of being smashed by a French round-shot. Truly the romance shines out but dimly when we insist on plain prose truth!

Only about ninety years ago Sheerness was covered by guns laid by angry mutineers, who had burst into rebellion after suffering wrong unspeakable. Had the sailors not held their hand and offered to hear reason, they might have laid the place in ruins, and opened the way to a foreign armada. They had reason enough for anger. Cheated of their pay, their food, their clothing, their liberty, imprisoned for years on pestilent foreign stations, crushed under savage discipline, they refused

SHEERNESS DOCKYARD, FROM THE RIVER.

longer to endure a bondage that the very brute beasts would have rejected. Then Sheerness saw her direst danger, and then England was near a disaster from which she might never have recovered. The whole grim story of the mutiny starts out vividly as we see the very place where the Admiralty messengers came in terror, and where the discarded officers were put on shore. Here and there we meet with a smart, well-looking seaman, and the very look of him reminds us that the bad days are gone. Jack is not like the scarecrows who clamoured for food and justice in the terrible times when Sheerness was panic-struck, and Gravesend Reach barricaded; he looks like a free, independent man; his rights as a citizen are recognised, and no petty tyrant can lay the lash on him.

The tendency to dwell on the past is almost irresistible as we move amid the stupendous evidences of modern ingenuity and resource. The clangour of hammers resounds in the dockyard. That monster, over whose iron ribs the swarming workmen clamber like midges, could have steamed quietly among Nelson's fleet and sent them all to the bottom in a couple of hours. Not one of them could have scratched her, not one could have run away from her, and, supposing that her ram were employed, she could have shorn through the *Victory* from bulwark to bulwark without even running the risk of being boarded. Out in the stream lies the rotting hulk which once was regarded as the prime work of the human hand and brain; in dock lies the iron monster that needs neither wind nor tide—the monster which could stand the brunt of the *Bellerophon's* broadside without suffering a dent. So the world changes.

It would take a month to describe the dockyard; indeed, in a single day's inspection, it is hardly possible to gain an idea of the magnitude of the place. It is a little world of industry, with a separate constitution, and separate laws. In such a vast organisation it is inevitable that blunders occur, and that woful waste goes on among the incredible masses of material that bewilder the senses. Nevertheless, a sight of Sheerness Dockyard gives a more definite idea of British power than reams of abstract declamation and shadowy description.

The town is marked off into strictly defined regions. Blue Town lies within the garrison limits, and is pervaded by the military. Mile Town faces toward the Nore, and lies within a strong line of fortifications. Banks Town and Marina front the open sea, and are clear of the atmosphere of business. The two last-named quarters form a merry little watering-place, and they are intensely modern. The sea rolls up the beach, pure and clean, and there are few signs of that dubious compound which makes the Southend flat a place of fear. There the children build their sand-castles, even as the children did in Homer's time by the blue eastern waters; there the enfranchised clerk carries out his peculiar system of enjoyment, and the usual happy, commonplace, invigorating life goes on during the season. If we described Marina, we should only describe the typical sea-side places into which the cities empty themselves in the autumn weather. We may leave Sheerness. The guide to the docks alone would occupy this book if we only indicated the points of interest. It is best merely to say that, alike to those who know the stirring history of our navy, and to those who are amused by Cockney jollity, the place is worth seeing.

The striped buoy rises and falls to the rhythm of the short seas, and the waving ball that surmounts the tall pole catches the eye at a long distance either riverward or eastward. This light is one of the marks that Englishmen think of wherever they may be on the surface of the globe. Not a passenger steamer goes past that light without a tremor of excitement running through all who are on board of her. It seems as though there were magic in the name, and whether for the sailor coming from the East Indies, or from round the Horn, or the coasters who have merely run down from the Tyne, the words, "Here we are, abreast the Nore!" have a sound that acts like a charm.

That worn and battered vessel that trails past you as though she were weary of infinite travel and incessant hard battering amid furious seas, has men aboard her who have chattered for hours together about all they would see and do when once they passed this point. When the seas are crashing down on the forecastle-head, and the falling water sounds like muffled drums; while the stinking lamp gutters in the foul atmosphere, then Jack, as he stretches on his squalid bunk, grumbles to his mates about the delight that will come when all this is over, and the buoy heaves up well within sight; and man-o'-warsmen who beat about in rickety gun-boats on the hideous coasts of East and West Africa, think with longing of the time when their cruel privations shall be over, and the magic announcement shall be made that sets the pulse of every mariner dancing.

To the mere landsmen, the whole stretch of the sea around the light is pleasant to sail over. The fishermen and bargemen say, "You are sure to get a breeze down by the Nore!" and there is hardly a day at any season of the year, or in any weather, on which this prophecy is not justified. The yachts that come lazily down, with their huge spinnakers spread like a swan's wing before cat's-paws that merely tremble over the surface of the Thames, immediately show signs that the captains are exercising caution when they cross that breezy band. You instinctively expect to see the spinnakers taken in, and to see the swift cutters lie hard over to the mainsail and foresail they sweep round the buoy. Once you have sailed to the eastward, you feel as though your craft were suspended between sea and land. On one side you have Southend, glittering in colour; on the other, you have the more sombre vision of Sheerness. Far up the river the rippling flood advances, as it were, in steady ranks upon you; and away to the south-west the marshes glitter, and the far-off hills look cold and blue. Between tide and tide a whole day of pleasure may be enjoyed by one who is content to watch merely the changes of sea and sky, and to speculate lazily as to the character of the vessels that pass in long procession. To men into whose spirits the charm of the Lower River has entered, there is no form of enjoyment dearer than merely to sail past the Nore, run outside near the Maplin sands, and there wait until the tide turns and the inward trip can be made with ease. In a small boat

MOUTH OF THE THAMES—LOW WATER.

it is best to keep slightly out of the track of the vessels that are running into the Swin, and to hit the happy medium between those that are going north about and those that are travelling south. The colliers go by flying light; the men on board are tolerably lazy; and as the dirty, rusty hulks lumber by, the seamen wave a kindly greeting. Smart, clean-built Scandinavian barques claw their way down, and the leisurely barges—which we have mentioned so often—pass by, laden to an extent that excites wonder at the temerity of the cool ruffians who man them.

Amid this unceasing panorama, every separate picture of which tells some fresh and strange story of far-off regions, of grimy labour, of storm and peril, it is easy for men who are content with a day of small things to sit for hours and hours, perhaps only exchanging monosyllabic comments on each new-comer. The most glorious sight of the many that may be seen at the Nore comes when some mighty sailing-ship looms to the northward of the marshes, and swims grandly on in the wake of her puffing, fussy little tug. As the two come on in their brief procession the tug represents mind; the vessel represents matter. That great ship that will so soon perhaps be sweeping down the league-long seas southward of the Horn trails meekly in the wake of a fat little screw, which could be placed on the deck of the convoy without causing a great deal of inconvenience. The ship is the embodiment of grace—the tug is the embodiment of ugliness; and yet until the river is clear the tug is master. But supposing a fine breeze springs up, then, of a sudden, there is a stir on board the vessel. From afar off you cannot tell exactly what is being done, but you know that presently her white wings will be shaken out; and, sure enough, as the vessel strikes the open, the sails fall—a cloud seems to spring from the water as if from the touch of a magician. Then the tug swings discreetly aside. Little by little, the wind lays its hand on each of the bellying sails and thrusts them out, till their broad bosoms glint in the sunshine and the hulk lunges and gathers way under their steady pull. The wind gains power, and the ship comes on with a creamy ripple of foam ringing her bows like some dainty ornament; and, with a sweep, she passes you by, leaving a billowy wake behind her; and before your last cheer has died into silence she is away on her journey. When one of the great four-masters glides out under her towers of canvas there is something in the sight that brings one's heart into his mouth. It never grows stale. You see the great hull with a line of wistful faces peering over the bulwark, and you know that you are only gazing upon a common-place emigrant ship. Yet the most prosaic of men comes to think of this majestic structure as a living being, and the poorest emigrant that ever wept over his farewells at Gravesend acquires a certain dignity from being carried away by her.

Shore-going folk often wonder at the contented impassivity of seamen who happen to have an idle hour in which to stare at the ships and the water. Observe an ordinary East Coast seaman spending his leisure. His eyes seem devoid of speculation, he stares sleepily seaward, and when he talks to a companion he uses brief, ill-formed sentences. But his mind is active, and if you listen to his low comments you will find that, in a quiet way, he studies the water and its passing burdens as men study a beloved book. If a ship is detained to wait for a tide or a pilot, the sailors find their pastime in contemplation and rude comment which the landsman does not understand. If that landsman only spent a while in a yacht on the Lower River they would learn that the solemn men who look so still and melancholy are probably feeling a placid pleasure, and the fixed silence is the expression of a sober contentment that cannot find expression in words. When our full-rigged vessel goes rolling away with the wind rushing hoarsely out of her courses the sailor feels acute delight; but he only grunts his admiration. The landsman may be excused if he breaks into unwonted ejaculations, and we have recognised our own right to the landsman's privilege. The present writer can never forget the shock of surprise with which he first saw a full-rigged ship slashing seaward from the Lower Hope. He rose as the dawn was painting the river with flashes of gold, and, lo! to leeward of the yacht, within forty yards, the monster ship was shouldering her way through the dappled flood. The smaller vessel was lying down till her copper gleamed to windward, and the swashing stream surged aft and rolled nearly up to the companion; but the little "floating chisel" could not long hold her own against the cloud-capped castle, and soon we watched her drawing proudly away on her long journey.

The trim gardens, the rich air of ordered beauty, the lovely song of birds —all the things that greet the senses on the Upper River are pleasant to the senses, but nothing in the gliding shallows that we love so well could equal the majesty, the strength, the glory, of that noble ship; and the sight of her was something to remember in happy nights when one cannot sleep for the delight of living.

Sometimes, when loitering northward of the Nore, we hear a sullen boom, and feel a tremor in the air. The artillerymen are at work on the ranges at Shoeburyness, and some tremendous piece of artillery is pitching masses of iron seaward. There is no danger, for it rarely happens that even an unwary bargeman ventures near the forbidden region. In our time we only remember one accident. The eighty-one ton gun had hurled a shrapnel shell over a distance of about six miles. On the landward side, the whole of the windows within a quarter of a mile were shivered from out the frames, and the officers' quarters were left desolate; while, to seaward, a great massacre took place among a flock of unwary gulls. But this is the only loss of life that has been caused by the projectiles which scream over the broad shallows. To

ARTILLERY PRACTICE AT SHOEBURYNESS.

persons of a military turn, Shoebury is a most interesting quarter. Everything is so trim, so business-like, so ineffably military; and the work goes on so calmly that no one would think that the groups of stern officers and dashing artillerymen were studying the art of destruction. In summer, when the volunteers are encamped, the whole place breaks into merriment as soon as the toilsome competitions are over, and the forts are well worth a visit from a tourist. The picturesque is lacking, but once more, the power—the immeasurable reserve force—of our nation strikes on the mind and wakens a feeling of pride.

GRAVESEND TO THE NORE.

Morning on the Upper River is joyous, and all through the bright summer days a sense of keen gladness grows with every hour. The sleepy afternoons, when the silence broods over the reaches like a voice, carry the day-long symphony of gladness through yet another movement; and in the evening, when the clear stars speak silence from their glimmering eyes, and wash the dusk with silver, everything grows beautiful, tender, and kindly to the thoughtful soul.

The Upper River is like a delicate lady, clad in all daintiness, and beaming with gentle beauty. The Lower River is like a burly man, who urges his way through his career with a sense of strength, with a disdain of obstacle, with a brutal persistence that keep up the masculine character. From the places where the ships curtsy at their creaking tiers, to the splendid stretch where the sea-breeze blows shrill, chilly with flecks of foam, every yard is vivid with interest.

We believe that no man ever grew tired of the Upper River. People haunt its reaches year after year, till it seems as though all the blessed summers were blended into one memory. We cannot think with joy of summer on the Lower River; but the bitter winter days, the scream of keen blasts, the monstrous procession that connects the world of the city with the great world of the outer sea—all these things are never-fading when once their impact has fairly gained the recesses of the soul.

Old sportsmen may still be found who shot over the saltings or glided round the forbidding points of the Lower Thames in their youth; the habit never leaves them, and, as the seasons roll, these men find their keenest delight from prowling among the shadowy marshes or facing the salt, shrill wind that pulses and beats around the Nore.

Sometimes a Cockney sceptic may be found who shudders and speaks of the Lower River as a place of horror. He sighs for the glades of Clieveden, for the mossy chestnuts of Hampton Court, for the sloping gardens of Sunbury. But let a wise sportsman take the sceptic's education in hand; let his wayward mind be disciplined by merry days among the swarming saltings, and he will acquire a taste that will be lasting. If he is judiciously taught he may come at last to feel the true ecstasy, the mysterious poetry, that touch the soul on shining nights when the moon-silvered roll of the water is gladsome, and the shadowy ships steal away to the sea. Then the sordid flats are touched into beauty by the cold gleam, and the winds, and the waters, and the sailing clouds, and the quiet ships pass like a mystic pageant, fleeting, fleeting, ever eastward. The veriest townsman that ever waked the echoes under Kingston Bridge with his clamour will own at such a time that few sights in England are finer than the noble outflow of our splendid river.

J. RUNCIMAN.

OUTWARD BOUND—PASSING THE NORE LIGHT.

INDEX.

Abbey Church at Dorchester, 72
—— of Abingdon, 63
Abingdon Bridge, 63
——, Abbatial parlours at, 64
——, Environs of, 68
——, Market Cross at, 65
——, Market-place of, 64
——, Shrine at, 66
—— to Streatley, 62 to 84
——, Town of, 62
Abney House, 134
Above Oxford, 1 to 32
Albert Bridge, Windsor, 169
—— Docks, 303, 320
—— Embankment, 266
Ancient Stone Cross at Abingdon, 65
"Angler's Rest," The, 176
Angling in the Thames, 13, 28, 29, 90, 107, 161, 170, 172, 186, 187, 193, 197, 218
Ankerwyke House, 173
Antiquity of Streatley and Goring, 87
Appleford Bridge, 70
Apps Court at Hampton, 200
Artillery Ranges at Shoeburyness, 360
Asiatic Home, 304
Aston Ferry, 119
Avon Canal, 99

B

Bablock Hythe, 24
Bampton, 20
Bankside, Southwark, 282
Barges on the Thames, 57
Barking Abbey, 326
—— Reach, 327
"Barley Mow," The, 71
Barn Elms, 253
Barnes Common, 248
——, The Village of, 248
Basildon Ferry, 88
——, Village of, 88
Battersea Bridge, 258
—— Park, 264
—— Reach, 257
Bear Garden, Southwark, 282
Beauchamp Tower, 300
Beefeaters and Guards at the Tower, 301
"Beetle and Wedge," The, 82

Bell Weir Lock, 173
"Bells of Ouseley" Tavern, 172
Belot, The Arctic Explorer, 316
Benedictine Novices at Oxford, 30
Benfleet Station, 312, 345
Bensington, The Village of, 77
—— Lock, 78
Bermondsey and Tooley Street, 298
Billingsgate Fish-market, 291, 293, 294
Binsey, The Village of, 32
Bird Life on the Thames, 18
Birthplace of River Thames, 2
Bisham and its Ghost, 126
—— Church, 127
—— House, 125
—— Priory, 126
—— Woods, 149
Bishopric of London and Fulham Palace, 255
"Black Cherry Fair," 187
Blackfriars Bridge, 277
Blackwall, 319
Blenheim Park, 27
Bolney Court, 104
Botanical Gardens at Battersea, 262
Boulter's Lock, 155
Bourne End, 133
Boveney Lock, 151
Boyle Farm and the "Dandies' Fête," 209
Boyne Hill, 142
Bray, 145
Bream Fishing at Dagenham, 325
Brentford, Town of, 246
Brethren of the Holy Rood and Abingdon, 65
Bridge at Richmond, 235
—— Ward Without, 282
Bridgwater Canal, 56
Burrow Marsh, 103
Bushey Park, 204, 210
Bye-laws of the Watermen's Company, 340

C

Canal from Birmingham to Oxford, 56
Canute's Country, 21
Canvey Island, 345
—— Point, 345, 346
Capital of Wessex, The, 73

Carp in the River, 111
Cassington Church, 30
Castle Eaton Bridge, 12
Caversham Bridge, 94
—— Warren, 94
"Cawsam Hill," 93
Chancer and Southwark, 282
Chelsea and Neighbourhood, 260
——, Cheyne Walk, 262
—— Don Saltero's Coffee-house, 262
——, Great Cheyne Row, 262
—— Hospital, 263
—— Suspension Bridge, 264
Chertsey, Angling at, 186 to 188
—— Bridge, 185
——, Bridge House at, 189
—— Church and Bell, 186, 188
—— Fairs, 187
—— Mead, 184
——, St. Ann's Hill at, 189
Cherwell Bridge at Oxford, 50
Cheyne Walk, Chelsea, 260, 262
Chiltern Hills, 86
Chilton Lodge, 98
Chinese Barges and Weirs, 56
Chiswick House, 250
Christ's Hospital at Abingdon, 67
Church of Old Windsor, 170
—— of St. Peter ad Vincula, 301
—— of St. Helen at Old Abingdon, 68
Churn, The, 7, 8
Clasper's Boathouse, 37
Cleeve Lock, 82
Cleopatra's Needle, 274
Cliefden House, 136, 138
—— Woods, 134, 138
Cliff Creek, Gravesend, 336
Clifton Bridge, 71
—— Hampden, 70
Coarse Fish in the Thames, 109
Commercial Docks, 307
Conservancy of the River, 179
Convent at Isleworth, 240
—— of Sheen, 237
Cookham Bridge, 135
—— Church, 136
Cooper's Hill, 175
Coronation Stone at Kingston, 214
Courses of the Thames, The, 34
Cowey Stakes at Walton, 195

THE THAMES.

Cowley's House at Chertsey, 108
Craven House, 254
Cricklade, The Town of, 10
Crockham Hill, 190
Cross Ness Sewage Works, 326
Crownmarsh Gifford, Parish of, 78
Culham Court, 119
—— Lock, 69
Cumnor, The Village of, 24
Custom House, 291, 296

D

Dace and Roach in the River, 111
Dagenham Breach, 325
—— Lake, 325
Danesfield House, 123
Dartford, 330
Datchet Mead and Falstaff, 168, 169
Day's Lock, 70
Denham's Description of the Thames, 175
Deptford Dockyard, 211
—— and Peter the Great, 315
Disused Weirpools, 23
Dock Labourers, 314
Dominican and Franciscan Teachers at Oxford, 39
Don Saltero's Coffee-house and Thomas Carlyle, 262
Dorchester Abbey, 70
Dorchester Abbey Church, 73
Dorchester, Capital of Wessex, 73
Duelling at Barn Elms, 254
Dufford Ferry, 21
Dumb-barges on the River, 312
Dyers' Company and the Swans, 133

E

Early Names of the Thames, 4
East India Docks, 310, 315
Eel Boats, 343
Eelbucks at Chasey Farm, 94
Eel Pie Island, 222
Egham, Town of, 178
English Frigate *Warspite*, The, 324
Environs of Abingdon, 68
Epitaphs in Bensington Churchyard, 78
Erith, 327, 329
Erkenwald's Monastery at Chertsey, 186
Essex Marshes, 342
Eton, 151
——, Chapel at, 164
—— College Buildings, 162
——, Founder of the Chapel, 165
—— Montem, 166
—— —— or Waterton, 164
——, Provosts at, 163
——, Rouse's Scholarships at, 163
——, Speech Day at, 165

Exciting Race from Cookham to Marlow, 147
Eynsham Bridge, 27
—— Cross, 27
——, The Village of, 27
—— Weir, 25

F

Fairford and the Coln, 13
"Fair Mile," The, 115
Famous Grotto at Twickenham, 222
Farringdon, The Town of, 20
Fawley Court and its History, 117, 118
First Lock, 14
Fish at Billingsgate, 294, 295
—— in the Thames, 13, 28, 29, 76, 99, 161, 170, 172, 193, 194, 197, 218
Fishers' Row, 34
Fishmongers' Hall, 288
—— Procession, 291
Fish Street Hill, 294
"Flash" Locks, 55
Flyfishers' Club at Hungerford, 98
Fly-fishing in the River, 111
Flooding the River, 56
Flowers, Some Thames, 16
Folly Bridge, 32
—— —— Lock, 55
Fords at Shefford, 21
Foreign Cattle Market at Deptford, 316
Four Noted Abbeys, 95
—— Streams, The, 35
Franciscans at Medmenham, 120, 121
—— and the Monkey, 121
Francis Rouse and Eton, 163
"French Horn" Inn, 100
Freshwater Wharf, 291, 292
Frogmore, 159, 167
Fulham and its Market Gardens, 256
—— and Putney, 254
—— Church, 256
—— Palace, 255

G

Garrick's Villa at Hampton, 198
Girls Rowing and Steering, 150
Godstow Nunnery, Ruins of, 31
——, The Village of, 30
Goring Church, 87
—— Mill, 86
—— Toll-gate, 87
—— Weir, 86
Great Marlow, 129
—— ——, The Church and its "Curiosities," 129
—— ——, The Deanery at, 130
—— —— Weir and Locks, 131
Gravesend Reach, 336
——, The Town of, 333
—— to the Nore, 337 to 362

Grayling in the Kennet, 98
Greenhithe, 327, 328, 330
Greenwich Hospital, 316
—— Observatory, 319
—— Park, 318, 319
—— Weir, 12
Grotto at Goring, 88
—— at Oatlands, 191
Groves at Bisham, 132
—— of Kew, The, 241
Gudgeon in the Thames, 176, 177
Guild of Pilots at Gravesend, 339

H

Hadleigh Castle, 348
Halliford Bridge, 194
Ham, Village of, 225
Hambledon Lock and its Islands, 119
Hammersmith Suspension Bridge, 251
Hampton Court, 201, 202
—— —— Bridge, 200
—— ——, Chestnut Avenue, 204
—— ——, Christmas at, 207
—— ——, English Brickwork at, 203
—— ——, Fountain Court, 206
—— ——, Great Hall, 204
—— ——, Painted Windows in Great Hall, 205
—— ——, Park at, 207
—— ——, Picture Gallery at, 206
—— —— to Richmond, 211 to 228
Hampton House, 198
—— Wick, 213
Hannington Bridge, 12
Hardwicke House, 91
Hart's Weir, 17
—— Wood, 88
Hedsor Park, 136
Hennerton Backwater, 104
Henley and its Natural Scenery, 113
—— Bridge and its Builder, 115
—— Church, 114
—— Races, 115
—— Regatta, 117
——, the Riparian Metropolis, 107
—— to Maidenhead, 113 to 142
——, Town of, 113
Hinksey Stream, The, 37
History of Kew Gardens, 241
Hogarth's Frolic, 289
Holme Park Woods, 100
Houses of Parliament, The, 267
Howberry Park, 78
Hungerford, Famous for Trout, 98
Hurley and its Islands, 123
Hurlingham, 257
Hythe Bridge, 34

I

Iffley Lock, 55
—— Mill, 54

INDEX. 365

Inglesham Weir, 13
Ingress Abbey, 330
Inns at Rotherhithe, 307
Inscriptions in Bensington Church, 78
Invention of Rowing, 54
Isis, The Head of the, 3
—— and the Thames, The, 72
Isle of Dogs, 308, 310
—— of Osney, 37
—— of Thorns, The, 186
Isleworth, Town of, 239
Izaak Walton at Datchet, 169

J

Jesus Hospital at Bray

K

Kempsford, 12
Kennington, The Village of, 58
Kentish Marshes, 344
Keston Heath, 316
Kew Bridge, 246
 - - Churchyard, 246
 — - Foot Lane, 227
—— Gardens, 237, 239, 241
- — ——, Chinese Pagoda in, 245
- — ——, Palm-house in, 243
—— Green, 246
—— Observatory, 237, 239
—— Palace and the "Dutch House," 241, 242
Kingston-on-Thames, 212, 215
 · · · Bridge, 211
 -   Church, 206, 215
 ——, Coronation Stone at, 214
Kingston, Rowing Clubs at, 211
King's Weir, 30
Kitcat Club, The, 253

L

Lady Place and its History, 123
Laleham and Dr. Arnold, 182
—— Church, 183
—— Ferry, 182
—— House, 183
Lambeth, Borough of, 264
—— Palace, 266
- - ——, Archbishop Laud and, 267
Langley or Ridge's Weir, 23
Lechlade, The Town of, 14
Leigh, The Village of, 349
Limehouse Church, 308
Little Marlow, 132
Littlecote Park, 98
Littlemore, The Village of, 58
Lollards' Tower, Lambeth, 266, 267
London Bridge and its History, 284, 286
—— and its Traffic, 284
—— Docks, The, 303

"Long Reach" Tavern, 329
Long Walk at Windsor, 167
Lower Hope, The, 344
Ludgate Hill and the Temple of Diana, 278

M

Magna Charta Island, 174
Maidenhead or Maidenhithe, 141, 143, 145
—— Bridge, 145
Main Stream for Barges at Oxford, 37
Maplodurham House, 93
—— Mill and Weir, 92
Maplin Sands, 357
Market Cross of Abingdon, 65
Marking Swans, 133
Marlborough, The Town of, 99
Marlow Bridge, 139
Marsh Lock, 105
Medley Weir, 32
Medmenham Abbey, 120
Middle Temple Hall, 277
Millbank Prison, 264
Millwall Docks, 310, 312
Mongewell House, 80
Monkey Island, 151
Monument, The, 291
Morning on the Lower Thames, 337
Mortlake and its History, 247
Mouslsey Hurst, 199
Moulsford Bridge, 80
Mysteries of Fly-fishing, 111

N

Names of Boat Landings and Stairs, 307
National Society's Training College at Battersea, 260
Navigation of the Iffley, 55
Newbridge, 22
Newbury, The Town of, 99
Newton Murren, Church of, 80
New Battersea, 264
Noah's Ark Weir, 23
Nore, The, 357
Northfleet, 330
—— Hope, 339
North Woolwich Gardens, 324
—— —— Pier, 324
Nuneham Lock, 55
——, Heights of, 59
—— Reach, 61
Nunnery of Sion, The, 237

O

Oatlands and its History, 192
- —— Grotto, 191
- —— Palace and Park, 190, 191
"Ocean Tramps," 338
Ock Street at Abingdon, 66

Old Buscot, The Village of, 16
"Old Chelsea Bun-house," 262
—— —— Church, 260
"Old Crabtree Inn," 252
Old London Bridge and its History, 285
—— ——, Nonsuch House on, 286
—— —— ——, St. Thomas's Chapel on, 286
—— —— —— Traitors' Gate on, 286
—— —— —— Theatres, 282
Old Navigation Stream, 35
Old St. Paul's Church, 280
Old Windsor, 170
Oldest Stone Bridge, The, 22
Origin of "The Merry Wives of Windsor," 159
Osier Farm on the Thames, 88
Osney Abbey, 30
Oxford to Abingdon, 33 to 61
——, Architectural Revival at, 47, 50
——, Austin Friars at, 39
——, Black Friars of, 38
—— Canal, 37
——, Carmelites or White Friars at, 39
——, Chapel of Brasenose at, 40
——, Church of St. Martin at Carfax at, 40
——, College of St. Mary Winton, 43
——, Colleges at, 42
- —, Degrees at, 42
- · ·, Disputations at, 48
——, Early Colleges at, 43
——, Eleanor Cross at, 51
- ——, Grey Friars of, 38
——, Halls at, 43
——, Houses of the Friars at, 34
——, Lincoln College of Priests at, 46
——, Magdalen Bridge, 50
——, —— College at, 44
——, Mendicant Order of Friars at, 38
——, Midnight at, 54
——, Oriel Barge at, 57
- ——, Preacher's Pool at, 38
- ——, Sheldonian Theatre at, 40
- ——, The Buildings of, 37
- ——, The Town of, 33
- ——, Trill Mill Stream at, 38
——, Undergraduate Revival at, 52
—— University and Parish Churches, 38

P

Pacey's Bridge, 35
Palace of Richmond and its History, 233, 234
—— of Westminster, 268, 270
Pangbourne, Village of, 80
Park Place and its History, 105
Paul's Walk, 279
"Pedlar's Acre," Lambeth, 266

THE THAMES.

Penton Hook Lock, 181
"Perdita's" Grave, 169
Petersham Park, 226
Phillimore Island, 104
Picnic Island, 175, 176
Pike and Perch Fishing, 110
Pilots and Watermen, 339, 340
Pimlico, District of, 264
Pink Hill Lock, 27
— — Weir, 27
Pleasant Pictures at Bray, 150
Plumstead Marshes, 326
"Pool," The, 301, 302
Pope's Villa at Twickenham, 222
Porch House, Chertsey, 188
Port Meadows, 32
Prisoners at Windsor, 158
Prize-fighting at Moulsey, 199
Pudding Lane, 294
"Puppy Pie" at Marlow Bridge, 130
Purfleet, Town of, 328
Purley Hall, 93
——, The Village of, 93
Putney and Fulham, 254
—— Bridge, 255
—— Heath, 255

Q

Quarry Woods, 131
Queen Bess and the Custom House, 297
"Queen's Tobacco Pipe," The, 305
Queenhithe and the Tax on Fish, 294

R

Racing Yachts on the Thames, 338
Radcot Bridge, 18
Radley, The Village of, 58
Ramsbury Manor, 98
Ranelagh Club, Barn Elms, 254
—— Gardens, 263
Ratcliff Highway, 305
Ray Mead, 144
Reaches on the River, 309
Reading Abbey, 95
— at the time of the Plague, 96
— —, The Town of, 95
Rearing Ponds at Sunbury, 197
"Red Lion" at Henley, 115
Regatta Island, 117
Relics of "Roman London," 292
Remenham Hill, 114, 119
Rewley Abbey, 39
Richmond, 227
— — —, Asgill House, 230, 232
—— Bridge, 229
—— Church, 234
—— Green, 232, 234
——, Mansfield House at, 228
—— Old Deer Park, 237
——, Old Palace of Sheen, 230
—— Palace, 231

Richmond, Queensberry House, 232, 234
— —, "Star and Garter" Hotel, 227
| ——, The Hill at, 227
— , Trumpeters' House, 232
— —, Water Supply of, 235
River Churn, The, 8
—— Cole, The, 13
—— Coln, The, 13
—— Cray, The, 329
· Darent, The, 329
Evenlode, The, 27
· · · — Iffley, 54
—— Isis, The, 11
—— Kennet, The, 95, 97
—— Lea, The, 319
- Leeh, The, 16
—— Loddon, The, 102
— — Ravensbourne, The, 316
——— Windrush, The, 21
——— Witham, The, 34
Riverside Amusements, 115
| — — Inns, 180
——— Solitude, 72
Roach, Dace, and Gudgeon in the River, 111
Romantic Episodes, 154, 157
Roman Fortifications at Dorchester, 73
Romney Island, 165
Rose Isle, 58
Rosherville Gardens, 333
Rotherhithe, 306
—— Railway Station, 307
Round House at Inglesham, 13
Rowdyism on the Thames, 144
Rowing Clubs at Kingston, 211
—— on the River, 54, 146
Royal Albert Dock, 320
—— Gardens at Kew, 239, 243
—— Naval College, Greenwich, 317
—— River, The, 241
Rules of the Company of Watermen, 339
Runnymede, 174, 175
Ruscombe, The Village of, 105
Rushy Lock, The, 20

S

Sailors and their Ways, 306
Sailors' Home, 304
Salesmen's Cries at Billingsgate, 295
Salmon in the Thames, 109
Salt Hill, 166
Sandford Mill, 58
St. Helen's Church at Abingdon, 66
St. Katharine's Docks, 303
St. Magnus' Church, Billingsgate, 291
St. Nicholas' Church at Abingdon, 64
St. Olave's Church, Bermondsey, 298
St. Patrick's Stream, 103
St. Paul's Cathedral, 277, 278
—— "Paul's Walk" in, 280
St. Saviour's Church, Southwark, 283

St. Stephen's Chapel at Westminster, 270
St. Thomas's Hospital, Lambeth, 266
Scenery at Pangbourne, 90
Sewage of London, 316
—— Pumping Stations, 316
Sewer Works at Barking, 326
Seven Springs, The, 6
Severn Canal and Thames, 9
Shakspeare and Windsor, 159
Shelley and Great Marlow, 129
Sheerness and its History, 353
—— Dockyard, 356
Shepperton, Village of, 194
Shillingford Bridge, 76, 77
Shiplake Lock, 104
Shipping in the Pool, 290, 291
Shoeburyness, 360
Shrimps at Gravesend, 334
Shrimp Hampers at Leigh, 350
Shrine at Abingdon, 66
Skinner's Weir, 27
Silvertown, The Colony of, 322
Sinking of the Cambridge Boat, 250, 251
Sinodun Hill, 72, 76
Sion House, 239
Somerford Keynes, 10
Somerset House, 274
Sonning Bridge, 100
Sonning-on-Thames, 100
Source of the Thames, 1
Southend, The Town of, 350
— - Pier, 350
Southley, 167
Southwark Bridge, 280
Spawning Season, 109
Spring Well at Goring, 87
Staines "Deep," 179
— —, Linoleum Works at, 178
——, Village of, 178
Stanton Harcourt, 25
State Barge at Teddington, 218
Steam Launches on the Thames, 29, 110, 144, 167, 170
Stepney, 308
Stoke Ferry, 81
Stone Church at Greenhithe, 330
Strange Names of Places, 208
Strathfieldsaye Park, 102
Strawberry Hill at Twickenham, 220
Streatley Hills, 80
—— Mill, 83, 86
—— the Mecca of Painters, 85
—— to Henley, 85 to 112
—— Tower, 84
—— Weirs, 83
Sundial at Kew, 243
Supposed Site of Saxon Palace, 171
Surbiton, 211
Surley Hall, 151
Surrey Docks, 308
Swallowfield, 103

INDEX.

"Swan," The, Thames Ditton, 209
Swans on the Thames, 133
Swashway, The, 353

T

"Tabard" Inn, Southwark, 282
Tadpole Bridge, 20
Tapestry Works at Windsor, 171
Taplow Court, 141
——— Woods, 141
Teddington, 215, 218
———, State Barge at, 218
———, "The Anglers" at, 216
———, Tombs in Church at, 219
——— Weir, 14, 218
Temple, The, 276
Temple Bar, 167
——— Gardens, 276
Temple's (Sir W.) Orangery at Sheen, 239
Tenfoot Weir, 21
Thames Angling, 107
——— ——— Preservation Society, 108, 197
——— and Severn Canal, 8, 9
——— and the Isis, 72
——— at Windsor, 161
———, Birthplace of, 2
——— Commission, 56
——— Conservancy Board, 179
——— Ditton, 207
———, Early Names of, 4
——— Embankment, 277
——— Flowers, 16
——— Haven, 344
——— Head, 2
——— Parade, 101
——— Sailing Club, 211
———, Source of the, 1
——— Street, 294
——— Subway, 299
———Swans, 133
———, Trout, 109
——— Tunnel, 306
——— Valley, The, 62
——— ——— Church, 209, 210
——— Watermen, 288
The Upper and Lower River, a Comparison, 361
Tilbury and its Memories, 331, 341
——— Docks, 330
——— Fort, 332, 341
——— Marshes, 330, 342
Tithing Saloon, 186
Tooley Street and the Three Tailors, 298

Tower Hill, 299
——— of London, 299
Training Ships on the Thames, 329
Traitor's Gate, 286
"Tramps," The, of the Ocean, 338
Trewsbury Mead, 8
Trill Mill Stream at Oxford, 38
Trinity House, 299
Tripcock Reach, 327
Trout-fishing on the Thames, 13, 90, 109
"Trout" Inn at Lechlade, 15
——— Stream at Pangbourne, 90
Twickenham, 220
———, Orleans House, 223
——— Parish Church, 222
———, Pope's Tomb, 223
———, Pope's Villa at, 222
———, Strawberry Hill at, 220
———, York House, 223

U

University Barge, The, 57
——— Boat Club at Oxford, 55
——— Boat Race, The First, 118
——— ——— ——— on the Thames, 250
Upper, Lower, and Middle Pools, 362

V

Vale of the Kennet, The, 99
Vauxhall, 264
Victoria Bridge at Pimlico, 264
——— Dock at Blackwall, 303, 320, 321
——— Embankment, 274
Vintners' Company and the Swans, 133

W

Walker's Picture of the "Harbour of Refuge," 146
Wallingford Castle and Museum, 79
———, The Town of, 78, 79
Walton-on-Thames, 195
———, Ashley Park at, 196
———, Cowey Stakes at, 195
———, The Bridge at, 194
———, The Scold's Bridle at, 196
Wandsworth Brewery, 257
Wapping, 308
——— Old Stairs, 301, 308
Wargrave Hill, 104
Water-fowl at Penton Hook, 181
Water Gate of Inigo Jones, 275
Water Hay Bridge, 10

Waterloo Bridge, 275
Watermen's Company. Rules of, 339, 340
West India Docks, 314
——— Mill at Somerford, 10
Westminster, City of, 267
——— Abbey, 269, 270, 271
——— Bridge, 273
——— Hall, 269, 270, 272
———, History of, 270
——— Palace, 268, 270
Weybridge, 190
——— Green, 192
Whitchurch Lock, 89
Whitebait Fishing at Gravesend, 334, 335
White Friars or Carmelites at Oxford, 39
White Place, near Cliefden, 139
Wild-fowl Shooting, 342, 344
Willow-peeling, 89
Windsor, Albert Bridge at, 168
——— Castle and its History, 152
——— Castle as a Palace and as a Prison, 158
———, Great Park at, 159, 160, 166
———, Herne's Oak at, 166
———, Home Park at, 166
———, Long Walk at, 167
———, Queen's Walk at, 167
———, Round Tower at, 158
———, St. George's Chapel at, 154
———, The Merry Wives of, 159
———, Town of, 159, 160
———, Victoria Bridge near, 169
Wine Cellars of the London Docks, 304
Witham Hill, 28
Wittenham Clump, 70
Wolsey's, Cardinal, College at Oxford, 66
——— Tower at Henley, 114
Wolvercott, Village of, 31
Woodstock, The Town of, 27
Woolwich Arsenal, 323
——— Barracks, 322
——— Church, 322
——— Dockyard, 323
——— Reach, 322
Wraysbury Church, 173
———, Village of, 174

Y

Yachting at Gravesend, 333
York House, 274

TABLE OF DISTANCES a

Taken, by permission of the Author, from Taunt's "Guide to the Thames."

		m. f. yds.
Thames Head to Cricklade		11 6 2
Seven Springs	,,	20 4 0

ABOVE OXFORD.

		m. f. yds.
Cricklade to Oxford		43 7 2
Water Eaton Bridge to Oxford		41 7 152
Castle Eaton Bridge	,,	39 3 44
Kempsford	,,	37 7 0
Hannington Bridge	,,	36 5 88
Inglesham Round House	,,	33 2 0
Lechlade Bridge	,,	32 4 88
St. John's	,,	31 7 0
Buscot Lock	,,	30 5 170
Hart's Weir	,,	29 3 20
Radcot Bridge	,,	26 2 100
Old Man's Bridge	,,	25 1 110
Rushy Lock	,,	23 0 110
Tadpole Bridge	,,	22 1 54
Tenfoot Bridge	,,	20 3 0
Duxford Ferry	,,	18 4 50
New Bridge	,,	15 1 160
Ridge's Weir	,,	14 0 104
Bablock Hythe Ferry	,,	11 4 34
Skinner's Weir	,,	9 7 0
Pinkle Lock	,,	8 7 0
Eynsham Bridge	,,	7 2 193
King's Weir	,,	4 4 32
Godstow Lock	,,	3 2 99
Medley Weir	,,	1 7 36
Osney Lock	,,	0 7 0

OXFORD TO PUTNEY.

	From Place to Place	From Oxford (Folly Bridge)	From London (Putney Br.)
	m. f. yds.	m. f. yds.	m. f. yds.
Oxford Bridge	0 0 0	0 0 0	104 3 66
Iffley Lock	1 3 150	1 3 150	102 7 136
Rose Island	0 6 124	2 2 54	102 1 12
Sandford Lock	0 6 166	3 1 0	101 2 66
Nuneham Bridge	2 5 160	5 6 160	98 4 126
Abingdon Lock	1 7 60	7 6 0	96 5 66
Abingdon Bridge	0 3 211	8 1 211	96 1 75
Culham Lock	2 0 0	10 1 211	94 1 75
Appleford Railway Bridge	1 2 76	11 4 67	92 6 219
Clifton Lock	1 4 54	13 0 121	91 2 165
Clifton Bridge	0 3 140	13 4 41	90 7 23
Day's Lock	2 4 40	16 0 81	88 2 203
Junction of River Thame	0 6 180	16 7 41	87 4 25
Keen Edge Ferry	1 0 140	17 7 181	86 3 105
Shillingford Bridge	0 6 100	18 6 61	85 5 5
Benson Lock	1 2 30	20 0 91	84 2 105
Wallingford Bridge	1 2 0	21 2 91	83 0 195
Nuneham Ferry	0 4 70	21 6 161	82 4 125
Stoke Ferry	2 1 0	23 7 161	80 3 125
Moulsford Railway Bridge	0 5 46	24 4 207	79 6 70
Moulsford Ferry	0 5 64	25 2 51	79 1 15
Cleeve Lock	1 2 78	26 4 129	77 6 157
Goring Lock	0 5 0	27 1 129	77 1 157
Basildon Railway Bridge	1 2 61	28 3 190	75 7 0
Gate-Hampton Ferry	0 2 66	28 6 36	75 5 30
Whitchurch Lock	2 4 33	31 2 69	73 0 217
Mapledurham Lock	2 2 70	33 4 139	70 6 147
The "Roebuck"	0 7 145	34 4 60	69 7 2
Caversham Bridge	2 6 206	37 3 50	67 0 16
Caversham Lock	0 4 120	37 7 170	66 3 116
River Kennet's Mouth	0 5 120	38 5 70	65 5 216
Sonning Lock	1 7 28	40 4 98	63 6 188
Sonning Bridge	0 2 60	40 6 158	63 4 128
Shiplake Lock	2 4 66	43 3 4	61 0 62
Shiplake Ferry	1 0 38	44 3 42	60 0 24
Boulney Ferry	1 0 44	45 3 86	58 7 200
Marsh Lock	0 4 78	47 5 164	58 4 122
Henley Bridge	0 7 109	46 7 53	57 4 13
Hambledon Lock	2 2 35	49 1 88	55 1 108

	From Place to Place	From Oxford (Folly Bridge)	From London (Putney Br.)
	m. f. yds.	m. f. yds.	m. f. yds.
Medmenham Ferry	2 0 66	51 1 154	53 1 132
Hurley Lock	1 4 168	52 6 102	51 4 184
Temple Lock	0 5 23	53 3 125	50 7 101
Marlow Bridge	1 3 201	54 7 106	49 3 180
Marlow Lock	0 1 107	55 0 213	49 2 73
Spade Oak Ferry	2 0 205	57 1 198	47 1 88
Cookham Bridge	1 5 66	58 7 44	45 4 22
Cookham Lower Ferry	0 4 110	59 3 154	44 7 132
Cliefden Ferry	0 3 44	59 6 198	44 4 88
Boulter's Lock	1 3 178	61 2 156	43 0 130
Maidenhead Bridge	0 5 70	62 0 6	42 3 60
Bray Lock	1 3 152	63 3 158	40 7 128
Monkey Island	0 4 128	64 0 66	40 3 0
Boveney Lock	2 5 0	66 5 66	37 6 0
Windsor Bridge	1 7 90	68 4 156	35 6 130
Romney Lock	0 3 96	69 0 32	35 3 34
Victoria Bridge	0 6 34	69 6 66	34 5 0
Albert Bridge	1 6 0	71 1 72	33 1 214
Old Windsor Lock	0 6 214	72 0 66	32 3 0
Magna Charta Island	1 3 0	73 3 66	31 0 0
Bell Weir Lock	1 3 157	74 7 3	29 4 63
Staines Bridge	0 7 105	75 6 108	28 4 88
Penton Hook Lock	1 3 168	77 5 146	26 5 140
Laleham Ferry	0 6 140	78 4 66	25 7 0
Chertsey Lock	1 1 4	79 5 70	24 5 216
Shepperton Lock	1 7 183	81 5 33	22 6 33
Halliford Point	1 2 33	82 7 66	21 4 0
Walton Bridge	0 6 156	83 6 2	20 5 64
Sunbury Lock	1 5 130	85 3 132	18 7 154
Hampton Ferry	2 0 110	87 4 22	16 7 44
Moulsey Lock	0 6 110	88 2 132	16 0 154
Thames Ditton	0 209	89 3 121	14 7 165
Kingston Bridge	1 7 55	91 2 170	13 0 110
Teddington Lock	1 6 88	93 1 88	11 2 22
Eel Pie Island	1 1 22	94 2 66	10 1 0
Richmond Bridge	1 4 140	95 6 206	8 4 80
Kew Bridge	2 7 124	98 6 110	5 4 176
Barnes Railway Bridge	2 0 178	100 7 68	3 3 218
Hammersmith Bridge	1 5 106	102 5 44	1 6 22
Putney Bridge	1 6 22	104 3 66	0 0 0

BETWEEN PUTNEY BRIDGE AND LONDON BRIDGE.

		m. f.
Putney Bridge	to London Bridge	7 3½
Battersea Railway Bridge	,,	5 5
Battersea Bridge	,,	5 0
Chelsea Bridge	,,	4 0
Vauxhall Bridge	,,	2 7½
Lambeth Bridge	,,	2 3
Westminster Bridge	,,	1 7½
Charing Cross Railway Bridge	,,	1 4
Waterloo Bridge	,,	1 2½
Blackfriars Bridge	,,	0 6
Southwark Bridge	,,	0 2½
Cannon Street Railway Bridge	,,	0 1½

BELOW LONDON BRIDGE.

		m. f.
Thames Tunnel to London Bridge		1 4
Deptford Dockyard	,,	3 5
Deptford Creek	,,	4 2
Blackwall Pier	,,	5 7
Woolwich Arsenal	,,	9 3½
Barking Creek	,,	11 1
Erith	,,	15 5
Dartford Creek	,,	17 3
Greenhithe	,,	20 4
Grays Thurrock	,,	22 3½
Gravesend	,,	25 1
Mucking Creek	,,	30 5
Yantlet Creek	,,	40 3½
Yantlet Creek to the Nore		5 miles (nautical)

PRINTED BY CASSELL & COMPANY, LIMITED, LA BELLE SAUVAGE, LONDON, E.C.

www.ingramcontent.com/pod-product-compliance
Lightning Source LLC
Chambersburg PA
CBHW030405230426
43664CB00007BB/756